Analytic Solutions of Functional Equations

Analytic Solutions of Functional Equations

Sui Sun Cheng
National Tsing Hua University, R. O. China

Wenrong Li
Binzhou University, P. R. China

NEW JERSEY · LONDON · SINGAPORE · BEIJING · SHANGHAI · HONG KONG · TAIPEI · CHENNAI

Published by

World Scientific Publishing Co. Pte. Ltd.
5 Toh Tuck Link, Singapore 596224
USA office: 27 Warren Street, Suite 401-402, Hackensack, NJ 07601
UK office: 57 Shelton Street, Covent Garden, London WC2H 9HE

British Library Cataloguing-in-Publication Data
A catalogue record for this book is available from the British Library.

ANALYTIC SOLUTIONS OF FUNCTIONAL EQUATIONS

Copyright © 2008 by World Scientific Publishing Co. Pte. Ltd.

All rights reserved. This book, or parts thereof, may not be reproduced in any form or by any means, electronic or mechanical, including photocopying, recording or any information storage and retrieval system now known or to be invented, without written permission from the Publisher.

For photocopying of material in this volume, please pay a copying fee through the Copyright Clearance Center, Inc., 222 Rosewood Drive, Danvers, MA 01923, USA. In this case permission to photocopy is not required from the publisher.

ISBN-13 978-981-279-334-8
ISBN-10 981-279-334-8

Printed in Singapore.

Preface

Functions are used to describe natural processes and forms. By means of finite or infinite operations, we may build many types of 'derived' functions such as the sum of two functions, the composition of two functions, the derivative function of a given function, the power series functions, etc.

Yet a large number of natural processes and forms are not explicitly given by nature. Instead, they are 'implicitly defined' by the laws of nature. Therefore we have functional equations (or more generally relations) involving our unknown functions and their derived functions.

When we are given one such functional equation as a mathematical model, it is important to try to find some or all solutions, since they may be used for prediction, estimation and control, or for suggestion of alternate formulation of the original physical model. In this book, we are interested in finding solutions that are 'polynomials of infinite order', or more precisely, power series functions.

There are many reasons for trying to find such solutions. First of all, it is sometimes 'obvious' from experimental observations that we are facing with natural processes and forms that can be described by 'smooth' functions such as power series functions. Second, power series functions are basically 'generated by' sequences of numbers, therefore, they can easily be manipulated, either directly, or indirectly through manipulations of sequences. Indeed, finding power series solutions are not more complicated than solving recurrence relations or difference equations. Solving the latter equations may also be difficult, but in most cases, we can 'calculate' them by means of modern digital devices equipped with numerical or symbolic packages! Third, once formal power series solutions are found, we are left with the convergence or stability problem. This is a more complicated problem which is not completely solved. Fortunately, there are now several standard techniques which have been proven useful.

In this book, basic tools that can be used to handle power series functions and analytic functions will be given. They are then applied to functional equations in which derived functions such as the derivatives, iterates and compositions of the unknown functions are involved. Although there are numerous functional equations in the literature, our main objective is to show by introductory examples how analytic solutions can be derived in relatively easy manners.

To accomplish our objective, we keep in mind that this book should be suitable for the senior and first graduate students as well as anyone who is interested in a quick introduction to the frontier of related research. Only basic second year advanced engineering mathematics such as the theory of a complex variable and the theory of ordinary differential equations are required, and a large body of seemingly unrelated knowledge in the literature is presented in an integrated and unified manner.

A synopsis of the contents of the various chapters follows.

- The book begins with an elementary example in Calculus for motivation. Basic definitions, symbols and results are then introduced which will be used throughout the book.
- In Chapter 2, various types of sequences are introduced. Common operations among sequences are then presented. In particular, scalar, term by term, convolution and composition products and their properties are discussed in detail. Algebraic derivation is also introduced.
- Power series functions are treated as generating functions of sequences and their relations are fully discussed. Stability properties are discussed and Cauchy's majorant method is introduced. The Siegel's lemma is an important tool in deriving majornats.
- In Chapter 4, the basic implicit function theorem for analytic functions is proved by Newton's binomial expansion theorem. Schröder and Poincaré type implicit functions together with several others are discussed. Application of the implicit theorems for finding power series solutions of polynomial or rational type functional equations are illustrated.
- In Chapter 5 analytic solutions for several classic ordinary differential equations or systems are derived. The Cauchy-Kowalewski existence theorem for partial differential equations is treated as an application. Then several selected functional differential equations are discussed and their analytic solutions found.
- In Chapter 6 analytic solutions for functional equations involving iterates of the unknown functions (or more general composition with other known functions) are treated. These equations are distinguished by whether derivatives of the unknown functions are involved. The last section is concerned with the existence of power solutions.

Some of the material in this book is based on classical theory of analytic functions, and some on theory of functional equations. However, a large number of material is based on recent research works that have been carried out by us and a number of friends and graduate students during the last ten years.

Our thanks go to J. G. Si, X. P. Wang, T. T. Lu and J. J. Lin for their hard works and comments. We would also like to remark that without the indirect help of many other people, this book would never have appeared.

We tried our best to eliminate any errors. If there are any that have escaped our attention, your comments will be much appreciated. We have also tried our best to rewrite all the material that we draw from various sources and cite them in our notes sections. We beg your pardon if there are still similarities left unattended or if there are any original sources which we have missed.

Sui Sun Cheng and Wenrong Li

Contents

Preface   v

1. Prologue   1
   - 1.1 An Example 1
   - 1.2 Basic Definitions 2
   - 1.3 Notes 9

2. Sequences   11
   - 2.1 Lebesgue Summable Sequences 11
   - 2.2 Relatively Summable Sequences 18
   - 2.3 Uniformly Summable Sequences 21
   - 2.4 Properties of Univariate Sequences 25
     - 2.4.1 Common Sequences 25
     - 2.4.2 Convolution Products 26
     - 2.4.3 Algebraic Derivatives and Integrals 32
     - 2.4.4 Composition Products 34
   - 2.5 Properties of Bivariate Sequences 42
   - 2.6 Notes 47

3. Power Series Functions   49
   - 3.1 Univariate Power Series Functions 49
   - 3.2 Univariate Analytic Functions 56
   - 3.3 Bivariate Power Series Functions 63
   - 3.4 Bivariate Analytic Functions 67
   - 3.5 Multivariate Power Series and Analytic Functions 68
   - 3.6 Matrix Power Series and Analytic Functions 71
   - 3.7 Majorants 72
   - 3.8 Siegel's Lemma 77
   - 3.9 Notes 82

4. Functional Equations without Differentiation   83

    4.1 Introduction 83
    4.2 Analytic Implicit Function Theorem 86
    4.3 Polynomial and Rational Functional Equations 90
    4.4 Linear Equations 100
        4.4.1 Equation I 100
        4.4.2 Equation II 102
        4.4.3 Equation III 103
        4.4.4 Equation IV 105
        4.4.5 Equation V 107
        4.4.6 Schröder and Poincaré Equations 110
    4.5 Nonlinear Equations 114
    4.6 Notes 121

5. Functional Equations with Differentiation   123

    5.1 Introduction 123
    5.2 Linear Systems 124
    5.3 Neutral Systems 128
    5.4 Nonlinear Equations 133
    5.5 Cauchy-Kowalewski Existence Theorem 139
    5.6 Functional Equations with First Order Derivatives 141
        5.6.1 Equation I 142
        5.6.2 Equation II 143
        5.6.3 Equation III 145
        5.6.4 Equation IV 147
        5.6.5 Equation V 148
        5.6.6 Equation VI 150
    5.7 Functional Equations with Higher Order Derivatives 152
        5.7.1 Equation I 153
        5.7.2 Equation II 154
        5.7.3 Equation III 156
        5.7.4 Equation IV 166
    5.8 Notes 170

6. Functional Equations with Iteration   175

    6.1 Equations without Derivatives 175
        6.1.1 Babbage Type Equations 176
        6.1.2 Equations Involving Several Iterates 182
        6.1.3 Equations of Invariant Curves 190
    6.2 Equations with First Order Derivatives 197
        6.2.1 Equation I 198
        6.2.2 Equation II 202

		6.2.3	Equation III	206

        6.2.3   Equation III 206
        6.2.4   Equation IV 212
        6.2.5   First Order Neutral Equation 214
  6.3   Equations with Second Order Derivatives 222
        6.3.1   Equation I 223
        6.3.2   Equation II 230
        6.3.3   Equation III 235
        6.3.4   Equation IV 240
  6.4   Equations with Higher Order Derivatives 244
        6.4.1   Equation I 247
        6.4.2   Equation II 249
  6.5   Notes 257

Appendix A Univariate Sequences and Properties 259

  A.1   Common Sequences 259
  A.2   Sums and Products 260
  A.3   Quotients 261
  A.4   Algebraic Derivatives and Integrals 261
  A.5   Tranformations 262
  A.6   Limiting Operations 263
  A.7   Operational Rules 263
  A.8   Knowledge Base 266
  A.9   Analytic Functions 267
  A.10 Operations for Analytic Functions 267

Bibliography 271

Index 283

Chapter 1

Prologue

1.1 An Example

As an elementary but motivating example, let $y(t)$ be the cash at hand of a corporation at time $t \geq 0$. Suppose the corporation invests its cash into a project which guarantees a positive interest rate r so that

$$\frac{dy}{dt} = ry, \ t \geq 0. \tag{1.1}$$

What is the cash at hand of the corporation at any time $t > 0$ given that $y(0) = 1$?

One way to solve this problem in elementary analysis is to assume that $y = y(t)$ is a "power series function" of the form

$$y(t) = a_0 + a_1 t + a_2 t^2 + a_3 t^3 + \cdots,$$

then we have

$$a_0 = y(0) = 1.$$

By formally operating the power series $y(t)$ term by term, we further have

$$y'(t) = a_1 + 2a_2 t + 3a_3 t^2 + \cdots,$$

and

$$ry(t) = ra_0 + ra_1 t + ra_2 t^2 + \cdots.$$

In view of (1.1), we see that

$$a_1 + 2a_2 t + 3a_3 t^2 + \cdots \equiv ra_0 + ra_1 t + ra_2 t^2 + \cdots.$$

By comparing coefficients on both sides, we may proceed formally and write

$$a_1 = r, \ 2a_2 = ra_1, \ 3a_3 = ra_2, \ ...,$$

This yields

$$a_1 = r, \ a_2 = \frac{r^2}{2}, \ a_3 = \frac{r^3}{3 \cdot 2}, \ ..., \ a_n = \frac{r^n}{n!}, \ ...,$$

so that

$$y(t) = 1 + rt + \frac{r^2}{2!} t^2 + \frac{r^3}{3!} t^3 + \cdots, \tag{1.2}$$

which is a "formal power series function".

In order that the formal solution (1.2) is a true solution, we need either to show that $y(t)$ is meaningful on $[0, \infty)$ and that the operations employed above are legitimate, or, we may show that $y(t)$ is equal to some previously known function and show that this function satisfies (1.1) and $y(0) = 1$ directly. If these can be done, then a power series solution exists and is given by (1.2).

Such solutions often reveal important quantitative as well as qualitative information which can help us understand the complex behavior of the physical systems represented by these equations.

In this book, we intend to provide some elementary properties of power series functions and its applications to finding solutions of equations involving unknown functions and/or their associated functions such as their iterates and derivatives.

1.2 Basic Definitions

Basic concepts from real and complex analysis and the theory of linear algebra will be assumed in this book. For the sake of completeness, we will, however, briefly go through some of these concepts and their related information. We will also introduce here some common notations and conventions which will be used in this book.

First of all, sums and products of a set of numbers are common. However, empty sums or products may be encountered. In such cases, we will adopt the convention that an empty sum is taken to be zero, while an empty product will be taken as one.

The union of two sets A and B will be denoted by $A \cup B$ or $A + B$, their intersection by $A \cap B$ or $A \cdot B$, their difference by $A \backslash B$, and their Cartesian product by $A \times B$. The notations $A^2, A^3, ...$, stand for the Cartesian products $A \times A$, $A \times A \times A$, ..., respectively. It is also natural to set $A^1 = A$. The number of elements in a set Ω will be denoted by $|\Omega|$.

The set of real numbers will be denoted by $\mathbf{R}$, the set of all complex numbers by $\mathbf{C}$, the set of integers by $\mathbf{Z}$, the set of positive integers by $\mathbf{Z}^+$, and the set of nonnegative integers by $\mathbf{N}$. We will also use $\mathbf{F}$ to denote either $\mathbf{R}$ or $\mathbf{C}$.

It is often convenient to extend the real number system by the addition of two elements, ∞ (which may also be written as $+\infty$) and $-\infty$. This enlarged set $[-\infty, \infty]$ is called the set of extended real numbers. In addition to the usual operations involving the real numbers, we will also require $-\infty < x < \infty$, $x + \infty = \infty$, $x - \infty = -\infty$ and $x/\infty = 0$ for $x \in \mathbf{R}$; $x \cdot \infty = \infty$ and $x \cdot -\infty = -\infty$ for $x > 0$; and

$$\infty + \infty = \infty, \ -\infty - \infty = -\infty, \ \infty \cdot (\pm\infty) = \pm\infty, \ -\infty \cdot (\pm\infty) = \mp\infty, \ 0 \cdot \infty = 0.$$

In the sequel, the equation

$$\frac{1}{u} = v$$

will be met where $v \in [0, \infty]$. The solution u will be taken as ∞ if $v = 0$ and as 0 if $v = \infty$.

The imaginary number $\sqrt{-1}$ in $\mathbf{C}$ will be denoted by $\mathbf{i}$. The symbols $0!$ and 0^0 will be taken as 1. Given a complex number z and an integer n, the n-th power of z is defined by $z^0 = 1$, $z^{n+1} = z^n z$ if $n \geq 0$ and $z^{-n} = (z^{-1})^n$ if $z \neq 0$ and $n > 0$.

Recall also that for any complex number $z = x + \mathbf{i}y$ where $x, y \in \mathbf{R}$, its real part is $\Re(z) = x$, its imaginary part is $\Im(z) = y$, its conjugate is $z^* = x - \mathbf{i}y$ and its modulus or absolute value is $|z| = (x^2 + y^2)^{1/2}$. We have $|z+w| \leq |z| + |w|$, $|zw| = |z||w|$ and $(zw)^* = z^* w^*$ for any $z, w \in \mathbf{C}$.

Given a nonzero $z = x + \mathbf{i}y \in \mathbf{C}$, if we let θ be the angle measured from the positive x-axis to the line segment joining the origin and the point (x, y), then we see that
$$z = |z|(\cos\theta + \mathbf{i}\sin\theta).$$
We define an argument of the nonzero z to be any angle $t \in \mathbf{R}$ (which may or may not lie inside $[0, 2\pi)$) for which
$$z = |z|(\cos t + \mathbf{i}\sin t),$$
and we write $\arg z = t$. A concrete choice of $\arg z$ is made by defining $\arg_0 z$ to be that number t_0, called the principal argument, in the range $(-\pi, \pi]$ such that
$$z = |z|(\cos t_0 + \mathbf{i}\sin t_0).$$
We may then write
$$\arg_0(zw) = \arg_0 z + \arg_0 w \pmod{2\pi}.$$

It is also easy to show that for any $z \neq 0$, given any positive integer n, there are exactly n distinct complex numbers $z_0, z_1, ..., z_{n-1}$ such that $z_i^n = z$ for each $i = 0, 1, ..., n-1$. The numbers $z_0, z_1, ..., z_{n-1}$ are called the n-th roots of z. The geometric picture of the n-th roots is very simple: they lie on the circle centered at the origin of radius $|z|^{1/n}$ and are equally spaced on this circle with one of the roots having polar angle $\frac{1}{n} \arg_0 z$.

Given a real or complex number α, and any real or complex valued functions f and g, we define $-f$, αf, $f \cdot g$, and $f + g$ by $(-f)(z) = -f(z)$, $(\alpha f)(z) = \alpha f(z)$, $(f \cdot g)(z) = f(z)g(z)$ and $(f + g)(z) = f(z) + g(z)$ as usual, while $|f|$ is defined by $|f|(z) = |f(z)|$. If no confusion is caused, the product $f \cdot g$ is also denoted by fg.

The zeroth power of a function, denoted by f^0, is defined by $f^0(z) = 1$, while the n-th power, denoted by f^n, is defined by $f^n(z) = (f(z))^n$.

The composition of f and g is denoted by $f \circ g$. The iterates of f are formally defined by $f^{[0]}(z) = z$, $f^{[1]}(z) = f(z)$, $f^{[2]}(z) = f(f(z))$, ..., and $f^{[n]}$ is called the n-th iterate of f. Note that $f^{[n]}$ may not be defined if the range of $f^{[n-1]}$ does not lie inside the domain of f.

The n-th derivative of a function is defined by
$$f'(z) = f^{(1)}(z) = \lim_{w \to 0} \frac{f(z+w) - f(z)}{w}$$

and $f^{(k)}(z) = (f^{(k-1)})'(z)$ for $k \geq 2$. As is customary, we will also define $f^{(0)}(z) = f(z)$.

Example 1.1. Recall that the identity function $f : \mathbf{F} \to \mathbf{F}$ defined by $f(t) = t$ for each $t \in \mathbf{F}$ is a polynomial function, so is any constant function $g : \mathbf{F} \to \mathbf{F}$ defined by $g(t) = c \in \mathbf{F}$. Any finite addition or multiplication of polynomial functions is also a polynomial function. For instance,

$$p(t) = c_0 + c_1 t + c_2 t^2 + \cdots + c_m t^m, \quad c_0, ..., c_m \in \mathbf{F},$$

is a polynomial. In case a polynomial is obtained by finite addition or multiplication of the identity function and nonnegative (positive) constant functions, it is called a polynomial with nonnegative (positive) coefficients.

Example 1.2. The previous example defines polynomials with real or complex independent variable. Polynomials with a function as the independent variable can also be defined. More specifically, let f be a complex valued function. Given a polynomial $p(t)$, formally 'replacing' each t^i by the i-th power f^i of f will result in a polynomial in f, which is denoted by $p(f)$. For instance, given

$$p(t) = c_0 + c_1 t + c_2 t^2 + \cdots + c_m t^m, \quad c_0, ..., c_m \in \mathbf{F},$$

we have

$$p(f) = c_0 f^0 + c_1 f^1 + c_2 f^2 + \cdots + c_m f^m, \quad c_0, ..., c_m \in \mathbf{F}.$$

Note that $p(f)$ is a function such that

$$p(f)(z) = c_0 f^0(z) + c_1 f^1(z) + c_2 f^2(z) + \cdots + c_m f^m(z)$$
$$= c_0 + c_1 f(z) + c_2 (f(z))^2 + \cdots + c_m (f(z))^m.$$

Another way to generate polynomials in f is to formally replace each t^i by the i-th iterate $f^{[i]}$ of f, resulting in $p[f]$. For instance, let p be the same polynomial above, then

$$p[f] = c_0 f^{[0]} + c_1 f^{[1]} + \cdots + c_m f^{[m]}, \quad c_0, ..., c_m \in \mathbf{F}.$$

As an example, let M be an n by n complex matrix, and $f(u) = Mu$ where $u \in \mathbf{C}^n$, then $f^{[0]} u = u$, $f^{[k]}(u) = M^k u$ for $k = 1, 2, ..., m$. Hence

$$p[f] = c_0 I + c_1 M + c_2 M^2 + \cdots + c_m M^m.$$

Example 1.3. Polynomials in several real or complex variables can also be defined in similar manners. More specifically, for each $i = 1, ..., \kappa$, let the projection function $f_i : \mathbf{F}^\kappa \to \mathbf{F}$ be defined by $f_i(t_1, t_2, ..., t_\kappa) = t_i$. Projection functions and constant functions are polynomials. Any finite addition or multiplication of polynomial functions is also a polynomial function. For instance,

$$p(t_1, t_2) = c_{00} + c_{10} t_1 + c_{01} t_2 + c_{20} t_1^2 + c_{11} t_1 t_2 + c_{02} t_2^2 + \cdots + c_{0m} t_2^m$$

is a polynomial in (t_1, t_2).

Example 1.4. The quotient of two polynomials is a rational function and is defined whenever its denominator is not zero. Any finite linear combination, products or quotients of rational functions are also rational functions.

Example 1.5. The exponential function exp of *a complex variable* is defined by
$$\exp(z) = e^x(\cos y + \mathbf{i}\sin y)$$
for each $z = x+\mathbf{i}y \in \mathbf{C}$. The value $\exp(z)$ is also written as e^z. Note that $e^x = \exp(x)$ for $x \in \mathbf{R}$ and $e^{\mathbf{i}y} = \cos y + \mathbf{i}\sin y$ for $y \in \mathbf{R}$. Furthermore, the function exp is $2\pi\mathbf{i}$-periodic and maps the strip $\{z \in \mathbf{C}|\ -\pi < \Im(z) \leq \pi\}$ one-to-one onto $\mathbf{C}\backslash\{0\}$.

Example 1.6. The logarithm function of a real variable is
$$\ln(x) = \int_1^x \frac{1}{t}dt,\ x > 0,$$
and the exponential function exp of *a real variable* is defined to be the inverse function of log. Thus $y = \exp(x)$ if $x = \ln(y)$. If z is a nonzero complex number, then there exist complex numbers w such that $e^w = z$. We define $\log z$ to be any number w such that $e^w = z$. Therefore
$$\log z = \ln|z| + \mathbf{i}\arg z,\ z \neq 0.$$
Note that one such w is the complex number $w = \ln(|z|) + \mathbf{i}\arg_0(z)$ and any other such w must have the form
$$\ln(|z|) + \mathbf{i}\arg_0(z) + 2\pi n\mathbf{i},\ n \in \mathbf{Z}.$$
The complex number $\ln(|z|) + \mathbf{i}\arg_0(z)$ will be called the principal logarithm of z and denoted by $\log_0(z)$. Thus the function $\log_0$ defined on $\{z \in \mathbf{C}|\ -\pi < \Im(z) \leq \pi\}$ is the inverse of exp.

Example 1.7. If $z, w \in \mathbf{C}$ and $z \neq 0$, we define
$$z^w = e^{w \log_0(z)}.$$
Note that if $n \in \mathbf{Z}$, then $z^0 = e^0 = 1$ and $z^{n+1} = e^{(n+1)\log_0(z)} = e^{n\log(z)}e^{\log_0(z)} = z^n z$ so that our definition here is compatible with the definition of the n-th power of z. Also, since
$$(z^{1/n})^n = \left(e^{\frac{1}{n}\log_0(z)}\right)^n = e^{\log_0(z)} = z,\ z \neq 0, n \in \mathbf{Z}^+,$$
$z^{1/n}$ is an n-th root of z.

Example 1.8. Some elementary functions are defined in terms of the exponential function:
$$\sin z = \frac{1}{2\mathbf{i}}\{e^{\mathbf{i}z} - e^{-\mathbf{i}z}\},\ \cos z = \frac{1}{2}\{e^{\mathbf{i}z} + e^{-\mathbf{i}z}\},$$
$$\sinh z = \frac{1}{2}\{e^z - e^{-z}\},\ \cosh z = \frac{1}{2}\{e^z + e^{-z}\}.$$
Note that when z is real, these functions coincide with the usual definitions of cosine, sine, hyperbolic sine and hyperbolic cosine. Basic properties of these functions can be found in standard text books.

A (univariate) sequence is a function defined over a set S of (usually consecutive) integers, and can be denoted by $\{u_k\}_{k \in S}$ or $\{u(k)\}_{k \in S}$. When S is finite and, say, equals $\{1, 2, ..., n\}$, a sequence is also denoted by $\{u_1, ..., u_n\}$. Bivariate or multivariate sequences are functions defined on subsets Ω of $\mathbf{Z}^2$ or $\mathbf{Z}^\kappa$ respectively. There are many different ways to denote bivariate or double sequences. One way is to denote a bivariate sequence by $\{u_{i,j}\}$. However, we may also denote it by $\{u_{ij}\}$ if no confusion is caused. Another way is by $\{u_i^{(j)}\}$. In general, when the independent variables have different interpretations, the latter notation is employed. For instance, $u_i^{(t)}$ may represent the temperature of a mass placed at the integral position i and in the time period t. For multivariate sequences, it is cumbersome to denote them by writing $\{u_{i,j,...,k}\}$. For this reason, we may employ the following device. First, an element in a subset of $\Omega \subseteq \mathbf{Z}^\kappa$ has the form $v = (v_1, v_2, ..., v_\kappa)$. Therefore, we may write $\{u_v\}_{v \in \Omega}$ for a multivariate sequence, and v is naturally called a *multi-index*. When v is treated as a multi-index, it will be convenient to use the standard notation $|v|_1 = v_1 + v_2 + \cdots + v_\kappa$, and $v! = v_1! v_2! \cdots v_\kappa!$. $|v|_1$ is usually called the order of v.

It will be necessary to list the components of a sequence in a linear order. For this purpose, we will order the multi-indices in a linear fashion. We say that a mapping $\Psi : \mathbf{N} \to \Omega \subseteq \mathbf{Z}^\kappa$ is an ordering for Ω if Ψ is one to one and onto. For example, let $\Omega = \mathbf{N} \times \mathbf{N}$, a well known ordering for Ω is the mapping $\tilde{\Psi}$ defined by

$$\tilde{\Psi}(0) = (0,0), \ \tilde{\Psi}(1) = (1,0), \ \tilde{\Psi}(2) = (0,1), \ \tilde{\Psi}(3) = (2,0), \\ \tilde{\Psi}(4) = (1,1), \ \tilde{\Psi}(5) = (0,2), \ ... \tag{1.3}$$

In terms of an ordering Ψ for Ω, a rearrangement or enumeration of a multivariate sequence $\{f_v\}_{v \in \Omega}$ is the sequence $\{g_i\}_{i \in \mathbf{N}}$ such that $g_i = f_{\Psi(i)}$.

The notation l^Ω will denote the set of all real or complex sequences defined on Ω. In particular, $l^\mathbf{N}$ denotes the set of all real or complex sequences of the form $\{f_k\}_{k \in \mathbf{N}}$. We will call f_k the k-th term of the sequence f. There are several common sequences in l^N which will be useful. First, for each $m \in \mathbf{N}$, $\hbar^{\langle m \rangle} \in l^\mathbf{N}$ denotes the Dirac sequence defined by

$$\hbar_k^{\langle m \rangle} = \begin{cases} 1 & k = m \\ 0 & k \neq m \end{cases},$$

and $\mathbf{H}^{(m)} \in l^\mathbf{N}$ denotes the jump (or Heaviside) sequence defined by

$$\mathbf{H}_k^{(m)} = \begin{cases} 0 & 0 \leq k < m \\ 1 & k \geq m \end{cases}.$$

Let $\alpha \in \mathbf{F}$, the sequence $\{\alpha, 0, 0, ...\}$ will be denoted by $\overline{\alpha}$ and is called a *scalar sequence*, and the geometric sequence $\{1, \alpha, \alpha^2, \alpha^3, ...\}$ will be denoted by $\underline{\alpha}$. Thus $\underline{z}_n = z^n$ for $n \in \mathbf{N}$. The sequence $\{0, 0, ...\}$ can be denoted by $\overline{0}$ (but it is also commonly denoted by 0), and $\{1, 0, 0, ...\}$ can be denoted by $\overline{1}$ or $\hbar^{\langle 0 \rangle}$. The 'summation' sequence $\{1, 1, 1, ...\}$ will be denoted by $\boldsymbol{\sigma}$ which is equal to $\mathbf{H}^{(0)}$, and the 'difference' sequence $\{1, -1, 0, 0, ...\}$ by δ. The sequence $\{1/0!, 1/1!, 1/2!, 1/3!, 1/4!, ...\}$ will be denoted by ϖ. It is also convenient to write $\hbar$ instead of $\hbar^{\langle 1 \rangle}$ and this practice will be assumed for similar situations in the sequel.

For any $z \in \mathbf{F}$, the sequence $\lfloor z \rfloor \in l^{\mathbf{N}}$ is defined by

$$\lfloor z \rfloor = \{1, z, z(z-1), z(z-1)(z-2), \ldots\}.$$

Thus $\lfloor z \rfloor_0 = 1$ and

$$\lfloor z \rfloor_m = z(z-1)(z-2)\cdots(z-m+1), \ m \in \mathbf{Z}^+.$$

Note that $\lfloor n \rfloor_n = n!$, $\lfloor 0 \rfloor_0 = 0! = 1$, and $0 = \lfloor n \rfloor_{n+1} = \lfloor n \rfloor_{n+2} = \cdots$ for $n \in \mathbf{N}$. Therefore, the sequence $\{1, -3, 3, -1, 0, 0, \ldots\}$ can be written as $\{(-1)^k \lfloor 3 \rfloor_k / k!\}_{k \in \mathbf{N}}$, and the sequence $\{1, 2, 3, \ldots\}$ as $\{\lfloor k+1 \rfloor_1\}_{k \in \mathbf{N}}$.

For any $z \in \mathbf{F}$, the binomial sequence $C^{(z)} \in l^{\mathbf{N}}$ is defined by

$$C^{(z)} = \lfloor z \rfloor \cdot \varpi = \left\{\frac{1}{0!}, \frac{z}{1!}, \frac{z(z-1)}{2!}, \frac{z(z-1)(z-2)}{3!}, \ldots\right\}$$

so that $C_0^{(z)} = 1$ and

$$C_m^{(z)} = \frac{z(z-1)\cdots(z-m-1)}{m!}, \ m \in \mathbf{N}, z \in \mathbf{C}.$$

In particular, for $i, j \in \mathbf{N}$ such that $j \leq i$, $C_j^{(i)}$ is the usual binomial coefficient.

A real function (including a real sequence, a real matrix, etc.) f is said to be nonnegative if $f(x) \geq 0$ for each x in its domain of definition. In such a case, we write $f \geq 0$. Similarly, given two real functions with a common domain of definition Ω, we say that f is less than or equal to g if $f(x) \leq g(x)$ for each $x \in \Omega$. The corresponding notation is $f \leq g$. Other monotonicity concepts for real functions (such as $f < g$, $f > 0$, etc.) are similarly defined.

The product set $\mathbf{F}^\kappa$, where κ is a positive integer, is assumed to be equipped with the usual vector operations and the usual Euclidean topology. In particular, the distance between two points $w = (w_1, \ldots, w_\kappa)$ and $z = (z_1, \ldots, z_\kappa)$ in $\mathbf{F}^\kappa$ is defined by

$$|w - z| = \left\{|w_1 - z_1|^2 + \cdots + |w_\kappa - z_\kappa|^2\right\}^{1/2}.$$

If $r > 0$ and $c = (c_1, \ldots, c_\kappa) \in \mathbf{F}^\kappa$, we will set

$$B(c; r) = \{z \in \mathbf{F}^\kappa |\ |z - c| < r\}$$

$$\bar{B}(c; r) = \{z \in \mathbf{F}^\kappa |\ |z - c| \leq r\},$$

and

$$B'(c; r) = \{z \in \mathbf{F}^\kappa |\ 0 < |z - c| < r\}.$$

They are usually called the open ball, the closed ball and the punctured ball respectively with center at c and radius r. It is well known that the set of all open balls can be used to generate the Euclidean topology for $\mathbf{F}^\kappa$. In particular, a subset Ω of

$\mathbf{F}^\kappa$ is said to be open if every point in Ω is the center of an open ball lying inside Ω.

Besides the open balls, polycylinders are also natural in future considerations. By a polycylinder of polyradius $\rho = (\rho_1, \rho_2, ..., \rho_\kappa)$, where $\rho_1, ..., \rho_\kappa > 0$, and polycenter $w = (w_1, w_2, ..., w_\kappa) \in \mathbf{F}^\kappa$, we mean the set

$$\{(z_1, ..., z_\kappa) \in \mathbf{F}^\kappa |\ |z_j - w_j| < \rho_j,\ 1 \leq j \leq \kappa\}.$$

We remark that the boundary of the above polycylinder is described by the set of inequalities

$$|z_j - w_j| \leq \rho_j,\ 1 \leq j \leq \kappa,$$

whereby at least one equality must hold. Thus for $\kappa = 2$, the boundary consists of those (z_1, z_2) for which

$$|z_1 - w_1| = \rho_1,\ |z_2 - d_2| \leq \rho_2,$$

and those for which

$$|z_1 - w_1| \leq \rho_1,\ |z_2 - d_2| = \rho_2.$$

A subset Ω of $\mathbf{F}^\kappa$ is said to be a domain if it is nonempty, open and pathwise connected (i.e., a nonempty open set such that any two points of which can be joined by a path lying in the set). We remark that a path in Ω from w to z is a continuous function γ from a real interval $[s,t]$ into Ω with $\gamma(s) = w$ and $\gamma(t) = z$. In this case, w and z are the initial and final points of the path.

In terms of the distance d and the open balls, we can then define as usual limits and continuity for complex-valued functions $f = f(z_1, z_2, ..., z_\kappa)$ defined on a domain Ω or a more general subset of $\mathbf{F}^\kappa$, we can also define partial derivatives, etc. More precisely, the limit

$$\lim_{h \to 0} \frac{f(c_1, ..., c_{i-1}, c_i + h, c_{i+1}, ..., c_\kappa) - f(c_1, ..., c_\kappa)}{h},$$

if it exists, is called the i-th partial derivative of f at $(c_1, ..., c_\kappa)$ and is denoted by

$$\frac{\partial f(c_1, ..., c_\kappa)}{\partial z_i}.$$

Higher partial derivatives of the form

$$\frac{\partial^{v_1}}{\partial z_1^{v_1}} \frac{\partial^{v_2}}{\partial z_2^{v_2}} \cdots \frac{\partial^{v_\kappa}}{\partial z_\kappa^{v_\kappa}} f(c_1, ..., c_\kappa)$$

are defined in recursive manners. Multi-indices can also be used to simplify such notations. Such simplifications are convenient and can be seen in our later sections.

We will need to define integrals for functions $f: \mathbf{F} \to \mathbf{F}$. One such integral is the Cauchy (line) integral

$$\int_\Gamma f(z) dz$$

where Γ is a well behaved path. In this book, it suffices to consider paths Γ that are representable by 'piecewise smooth' functions $\gamma : [a,b] \to \mathbf{F}$, that is, there are points $t_0, t_1, ..., t_n$ with $a = t_0 < t_1 < \cdots < t_n = b$ such that γ' is continuous on each $[t_k, t_{k+1}]$ for $k = 0, ..., n-1$. Then by the standard theory of Riemann-Stieltjes integral, when f is continuous on the image $\Gamma([0,1]) \subset \mathbf{F}$,
$$\int_\Gamma f(z)dz = \int_0^1 f(\gamma(t))\gamma'(t)dt.$$
In case Γ is the straight line segment joining the point $u = \gamma(a)$ to $v = \gamma(b)$, we will also write
$$\int_\Gamma f(z)dz = \int_u^v f(z)dz.$$
Note that when $\mathbf{F} = \mathbf{R}$, the above line integral is compatible with the usual Riemann integral of a real function.

Recall that Ω is a metric space if there is a metric $d : \Omega \times \Omega \to [0, \infty)$ which satisfies (i) for every pair of $x, y \in \Omega$, $d(x,y) = 0$ if, and only if, $x = y$, (ii) $d(x,y) = d(y,x)$ for $x, y \in \Omega$, and (iii) $d(x,z) \le d(x,y) + d(y,x)$ for $x, y, z \in \Omega$. Ω is said to be complete if every Cauchy sequence in Ω converges to a point in Ω. $T : \Omega \to \Omega$ is a contraction if there is number λ in $[0,1)$ such that $d(Tx, Ty) \le \lambda d(x,y)$ for all $x, y \in \Omega$.

A large number of metric spaces are normed linear spaces, that is, linear spaces whose metrics are induced by norms. Recall that a norm $\|\cdot\|$ on a linear space Ω is a function that maps Ω into $[0, \infty)$ such that (i) for every $x \in \Omega$, $\|x\| = 0$ if, and only if, $x = 0$, (ii) $\|\alpha x\| = |\alpha|\|x\|$ for any scalar α and $x \in \Omega$, and (iii) $\|x + y\| \le \|x\| + \|y\|$ for $x, y \in \Omega$. When a normed linear space is also a complete metric space, it is called a Banach space.

A well known result for mappings defined on complete metric spaces is the Banach contraction mapping theorem: If Ω is a nonempty complete metric space and $T : \Omega \to \Omega$ a contraction mapping, then T has a fixed point in Ω.

1.3 Notes

There are several standard reference books on functional equations, see for examples, the books by Aczel [1], Aczel and Dhombres [2], Kuczma [104], Kuczma and Choczewshi [107], and the survey papers by Cheng [29], Kuczma [102], Li and Si [126], Zhang et al. [232]. In this book, we also treat differential equations as functional equations. The corresponding references are too many to list. The books by Bellman and Cooke [16], Coddington and Levinson [40], Driver [52], Friedrichs [66], Hale [73], Hille [78], Kamke [92], Sansone [167], etc., are related to some of our discussions.

There are also several text books which emphasize on analytic functions, see for examples, Balser [13], Krantz and Parks [99], Krantz [100], Smith [211], Sneddon [212], Valiron [216].

In this book, we do not use sophisticated mathematics beyond the usual material taught in courses such as Advanced Engineering mathematics. The reader may also consult text books in real and complex analysis such as Apostol [5], Fichtenholz [62, 63], Kaplan [94], Watson [223], Whittaker and Watson [224], etc.

We have introduced univariate sequences and discussed some of their properties. Further properties will be discussed in later chapters. A summary of their properties can be found in the Appendix.

Chapter 2

Sequences

2.1 Lebesgue Summable Sequences

Note that a power series appears to be a 'sum' of infinitely many terms. For this reason, we need to introduce means to deal with infinite sums.

Let Ω be a (finite or infinite) subset of Z^κ where κ is a positive integer. Each member in the set l^Ω of all functions defined on Ω is then a multiply indexed sequence of the form $\{f_k|\ k \in \Omega\}$. Such a sequence will be denoted by f or $\{f_k\}$ or $\{f_k\}_{k\in\Omega}$ instead of $\{f_k|k \in \Omega\}$ if no confusion is caused.

For any $\alpha \in \mathbf{C}$ and $f = \{f_k\}, g = \{g_k\}$ in l^Ω, we define $-f, \alpha f, |f|$ and $f + g$ respectively by $\{-f_k\}, \{\alpha f_k\}, \{|f_k|\}$ and $\{f_k + g_k\}$ as usual. The termwise product $f \cdot g$ is defined to be $\{f_k g_k\}$. The products $f \cdot f, f \cdot f \cdot f, \ldots$ will be denoted by $f^2, f^3, \ldots$ respectively. We define $f^1 = f$ and $f^0 = \{1\}$. The sequence f^p is called the p-th termwise (product) power of f. If $f_k \neq 0$ for all k, then there is a unique sequence $x \in l^\Omega$ such that $x \cdot f = \{1\}$. This unique sequence will be denoted by f^{-1}.

For any $f, g \in l^\Omega$, if $f_k \leq g_k$ for all $k \in \Omega$, then we write $f \leq g$. The notation $f < g$ is similarly defined.

Any sequence with zero values only will be denoted by 0. The sequence in $l^\mathbf{N}$ whose i-th term is 1 and the other terms are 0 will be called the *Dirac delta sequence* and denoted by $\hbar^{\langle i \rangle}$.

For a given real sequence $f = \{f_k\}$, we can always write it in the form $f^+ - f^-$ for some nonnegative sequences f^+ and f^-. Indeed, the positive part f^+ is given by $(|f| + f)/2$, and the negative part by $(|f| - f)/2$. A sequence $f = \{f_k\}$ is said to have finite support if the number of nonzero terms of f is finite. The set $\Phi(f)$ of $k \in \Omega$ for which $f_k \neq 0$ will be called the support of f. When $\{f^{(j)}\}_{j\in\mathbf{N}}$ is a sequence of sequences in l^Ω, we say that $\{f^{(j)}\}_{j\in\mathbf{N}}$ converges (pointwise) to $f \in l^\Omega$ if

$$\lim_{j\to\infty} f_k^{(j)} = f_k,\ k \in \Omega.$$

Note that for any nonnegative sequence $f = \{f_k\} \in l^\Omega$, we can always find a

sequence $\{g^{(j)}\}_{j\in \mathbf{N}}$ of nonnegative sequences in l^Ω such that
$$0 \le g^{(0)} \le g^{(1)} \le \cdots \le f$$
and $g^{(j)}$ converges pointwise to f as $j \to \infty$. For instance, if $f \in l^{\mathbf{N}}$, we may pick
$$g^{(j)} = \sum_{k=0}^{j} f_k \hbar^{\langle k \rangle}, \ j \in \mathbf{N}.$$

The concept of a Lebesgue summable sequence will be needed in order to define a convergent series. This will be done in steps.

First of all, for a sequence f with finite support, we define its sum by the number
$$\sum_\Omega f = \sum_{k \in \Phi(f)} f_k.$$
For a nonnegative sequence $f = \{f_k\}$ in l^Ω, we define its sum by
$$\sup \sum_\Omega g,$$
where the supremum is taken over all sequences g with finite support such that $0 \le g \le f$, and denoted by
$$\sum_\Omega f \ \text{or} \ \sum_{k \in \Omega} f_k.$$
If the supremum on the right hand side is finite, we say that f is (Lebesgue) summable and denote this fact by
$$\sum_\Omega f < \infty.$$
Occasionally, it is convenient to allow the right hand side to be infinite and in such a case, we write
$$\sum_\Omega f = \infty.$$

Note that it easily follows from the definition that a finite linear combination of nonnegative Lebesgue summable sequences is Lebesgue summable and its sum is equal to the corresponding linear combination of the separate sums, that is, for nonnegative $\alpha, \beta \in \mathbf{R}$ and nonnegative $f, g \in l^\Omega$,
$$\sum_\Omega (\alpha f + \beta g) = \alpha \sum_\Omega f + \beta \sum_\Omega g,$$
and that if $0 \le f \le g$, then
$$0 \le \sum_\Omega f \le \sum_\Omega g. \tag{2.1}$$

We remark that the above definition of the sum of a nonnegative sequence in $l^{\mathbf{N}}$ can be simplified to
$$\sum_{\mathbf{N}} f = \lim_{j \to \infty} \sum_{k=0}^{j} f_k; \tag{2.2}$$

in $l^\mathbf{Z}$ to

$$\sum_\mathbf{Z} f = \sup_{m,j\in Z} \sum_{k=m}^{j} f_k = \lim_{m,j\to\infty} \sum_{k=-m}^{j} f_k; \qquad (2.3)$$

and in $l^{\mathbf{N}\times\mathbf{N}}$ to

$$\sum_{\mathbf{N}\times\mathbf{N}} f = \lim_{m,n\to\infty} \sum_{i=0}^{m}\sum_{j=0}^{n} f_{ij}. \qquad (2.4)$$

To see this, for $f \in l^\mathbf{N}$, since

$$\sum_{k=0}^{j} f_k = \sum_\mathbf{N} h,$$

where $h = \{f_0, ..., f_j, 0, 0, ...\}$ is a sequence with finite support, we have

$$\sum_{k=0}^{j} f_k = \sum_\mathbf{N} h \le \sum_\mathbf{N} f.$$

Conversely, for any g such that $0 \le g \le f$ and $\Phi(g)$ is finite, since $\Phi(g) \subseteq \{0, ..., m\}$ for some m, we see that $0 \le g \le u \le f$, where $u = \{f_0, ..., f_m, 0, 0, ...\}$, and

$$\sum_\mathbf{N} g \le \sum_\mathbf{N} u = \sum_{k=0}^{m} f_k \le \lim_{j\to\infty} \sum_{k=0}^{j} f_k.$$

For $f \in l^\mathbf{Z}$ or $f \in l^{\mathbf{N}\times\mathbf{N}}$, (2.3) or (2.4) are similarly proved.

We pause here to recall that for a sequence $f = \{f_k\}_{k\in\mathbf{N}}$ in $l^\mathbf{N}$, the sequence $\left\{\sum_{k=0}^{j} f_k\right\}_{j\in\mathbf{N}}$ is called the partial sum sequence generated by f. The limit $L = \lim_{j\to\infty}\sum_{k=0}^{j} f_k$, if it exists, is usually called the sum of the 'series' $\sum_{k=0}^{\infty} f_k$. For this reason, the (finite or infinite) limits on the right hand side of (2.2) and (2.3) will also be denoted by the conventional notations, that is,

$$\lim_{j\to\infty} \sum_{k=0}^{j} f_k = \sum_{k=0}^{\infty} f_k = \sum_\mathbf{N} f$$

and

$$\lim_{m,j\to\infty} \sum_{k=-m}^{j} f_k = \sum_{k=-\infty}^{\infty} f_k = \sum_\mathbf{Z} f$$

respectively. Limits of partial sum sequences will be discussed in details in the next section.

We remark also that our definition of a sum of infinite sequence is a special case of the Lebesgue integral for measurable functions. Thus standard results from the theory of Lebesgue integrals can be applied. In particular, Lebesgue's monotone convergence theorem holds.

Theorem 2.1 (Lebesgue Monotone Convergence Theorem). *Let $g \in l^\Omega$ and let $\{f^{(j)}\}_{j \in \mathbf{N}}$ be a sequence of nonnegative sequences $f^{(j)} \in l^\Omega$ such that*

$$0 \leq f_k^{(0)} \leq f_k^{(1)} \leq \cdots < \infty, \ k \in \Omega,$$

and

$$\lim_{j \to \infty} f_k^{(j)} = g_k, \ k \in \Omega,$$

then

$$\lim_{j \to \infty} \sum_\Omega f^{(j)} = \sum_\Omega g.$$

Indeed, since

$$0 \leq \sum_\Omega f^{(j)} \leq \sum_\Omega f^{(j+1)} \leq \sum_\Omega g,$$

thus

$$\lim_{j \to \infty} \sum_\Omega f^{(j)} \in [0, \infty]$$

and

$$\lim_{j \to \infty} \sum_\Omega f^{(j)} \leq \sum_\Omega g.$$

To see the converse, let u be a sequence with finite support that satisfies $0 \leq u \leq g$. Let c be a constant in $(0,1)$. Since $f^{(j)} \to g$, we have $f^{(j)} \geq cu$ for all large j. Hence

$$\sum_\Omega f^{(j)} \geq \sum_\Omega cu = c \sum_\Omega u$$

for all large j. Since c and u are arbitrary, we must have

$$\lim_{j \to \infty} \sum_\Omega f^{(j)} \geq \sup \sum_\Omega u = \sum_\Omega g.$$

The proof is complete.

As a corollary, if $\{g^{(j)}\}_{j \in \mathbf{N}}$ is a sequence of nonnegative sequences in l^Ω such that

$$\sum_{j=0}^\infty g_k^{(j)} < \infty, \ k \in \Omega,$$

then

$$0 \leq \sum_{j=0}^0 g_k^{(j)} \leq \sum_{j=0}^1 g_k^{(j)} \leq \sum_{j=0}^2 g_k^{(j)} \leq \cdots < \infty, \ k \in \Omega,$$

and

$$\lim_{m \to \infty} \sum_{j=0}^m g_k^{(j)} = \sum_{j=0}^\infty g_k^{(j)}, \ k \in \Omega.$$

Hence the Lebesgue monotone convergence theorem leads us to

$$\sum_{k\in\Omega}\left\{\sum_{j=0}^{\infty} g_k^{(j)}\right\} = \lim_{n\to\infty}\sum_{k\in\Omega}\left\{\sum_{j=0}^{n} g_k^{(j)}\right\} = \lim_{n\to\infty}\sum_{j=0}^{n}\sum_{\Omega} g^{(j)} = \sum_{j=0}^{\infty}\sum_{\Omega} g^{(j)}, \quad (2.5)$$

where we have used the linearity of the Lebesgue sum in the second equality.

As another corollary, we have Fatou's lemma: If $\{f^{(n)}\}_{n\in\mathbf{N}}$ is a sequence of nonnegative sequences in l^Ω such that

$$\liminf_{n\to\infty} f_k^{(n)} < \infty, \ k \in \Omega,$$

then

$$\sum_\Omega \liminf_{n\to\infty} f^{(n)} \le \liminf_{n\to\infty} \sum_\Omega f^{(n)}.$$

To see this, let $h^{(m)} = \left\{h_k^{(m)}\right\}_{k\in\Omega}$ be defined by $h_k^{(m)} = \inf_{n\ge m} f_k^{(n)}$ for each $k \in \Omega$ and $m \ge 0$. Then $0 \le h_k^{(0)} \le h_k^{(1)} \le \cdots \le h_k^{(m)} \le f_k^{(m)}$ for each $k \in \Omega$ and $m \ge 0$, and

$$\lim_{m\to\infty} h_k^{(m)} = \liminf_{n\to\infty} f_k^{(n)} < \infty, \quad k \in \Omega,$$

so that

$$\sum_\Omega \liminf_{n\to\infty} f^{(n)} = \sum_\Omega \lim_{m\to\infty} h^{(m)} = \lim_{m\to\infty} \sum_\Omega h^{(m)}$$

$$= \liminf_{m\to\infty} \sum_\Omega h^{(m)} \le \liminf_{m\to\infty} \sum_\Omega f^{(m)}.$$

We have mentioned that any discrete set Ω in $\mathbf{Z}^\kappa$ can be linearly ordered. Note however, that for each linear ordering, the corresponding sum of a sequence defined over Ω may be different from the one that arises from another linear ordering. Fubini's theorem states, however, that such cannot be the case. We will state Fubini's theorem for $\Omega = \mathbf{N} \times \mathbf{N}$, the general case being similar. Recall first that $\{g_k\}_{k\in\mathbf{N}}$ is called an enumeration or rearrangement of the sequence $\{f_v\}_{v\in\Omega}$ if there is a linear ordering $\Psi : \mathbf{N} \to \Omega$ such that $g_k = f_{\Psi(k)}$.

Theorem 2.2 (Fubini Theorem). *Suppose $\{g_k\}_{k\in\mathbf{N}}$ is any enumeration of the nonnegative doubly indexed sequence $\{f_{ij}\}_{i,j\in\mathbf{N}}$. Then $\{g_k\}_{k\in\mathbf{N}}$ is Lebesgue summable if, and only if,*

$$\sum_{j=0}^{\infty} f_{ij} < \infty \ \text{for } i \in \mathbf{N}, \ \text{and} \ \sum_{i=0}^{\infty}\left\{\sum_{j=0}^{\infty} f_{ij}\right\} < \infty; \quad (2.6)$$

moreover, if $\{g_k\}_{k\in\mathbf{N}}$ is Lebesgue summable, then

$$\sum_\mathbf{N} g = \sum_{i=0}^{\infty}\left\{\sum_{j=0}^{\infty} f_{ij}\right\}. \quad (2.7)$$

For a proof, let us first assume that (2.6) holds. Let M be any integer and choose integers I and J so large that $g_1, ..., g_M$ occur among $\{f_{ij}|\ 0 \le i \le I, 0 \le j \le J\}$. Then

$$\sum_{m=0}^{M} g_k \le \sum_{i=0}^{I} \sum_{j=0}^{J} f_{ij} \le \sum_{i=0}^{\infty} \left\{ \sum_{j=0}^{\infty} f_{ij} \right\} < \infty.$$

This shows that g is Lebesgue summable.

Conversely, assume that g is Lebesgue summable. Let J be an integer and, for a fixed $i \in \mathbf{N}$, choose the integer M so large that $f_{i1}, ..., f_{iJ}$ occur among $g_1, ..., g_M$. Then

$$\sum_{j=0}^{J} f_{ij} \le \sum_{k=0}^{M} g_k \le \sum_{k=0}^{\infty} g_k,$$

which implies

$$\sum_{j=0}^{\infty} f_{ij} < \infty,\ i \in \mathbf{N}.$$

Now let $\{w^{(n)}\}_{n \in \mathbf{N}}$ be a sequence of nonnegative sequences in $l^{\mathbf{N} \times \mathbf{N}}$ each of which has finite support and $0 \le w^{(0)} \le w^{(1)} \le \cdots \le f$ as well as $\lim_{n \to \infty} w_k^{(n)} = f_k$ for $k \in \mathbf{N}$. For each $n \in \mathbf{N}$, let $v^{(n)}$ be the corresponding enumeration of $w^{(n)}$. Then since

$$\sum_{\mathbf{N}} v^{(n)} = \sum_{i=0}^{\infty} \sum_{j=0}^{\infty} w_{ij}^{(n)},$$

and since

$$\sum_{j=0}^{\infty} w_{ij}^{(n)} \le \sum_{j=0}^{\infty} f_{ij} < \infty,\ i \in \mathbf{N},$$

we may apply Lebesgue's monotone convergence theorem to obtain

$$\sum_{\mathbf{N}} g = \lim_{n \to \infty} \sum_{\mathbf{N}} v^{(n)} = \lim_{n \to \infty} \sum_{i=0}^{\infty} \sum_{j=0}^{\infty} w_{ij}^{(n)} = \sum_{i=0}^{\infty} \lim_{n \to \infty} \sum_{j=0}^{\infty} w_{ij}^{(n)} = \sum_{i=0}^{\infty} \sum_{j=0}^{\infty} f_{ij},$$

which shows that (2.6) and (2.7) hold.

Let us denote by l_1^Ω the subset of all sequences $f \in l^\Omega$ for which $|f|$ is Lebesgue summable. Let us also denote by l_p^Ω the set of all sequences $f \in l^\Omega$ for which

$$\|f\|_p \equiv \left\{ \sum_\Omega |f|^p \right\}^{1/p} < \infty,\ p \in (0, \infty).$$

The number $\|f\|_p$ is called the l_p^Ω-norm of f, while the infinity norm of f is

$$\|f\|_\infty = \max_{k \in \Omega} \{|f_k|\}.$$

The set of all sequences $f \in l^\Omega$ for which $\|f\|_\infty < \infty$ will be denoted by l^Ω_∞.

Let $f \in l^\Omega_1$. We define its sum by

$$\sum_\Omega f = \sum_\Omega u^+ - \sum_\Omega u^- + \mathbf{i}\sum_\Omega v^+ - \mathbf{i}\sum_\Omega v^-,$$

where $f = u + \mathbf{i}v$, and u^+, v^+, u^-, v^- are the positive parts and negative parts defined before. Note that each of the four sums on the right hand side exists since $0 \le u^+, v^+, u^-, v^- \le |f|$.

If f is a real multivariate sequence (which may or may not be in l^Ω_1), we define its sum by

$$\sum_\Omega f = \sum_\Omega f^+ - \sum_\Omega f^-,$$

provided that at least one of the sums on the right hand side is finite. The left side is then a number in the extended real number system $[-\infty, \infty]$.

Note that it easily follows from the definition of l^Ω_1 that the sum of a finite linear combination of Lebesgue summable sequences in l^Ω_1 is equal to the corresponding linear combination of the separate sums, and that for any $f \in l^\Omega_1$,

$$\left|\sum_\Omega f\right| \le \sum_\Omega |f|. \tag{2.8}$$

Lebesgue's dominated convergence theorem also holds.

Theorem 2.3 (Lebesgue Dominated Convergence Theorem). *Suppose $\{f^{(n)}\}_{n \in \mathbf{N}}$ is a sequence of complex sequences in l^Ω such that $f = \lim_{n\to\infty} f^{(n)} \in l^\Omega$. If there is $g \in l^\Omega_1$ such that $\left|f^{(n)}\right| \le g$ for $n \in \mathbf{N}$, then $f \in l^\Omega_1$,*

$$\lim_{n\to\infty} \sum_\Omega \left|f^{(n)} - f\right| = 0, \tag{2.9}$$

and

$$\lim_{n\to\infty} \sum_\Omega f^{(n)} = \sum_\Omega f. \tag{2.10}$$

Indeed, since $|f| \le g$, and since $\left|f^{(n)} - f\right| \le 2g$, by Fatou's lemma, we see that

$$\sum_\Omega 2g = \sum_\Omega \liminf_{n\to\infty} \left(2g - \left|f^{(n)} - f\right|\right) \le \liminf_{n\to\infty} \sum_\Omega \left(2g - \left|f^{(n)} - f\right|\right)$$

$$= \sum_\Omega 2g - \limsup_{n\to\infty}\left(\sum_\Omega \left|f^{(n)} - f\right|\right).$$

Thus

$$\lim_{n\to\infty} \sum_\Omega \left|f^{(n)} - f\right| = \limsup_{n\to\infty} \sum_\Omega \left|f^{(n)} - f\right| = 0.$$

Finally, (2.10) follows from (2.9) in view of (2.8).

There is also a useful Fubini's theorem for l^Ω_1-sequences.

Theorem 2.4. *Suppose $\{g_k\}_{k\in\mathbf{N}}$ is any enumeration of the doubly indexed sequence $\{f_{ij}\}_{i,j\in\mathbf{N}}$. Then $\{g_k\} \in l_1^{\mathbf{N}}$ if, and only if,*

$$\sum_{j=0}^{\infty} |f_{ij}| < \infty \text{ for } i \in \mathbf{N}, \text{ and } \sum_{i=0}^{\infty} \left\{ \sum_{j=0}^{\infty} |f_{ij}| \right\} < \infty;$$

moreover, if $\{g_k\} \in l_1^{\mathbf{N}}$, then

$$\sum_{\mathbf{N}} g = \sum_{i=0}^{\infty} \left\{ \sum_{j=0}^{\infty} f_{ij} \right\}.$$

The proof is not difficult and follows from breaking f into real, complex, positive and negative parts and then applying Fubini's theorem for nonnegative sequences.

Example 2.1. If $f \in l_1^{\mathbf{N}}$ and $g \in l_1^{\mathbf{N}}$, then the bivariate sequence $h = \{f_i g_j\}_{i,j\in\mathbf{N}}$ belongs to $l_1^{\mathbf{N}\times\mathbf{N}}$ and $\sum_{\mathbf{N}\times\mathbf{N}} h = \sum_{\mathbf{N}} f \sum_{\mathbf{N}} g$.

Example 2.2. If $a, b \in \mathbf{F}$ and $|a| + |b| < 1$, then from

$$\sum_{k=0}^{n} C_k^{(n)} a^k b^{n-k} = (a+b)^n, \ n \in \mathbf{N},$$

we see that

$$\sum_{n=0}^{\infty} \sum_{k=0}^{n} C_k^{(n)} |a|^k |b|^{n-k} = \frac{1}{1 - |a| - |b|} < \infty.$$

Thus

$$\sum_{k=0}^{\infty} \sum_{n=k}^{\infty} C_k^{(n)} a^k b^{n-k} = \sum_{n=0}^{\infty} \sum_{k=0}^{\infty} C_k^{(n)} a^k b^{n-k} = \sum_{n=0}^{\infty} \sum_{k=0}^{n} C_k^{(n)} a^k b^{n-k} = \frac{1}{1 - a - b}.$$

2.2 Relatively Summable Sequences

In the previous section, we use suprema to define sums of sequences. Sums of sequences defined by limits of their partial sum sequences are also studied quite extensively. For this reason, we will recall some of the related information in this section. We say that a sequence $\{f_k\}$ in $l^{\mathbf{N}}$ (which is not necessarily nonnegative) is summable if the limit

$$\lim_{n\to\infty} \sum_{k=0}^{n} f_k \qquad (2.11)$$

exists, otherwise we say that f is not summable or the *infinite series* $\sum_{k=0}^{\infty} f_k$ *diverges*. In case it is summable, the corresponding limit s is called its sum and we write

$$\sum_{k=0}^{\infty} f_k = s.$$

In case f is nonnegative and its infinite series diverges, we will also write
$$\sum_{k=0}^{\infty} f_k = \infty.$$
Recall that for a **nonnegative** sequence $f = \{f_k\}$ in $l^{\mathbf{N}}$, its Lebesgue sum and the limit of partial sum sequence is equal:
$$\sum_{\mathbf{N}} f = \sum_{k=0}^{\infty} f_k \equiv \lim_{n \to \infty} \sum_{k=0}^{n} f_k.$$

For multivariate sequences, we may generalize the above concept as follows. Let Ω be a subset of $\mathbf{Z}^\kappa$ where κ is a positive integer and let $\Psi : \mathbf{N} \to \Omega$ be an ordering for Ω. Let $f = \{f_v\}_{v \in \Omega}$, we call
$$\sigma_i = \sum_{j=0}^{i} f_{\Psi(j)}$$
the (generalized) partial sum of index i relative to Ψ. If
$$\lim_{i \to \infty} \sigma_i = \lim_{i \to \infty} \sum_{j=0}^{i} f_{\Psi(j)} = s,$$
then we say that the sequence f is summable relative to the ordering Ψ and we say that s is the sum relative to Ψ. We also say that f is relatively summable if f is summable relative to some ordering Ψ when the specific form of the mapping Ψ is not important.

By means of standard analytic arguments, it is easily shown that if $f = \{f_v\}$ and $g = \{g_v\}$ have sums s and t relative to Ψ respectively, then the sum of $\alpha f + \beta g$ relative to Ψ is $\alpha s + \beta t$. In particular, f is summable relative to Ψ if, and only if, its real part and its imaginary part are summable relative to Ψ. Furthermore, for a nonnegative sequence $f = \{f_v\}_{v \in \Omega} \in l^\Omega$, if we take $u^{(k)}$ to be the sequence which is equal to the values of f when restricted to $\Psi(\{0, 1, ..., k\})$ and equal to 0 otherwise, then $0 \leq u^{(0)} \leq u^{(1)} \leq \cdots \leq f$ and $\lim_{k \to \infty} u^{(k)} = f$. By Lebesgue's monotone convergence theorem,
$$\sum_{\Omega} f = \lim_{k \to \infty} \sum_{\Omega} u^{(k)} = \lim_{k \to \infty} \sum_{i=0}^{k} f_{\Psi(i)}$$
for any ordering Ψ of Ω. In particular, for any sequence $f = \{f_v\} \in l^\Omega$ and any ordering Ψ of Ω,
$$\sum_{\Omega} |f| = \lim_{k \to \infty} \sum_{i=0}^{k} |f_{\Psi(i)}|.$$

In case $\lim_{k \to \infty} \sum_{i=0}^{k} |f_{\Psi(i)}|$ exists for any Ψ, we say that f is **absolutely summable** (relative to Ψ). Note that a sequence in l^Ω is absolutely summable relative to any Ψ if, and only if, it belongs to l_1^Ω. Furthermore, if $f \in l_1^\Omega$, then

$$\sum_\Omega f = \sum_\Omega u^+ - \sum_\Omega u^- + \mathbf{i}\sum_\Omega v^+ - \mathbf{i}\sum_\Omega v^-$$

$$= \lim_{k\to\infty} \sum_{i=0}^k u^+_{\Psi(i)} - \lim_{k\to\infty} \sum_{i=0}^k u^-_{\Psi(i)} + \mathbf{i}\lim_{k\to\infty} \sum_{i=0}^k v^+_{\Psi(i)} - \mathbf{i}\lim_{k\to\infty} \sum_{i=0}^k v^-_{\Psi(i)}$$

$$= \lim_{k\to\infty} \sum_{i=0}^k f_{\Psi(i)},$$

where $f = u + \mathbf{i}v$, that is, if f is absolutely summable, then its sum is independent of the ordering Ψ.

Theorem 2.5. *If $f = \{f_v\}_{v\in\Omega}$ is summable relative to an ordering Ψ for Ω, then f is bounded.*

Indeed, since

$$0 = \lim_{i\to\infty}\left\{\sum_{j=0}^{i+1} f_{\Psi(j)} - \sum_{j=0}^{i} f_{\Psi(j)}\right\} = \lim_{i\to\infty} f_{\Psi(i)},$$

thus $|f_{\Psi(i)}| \le M$ for i greater than some integer I. Thus $|f_{\Psi(i)}| \le \max\{|f_{\Psi(0)}|,\ldots|f_{\Psi(I)}|,M\}$ as required.

In the special case when $\Psi: \mathbf{N} \to \mathbf{N}$ is the identity mapping, the sum of $f \in l^{\mathbf{N}}$ relative to Ψ is the limit (2.11) defined above. *For the sake of convenience, we will say that $f \in l^{\mathbf{N}}$ is summable if it is summable relative to the identity mapping.*

There is a large collection of summability criteria such as the root test, integral test, etc., discussed in elementary analysis texts.

Example 2.3. If $f = \{f_k\}, g = \{g_k\} \in l^{\mathbf{N}}$ are summable and their sums are f and g respectively, then

$$\lim_{m,n\to\infty}\left(\sum_{i=0}^m f_i + \sum_{j=0}^n g_j\right) = \sum_{i=0}^\infty f_i + \sum_{j=0}^\infty g_j.$$

Conversely, if $\lim_{m,n\to\infty}\left(\sum_{i=0}^m f_i + \sum_{j=0}^n g_j\right)$ exists, then f and g are summable.

Example 2.4 (Dirichlet Test). Let $\{\sum_{k=0}^n a_k\}_{n\in\mathbf{N}}$ be a bounded sequence and $\{b_n\}_{n\in\mathbf{N}}$ be a decreasing sequence tending to 0. Then $\{a_n b_n\}_{n\in\mathbf{N}}$ is summable.

Example 2.5 (Abel Test). Let $\{a_n\}_{n\in\mathbf{N}}$ be summable and $\{b_n\}_{n\in\mathbf{N}}$ a monotonic and convergent sequence. Then $\{a_n b_n\}_{n\in\mathbf{N}}$ is summable.

2.3 Uniformly Summable Sequences

We first recall the concept of uniformly convergent sequence and series of functions in elementary analysis. Given a sequence of real functions $u_0(x), u_1(x), ...$, defined on a set X, if $u(x) = \lim_{n \to \infty} u_n(x)$ for every $x \in X$, and if given any $\varepsilon > 0$, there is an integer $I \geq 0$ such that $n > I$ implies
$$|u_n(x) - u(x)| < \varepsilon$$
for all $x \in X$, then the sequence is said to converge to f uniformly on X. In particular, if $u_n(x) = f_0(x) + \cdots + f_n(x)$ for each $n \in \mathbf{N}$ and $x \in X$, and if the sequence $\{u_n(x)\}$ is uniformly convergent to $u(x)$ on X, then we say that the series $\sum_{i=0}^{\infty} f_i(x)$ converges uniformly on X to the function $u(x)$. There are a large of number of properties of uniformly convergent sequence of functions and uniformly convergent functional series.

By means of the generalized partial sums introduced in the last section, we can carry some of these properties to functional series with multiple indices. Let Λ be a nonempty set in $\mathbf{F}^\kappa$ and let $f^{(\lambda)} \in l^\Omega$ for each $\lambda \in \Lambda$. We now have a family $\{f^{(\lambda)}\}_{\lambda \in \Lambda}$ of sequences in l^Ω. Since $f^{(\lambda)} = \{f_v^{(\lambda)}\}_{v \in \Omega}$, we may also look at our family as a sequence of functions $f_v = f_v(\lambda)$ defined on Λ. For this reason, we will write $f(\lambda)$ instead of $f^{(\lambda)}$ if no confusion is caused. For each $\lambda \in \Lambda$, if $f(\lambda)$ is summable relative to an ordering Ψ of Ω, then we may define a function $\widetilde{f} : \Lambda \to \mathbf{F}$ such that
$$\widetilde{f}(\lambda) = \sum_{j=0}^{\infty} f_{\Psi(j)}(\lambda).$$
Given an ordering Ψ for Ω, if for every $\varepsilon > 0$, there is $I \in \mathbf{N}$ such that $i > I$ implies
$$\left| \sum_{j=0}^{i} f_{\Psi(j)}(\lambda) - \widetilde{f}(\lambda) \right| < \varepsilon \tag{2.12}$$
for all $\lambda \in \Lambda$, then we say that the family $\{f(\lambda)\}_{\lambda \in \Lambda}$ is uniformly summable in Λ with respect to Ψ, and its sum function is $\widetilde{f}(\lambda)$.

Note that when Λ is a subset of $\mathbf{F}^\kappa$, $\Omega = \mathbf{N}$ and Ψ is the identity mapping on $\mathbf{N}$, then (2.12) reduces to
$$\left| \sum_{n=0}^{j} f_n(\lambda) - \widetilde{f}(\lambda) \right| < \varepsilon.$$
Therefore in this case, we are back to the usual uniform convergence of a functional series.

Theorem 2.6 (Cauchy's Test). *Let Ψ be an ordering for Ω. The family $\{f(\lambda)\}_{\lambda \in \Lambda}$ of sequences in l^Ω is uniformly summable on Λ relative to Ψ if, and only if, for every $\varepsilon > 0$, there is $I \in \mathbf{N}$ such that $m, n > I$ implies*
$$\left| \sum_{j=0}^{m} f_{\Psi(j)}(\lambda) - \sum_{j=0}^{n} f_{\Psi(j)}(\lambda) \right| < \varepsilon, \; \lambda \in \Lambda. \tag{2.13}$$

Indeed, suppose $\{f(\lambda)\}_{\lambda \in \Lambda}$ is uniformly summable in Λ relative to Ψ with sum function $\widetilde{f}(\lambda)$. Then

$$\left|\sum_{j=0}^{m} f_{\Psi(j)}(\lambda) - \widetilde{f}(\lambda)\right|, \left|\sum_{j=0}^{n} f_{\Psi(j)}(\lambda) - \widetilde{f}(\lambda)\right| < \frac{\varepsilon}{2}$$

for all large m and n. Thus (2.13) hold for all large m and n in view of the triangle inequality. Conversely, if (2.13) hold for all large m and n, then by Cauchy's convergence theorem for real sequences, $\left\{\sum_{j=0}^{i} f_{\Psi(j)}(\lambda)\right\}_{i \in \mathbf{N}}$ converges to some $\widetilde{f}(\lambda)$ for each $\lambda \in \Lambda$. If we now replace ε in (2.13) by $\varepsilon' \in (0, \varepsilon)$, fix m in (2.13) and take limits on both sides as $n \to \infty$, we see that

$$\left|\widetilde{f}(\lambda) - \sum_{j=0}^{m} f_{\Psi(j)}(\lambda)\right| \leq \varepsilon' < \varepsilon, \ \lambda \in \Lambda$$

as required.

Theorem 2.7 (Weierstrass Test). *Let Ψ be an ordering for Ω. Let $\{M_n\}_{n \in \Omega}$ be a sequence of nonnegative numbers such that*

$$0 \leq |f_n(\lambda)| \leq M_n, \ n \in \Omega, \ \lambda \in \Lambda.$$

Then the family $\{f(\lambda)\}_{\lambda \in \Lambda}$ of sequences in l^Ω is uniformly summable on Λ relative to Ψ if $\{M_n\}_{n \in \Omega}$ is summable relative to Ψ.

Indeed,

$$\left|\sum_{j=0}^{n+p} f_{\Psi(j)}(\lambda) - \sum_{j=0}^{n} f_{\Psi(j)}(\lambda)\right| \leq \sum_{j=n+1}^{n+p} M_{\Psi(j)}$$

for all $\lambda \in \Lambda$ and any $p \in \mathbf{Z}^+$. Since $\{M_n\}_{n \in \Omega}$ is summable relative to Ψ, in view of Cauchy's convergence criterion for real sequences, the right hand side can be made arbitrary small by requiring large n. Our previous theorem then yields our proof.

Theorem 2.8. *Assume that the family $\{f(\lambda)\}_{\lambda \in \Lambda}$ of sequences in l^Ω is uniformly summable on Λ relative to the ordering Ψ with sum function $\widetilde{f}(\lambda)$. Let μ be an accumulation point of Λ. Suppose each function $f_n = f_n(\lambda)$ satisfies*

$$\lim_{\lambda \to \mu} f_n(\lambda) = c_n, \ n \in \Omega.$$

Then $\{c_n\}$ is summable relative to Ψ and its sum C is given by

$$C = \lim_{\lambda \to \mu} \widetilde{f}(\lambda).$$

To see the proof, note that taking limits on both sides of (2.13) as $\lambda \to \mu$, we see that

$$\left| \sum_{j=0}^{m} c_{\Psi(j)} - \sum_{j=0}^{n} c_{\Psi(j)} \right| \leq \varepsilon.$$

By Cauchy's convergence criteria, we see that $\{c_n\}$ is summable relative to Ψ. Next, note that

$$\left| \widetilde{f}(\lambda) - C \right| \leq \left| \sum_{j=0}^{i} f_{\Psi(j)}(\lambda) - \sum_{j=0}^{i} c_{\psi(j)} \right| + \left| \widetilde{f}(\lambda) - \sum_{j=0}^{i} f_{\Psi(j)}(\lambda) \right| + \left| C - \sum_{j=0}^{i} c_{\Psi(j)} \right|.$$

If i is sufficiently large, the last two terms can be made arbitrary small and independent of λ, while if λ is sufficiently close to μ, the first term on the right hand side can be made arbitrary small. The proof is complete.

As an immediate consequence, assume that the family $\{f(\lambda)\}_{\lambda \in \Lambda}$ of sequence in l^Ω is uniformly summable on Λ relative to the ordering Ψ and each function $f_n = f_n(\lambda)$ is continuous at a point $\mu \in \Lambda$, then the corresponding sum function $\widetilde{f}$ is also continuous at μ.

Theorem 2.9. *Let $\Lambda = \overline{B}(a; \delta) \subset \mathbf{F}$. Let $\widetilde{f}$ be the sum function of the family $\{f(\lambda)\}_{\lambda \in \Lambda}$ of real sequences in $l^\mathbf{N}$ uniformly summable in Λ relative to an ordering Ψ for $\mathbf{N}$, where each $f_k(\lambda)$ is continuous at each point λ in Λ. Then the sequence $\left\{ \int_\Gamma f_j(\lambda) d\lambda \right\}_{j \in \mathbf{N}}$, where Γ is the straight line segment from a to $z \in B(a; \delta)$, is summable relative to Ψ and*

$$\int_\Gamma \widetilde{f}(\lambda) d\lambda = \sum_{j=0}^{\infty} \int_\Gamma f_{\Psi(j)}(\lambda) d\lambda.$$

To see this, assume without loss of generality that Ψ is the identity mapping on $\mathbf{N}$. Since $\widetilde{f}$ is continuous by the previous Theorem, its integral $\int_\Gamma \widetilde{f}(\lambda) d\lambda$ exists. Furthermore,

$$\left| \widetilde{f}(\lambda) - \sum_{j=0}^{n} f_j(\lambda) \right| < \frac{\varepsilon}{2\delta}, \quad \lambda \in \overline{B}(a; \delta),$$

for all sufficiently large n, thus

$$\left| \int_\Gamma \widetilde{f}(\lambda) d\lambda - \sum_{j=0}^{n} \int_\Gamma f_j(\lambda) d\lambda \right| \leq 2\delta \sup_{\lambda \in \overline{B}(a;\delta)} \left| \widetilde{f}(\lambda) - \sum_{j=0}^{n} f_j(\lambda) \right| < \varepsilon$$

for all large n.

Theorem 2.10. *Let $\Lambda = \overline{B}(a; \delta)$. Let $\{f(\lambda)\}_{\lambda \in \Lambda}$ be a family of real sequences in l^Ω such that each $f_n = f_n(\lambda)$ has a finite derivative f'_n in Λ. Suppose $\{f_n(c)\}$ is summable for $c \in \Lambda$ relative to an ordering Ψ and the family $\{f'(\lambda)\}_{\lambda \in \Lambda}$ is uniformly* summable in Λ relative to Ψ. Then $\{f(\lambda)\}_{\lambda\in\Lambda}$ is uniformly summable in Λ relative to Ψ to $\widetilde{f}(\lambda)$ and

$$\frac{d\widetilde{f}(z)}{d\lambda} = \sum_{j=0}^{\infty} f'_{\Psi(j)}(z), \ z \in \Lambda.$$

To see the sketch of the proof, choose two distinct points $\lambda, \mu \in \overline{B}(a;\delta)$. Without loss of any generality, we will also assume that Ψ is the identity mapping for $\mathbf{N}$. Let

$$g_n(\lambda) = \begin{cases} \frac{f_n(\lambda)-f_n(\mu)}{\lambda-\mu} & \lambda \neq \mu \\ f'_n(\mu) & \lambda = \mu \end{cases}. \tag{2.14}$$

Then the family

$$\{g(\lambda)\}_{\lambda\in\Lambda}$$

is uniformly summable in Λ. Indeed, given any $\varepsilon > 0$, there is a number $\mathbf{N}$ such that

$$\left| \sum_{k=n+1}^{n+m} f'_k(\lambda) \right| < \varepsilon, \ \lambda \in \Lambda$$

for all $n > N$ and $m \in \mathbf{Z}^+$ by the uniform summability of $\{f'_n(\lambda)\}_{\lambda\in\Lambda}$. Let

$$U(\lambda) = \sum_{k=n+1}^{n+m} f_k(\lambda)$$

where n and m are temporarily fixed. Then

$$\left| \sum_{k=n+1}^{n+m} \frac{f_k(\lambda)-f_k(\mu)}{\lambda-\mu} \right| = \left| \frac{U(\lambda)-U(\mu)}{\lambda-\mu} \right| \leq |U'(\xi)| = \left| \sum_{k=n+1}^{n+m} f'_k(\xi) \right| < \varepsilon$$

for all $\lambda \in \Lambda\setminus\{\mu\}$, where ξ is between λ and μ, and

$$\left| \sum_{k=n+1}^{n+m} g_k(\mu) \right| = \left| \sum_{k=n+1}^{n+m} f'_k(\mu) \right| < \varepsilon.$$

This shows that $\{g(\lambda)\}_{\lambda\in(a,b)}$ is uniformly summable in Λ.

Now take $\mu = c$ in (2.14), the corresponding family $\{g(\lambda)\}$ is uniformly summable. Thus $\{g(\lambda)(\lambda-c)\}$ is also uniformly summable since $|\lambda-c|$ is bounded on Λ. In turn, we see that $\{g(\lambda)(\lambda-c)+f(c)\} = \{f(\lambda)\}$ is uniformly summable. Finally,

$$\frac{d\widetilde{f}(z)}{d\lambda} = \lim_{\lambda\to z} \frac{\widetilde{f}(\lambda)-\widetilde{f}(z)}{\lambda-z} = \sum_{n=0}^{\infty} \lim_{\lambda\to z} \frac{f_n(\lambda)-f_n(z)}{\lambda-z} = \sum_{n=0}^{\infty} f'_n(z)$$

as desired.

2.4 Properties of Univariate Sequences

2.4.1 *Common Sequences*

As before, let $l^{\mathbf{N}}$ be the set of all real or complex sequences of the form $f = \{f_k\}_{k \in \mathbf{N}}$. We will call f_k the k-th term of the sequence f. Note that the first k terms of f are $f_0, ..., f_{k-1}$ respectively.

Let m be a nonnegative integer. Recall that $\hbar^{\langle m \rangle} \in l^{\mathbf{N}}$ denotes the Dirac sequence defined by

$$\hbar_k^{\langle m \rangle} = \begin{cases} 1 & k = m \\ 0 & k \neq m \end{cases}.$$

Besides the Dirac sequences, there are a number of common sequences in $l^{\mathbf{N}}$ which deserve special notations. First of all, let α be a complex number, the sequence $\{\alpha, 0, 0, ...\}$ is denoted by $\overline{\alpha}$ and is called a *scalar sequence*. In particular, the sequences $\{0, 0, ...\}$ and $\{1, 0, ...\}$ is denoted by $\overline{0}$ (or 0 if no confusion is caused) and $\overline{1}$ respectively. The sequence $\{1, 1, 1, ...\}$ is denoted by σ, and the sequence $\{1, -1, 0, 0, ...\}$ by δ. The sequence $\{1/0!, 1/1!, 1/2!, 1/3!, 1/4!, ...\}$ is called the *exponential sequence* and is denoted by ϖ. For $m \in \mathbf{N}$, $\mathbf{H}^{(m)}$ is the jump (or Heaviside) sequence defined by

$$\mathbf{H}_k^{(m)} = \begin{cases} 0 & 0 \leq k < m \\ 1 & k \geq m \end{cases}.$$

Note that we have also used $\hbar^{\langle 0 \rangle}$ to denote $\overline{1}$ and $\mathbf{H}^{(0)}$ for σ. It is also convenient to write $\hbar$ instead of $\hbar^{\langle 1 \rangle}$ and this practice will be assumed for similar situations in the sequel.

The sequence $\{f_1 - f_0, f_2 - f_1, ...\}$ obtained by taking the difference of the consecutive components of the sequence $\{f_k\}$ will be denoted by Δf and is called the *first difference* of f. The higher differences $\Delta^m f$, $m = 2, 3, ...$, are defined recursively by $\Delta^m f = \Delta(\Delta^{m-1} f)$. Thus

$$(\Delta f)_k = f_{k+1} - f_k,$$
$$(\Delta^2 f)_k = f_{k+2} - 2f_{k+1} + f_k,$$
$$(\Delta^3 f)_k = f_{k+3} - 3f_{k+2} + 3f_{k+1} - f_k,$$

etc., for $k \in \mathbf{N}$. We also define $\Delta^0 f = f$ and $\Delta^1 f = \Delta f$.

Example 2.6. The following is the well known telescoping property for the difference operations: for $f \in l^{\mathbf{N}}$,

$$\sum_{k=a}^{b} (\Delta f)_k = f_{b+1} - f_a, \ 0 \leq a \leq b.$$

The sequence $\{f_m, f_{m+1}, ...\}$ obtained by 'deleting' the first m terms of the sequence $\{f_0, ..., f_m, f_{m+1}, ...\}$ will be denoted by $E^m f$, and the sequence

$\{0, ..., 0, f_0, f_1, ..\}$ obtained by 'adding' m zeros to the front of the terms of f by $E^{-m}f$. The more precise definitions of $E^m f$ and $E^{-m} f$ are respectively

$$E^m f = \{f_{m+k}\}_{k \in \mathbb{N}},$$

and

$$(E^{-m} f)_k = \begin{cases} f_{-m+k} & k \geq m \\ 0 & 1 \leq k \leq m \end{cases}.$$

These definitions require $m \geq 1$. However, it is natural to define $E^0 f = f$ and $Ef = E^1 f$. Note that we have

$$E^m E^{-m} f = f, \ m \in \mathbb{Z}^+,$$

but for $m \in \mathbb{Z}^+$, $E^{-m} E^m f$ is not equal to f in general. The sequence $E^m f$ will be called a *translated* or *shifted sequence* of f.

Example 2.7. It is easy to see that $E^m(f+g) = E^m f + E^m g$, $E(Ef) = E^2 f$, and $\Delta f = Ef - f$ holds for any $f \in l^{\mathbb{N}}$. Thus,

$$\Delta^2 f = E(Ef - f) - (Ef - f) = E^2 f - 2Ef + f,$$

$$\Delta^3 f = E^3 f - 3E^2 f + 3Ef - f,$$

etc.

For any number $\lambda \in \mathbf{F}$, where $\lambda = 0$ is allowed, the geometric sequence $\{\lambda^n\}_{n \in \mathbb{N}}$, where $\lambda \in \mathbf{C}$, is denoted by $\underline{\lambda}$. By means of this notation, we see that $\underline{\lambda}_n = \lambda^n$ for $n \in \mathbb{N}$. The product sequence $\underline{\lambda} \cdot f = \{\lambda^k\} \cdot \{f_k\} = \{\lambda^k f_k\}$ is called an *attenuated sequence* of f. It is easily seen that $\underline{0} \cdot f = \overline{f_0}$, $\underline{1} \cdot f = f$, $\underline{\lambda} \cdot \underline{\mu} \cdot f = \underline{\lambda \mu} \cdot f$ and

$$\underline{\lambda} \cdot (f + g) = \underline{\lambda} \cdot f + \underline{\lambda} \cdot g, \quad f, g \in l^{\mathbb{N}}.$$

2.4.2 Convolution Products

For any $f = \{f_k\}$ and $g = \{g_k\}$ in $l^{\mathbb{N}}$, we define the *convolution product* $f * g$, by

$$(f * g)_k = \sum_{i=0}^{k} f_{k-i} g_i, \quad k \in \mathbb{N}.$$

Example 2.8. For any $f = \{f_k\}$ in $l^{\mathbb{N}}$, $\overline{1} * f = f$, $\overline{0} * f = \overline{0}$, $\overline{\alpha} * \overline{\beta} = \overline{\alpha\beta}$, $\overline{\alpha} * f = (\alpha\overline{1}) * f = \alpha(\overline{1} * f) = \alpha f$ and $\hbar * f = E^{-1} f$. The last equality means that the product $\hbar * f$ is equal to translating f 'one unit' to the right.

It is also easy to verify that for any $f = \{f_k\}$, $g = \{g_k\}$ and $h = \{h_k\}$ in $l^{\mathbb{N}}$,

$$f * (g + h) = f * g + f * h.$$

We will also denote the products $f*f$, $f*f*f, \ldots$ by $f^{\langle 2 \rangle}, f^{\langle 3 \rangle}, \ldots$ respectively. If no confusion is caused, the k-th term of the sequence $f^{\langle p \rangle}$ will be written as $f_k^{\langle p \rangle}$ instead of $\left(f^{\langle p \rangle}\right)_k$. Note that

$$f_k^{\langle p \rangle} = \sum_{v_1+\cdots+v_p=k;\, v_1,\ldots,v_p \in \mathbf{N}} f_{v_1} f_{v_2} \cdots f_{v_p}, \quad k \in \mathbf{N}. \tag{2.15}$$

Although p is implicitly defined to be greater than 1, the same formula holds for $p = 1$. Thus we will define $f^{\langle 1 \rangle} = f$. For the sake of convenience, we will also define $f^{\langle 0 \rangle} = \overline{1}$. The sequence $f^{\langle p \rangle}$ is called the p-th convolution (power) product of f.

Example 2.9. Recall $\hbar = \{0,1,0,0,\ldots\}$. We have $\hbar^{\langle 0 \rangle} = \{1,0,0\ldots\}$, $\hbar^{\langle 2 \rangle} = \{0,0,1,0,\ldots\} = \hbar * \hbar$, $\hbar^{\langle 3 \rangle} = \hbar * \hbar * \hbar$, etc.

Example 2.10. Let $f, g \in l^{\mathbf{N}}$. Then

$$(f+g)^{\langle k \rangle} = \sum_{i=0}^{k} C_i^{(k)} f^{\langle i \rangle} * g^{\langle k-i \rangle}, \quad k \in \mathbf{N}.$$

Theorem 2.11 (Merten's Theorem). *If $f = \{f_k\} \in l_1^{\mathbf{N}}$ and $g = \{g_k\} \in l^{\mathbf{N}}$ is summable, then $f*g$ is summable and*

$$\sum_{n=0}^{\infty}(f*g)_n = \left(\sum_{\mathbf{N}} f\right)\left(\sum_{k=0}^{\infty} g_k\right).$$

Proof. Let $h_n = (f*g)_n$, $F_n = \sum_{k=0}^{n} f_k$ and $G_n = \sum_{k=0}^{n} g_k$ for $n \in \mathbf{N}$. Let $\sum_{\mathbf{N}} f = \alpha$, $\sum_{\mathbf{N}} |f| = \gamma$ and $\sum_{k=0}^{\infty} g_k = \beta$. Then

$$\sum_{n=0}^{p} h_n = \sum_{n=0}^{p}\sum_{k=0}^{n} f_k g_{n-k} = \sum_{k=0}^{p}\sum_{n=k}^{p} f_k g_{n-k} = \sum_{k=0}^{p} f_k \sum_{m=0}^{p-k} g_m = \sum_{k=0}^{p} f_k G_{p-k}$$

$$= \sum_{k=0}^{p} f_k \beta - \sum_{k=0}^{p} f_k (\beta - G_{p-k}) = F_p \beta - \sum_{k=0}^{p} f_k (\beta - G_{p-k}).$$

To conclude our proof, it suffices now to show that

$$\lim_{p \to \infty} \sum_{k=0}^{p} f_k (\beta - G_{p-k}) = 0.$$

To see this, note first that $\lim_{n \to \infty}(\beta - G_n) = 0$. Thus we may choose $M > 0$ such that $|\beta - G_n| \leq M$ for $n \in \mathbf{N}$. Given $\varepsilon > 0$, choose P in N sufficiently large so that $n > P$ implies $|\beta - G_n| < \varepsilon/(2\gamma)$ and

$$\sum_{n=P+1}^{\infty} |f_n| < \frac{\varepsilon}{2M}.$$

Then for $p > 2P$, we have

$$\left|\sum_{k=0}^{p} f_k(\beta - G_{p-k})\right| \leq \sum_{k=0}^{P} |f_k||\beta - G_{p-k}| + \sum_{k=P+1}^{p} |f_k||\beta - G_{p-k}|$$

$$\leq \frac{\varepsilon}{2\gamma} \sum_{k=0}^{P} |f_k| + M \sum_{k=P+1}^{p} |f_k|$$

$$\leq \frac{\varepsilon}{2\gamma} \sum_{k=0}^{\infty} |f_k| + M \sum_{k=P+1}^{\infty} |f_k|$$

$$< \frac{\varepsilon}{2} + \frac{\varepsilon}{2}.$$

This proves

$$\lim_{p \to \infty} \sum_{n=0}^{p} h_n = \lim_{p \to \infty} F_p \beta = \left(\sum_{\mathbf{N}} f\right)\left(\sum_{k=0}^{\infty} g_k\right).$$

As an immediate corollary, if $f = \{f_k\}$, $g = \{g_k\} \in l_1^{\mathbf{N}}$, then $f * g \in l_1^{\mathbf{N}}$, and

$$\sum_{\mathbf{N}} f * g = \left(\sum_{\mathbf{N}} f\right)\left(\sum_{\mathbf{N}} g\right).$$

Indeed, let $h_n = (f * g)_n$ for $n \in \mathbf{N}$. Since

$$\sum_{k=0}^{n} |h_k| \leq \sum_{k=0}^{n} |f_k| \sum_{k=0}^{n} |g_k| \leq \sum_{k=0}^{\infty} |f_k| \sum_{k=0}^{\infty} |g_k|,$$

and since $\{\sum_{k=0}^{n} |h_k|\}$ is nondecreasing, thus $\sum_{k=0}^{\infty} |h_k| < \infty$. We have thus shown that $f * g \in l_1^{\mathbf{N}}$. Furthermore,

$$\sum_{\mathbf{N}} f * g = \sum_{k=0}^{\infty} (f * g)_k = \left(\sum_{\mathbf{N}} f\right)\left(\sum_{k=0}^{\infty} g_k\right) = \left(\sum_{\mathbf{N}} f\right)\left(\sum_{\mathbf{N}} g\right).$$

Several elementary facts related to the convolution product of sequences will be useful later. First of all, we may easily show that for any two sequences $f = \{f_k\}$ and $g = \{g_k\}$ in $l^{\mathbf{N}}$, $f * g = g * f$. Furthermore, if $f * g = \bar{0}$, then $f = \bar{0}$ or $g = \bar{0}$. Indeed, suppose that $f_0 = \cdots = f_{m-1} = 0$, $f_m \neq 0$, $g_0 = \cdots = g_{n-1} = 0$ and $g_n \neq 0$. Then we have

$$(f * g)_{m+n} = f_0 g_{m+n} + \cdots + f_m g_n + \cdots + f_{m+n} g_0 = f_m g_n \neq 0.$$

This shows that $f * g \neq \bar{0}$. It is also easily verified that under the above addition and convolution product, $l^{\mathbf{N}}$ is a commutative ring with no zero divisor, i.e. $f * g = \bar{0}$ implies $f = \bar{0}$ or $g = \bar{0}$, and the additive and multiplicative identities are $\bar{0}$ and $\bar{1}$ respectively.

Theorem 2.12. *Let $f = \{f_k\}$, $g = \{g_k\}$ be sequences in $l^{\mathbf{N}}$. If $g_0 \neq 0$, then there is a unique sequence $x = \{x_k\} \in l^{\mathbf{N}}$ such that $g * x = f$.*

Indeed, we simply note that the infinite linear system
$$g_0 x_0 = f_0,$$
$$g_0 x_1 + g_1 x_0 = f_1,$$
$$g_0 x_2 + g_1 x_1 + g_2 x_0 = f_2,$$
$$... = ...,$$
can be solved successively in the following unique manner: $x_0 = f_0/g_0$, $x_1 = (f_1 - g_1 x_0)/g_0, ...$.

In case $g = \{g_k\}$ satisfies $g_0 \neq 0$, the quotient f/g will denote the (unique) solution sequence of the equation
$$g * x = f.$$
One important question is how to find the explicit form of a quotient f/g. The algorithm just stated is one way to calculate f/g. However, it may also be found by other means.

Example 2.11. Let $f = \{f_k\} \in l^{\mathbf{N}}$. If $f_0 = 0$, then the first n terms of the convolution product $f^{\langle n \rangle}$ are equal to zero. Indeed, let
$$f^{\langle 1 \rangle} = f = \{0, f_1, f_2, f_3, ...\},$$
then
$$f^{\langle 2 \rangle} = \{0, 0, f_1^2, 2f_1 f_2, 2f_1 f_3 + f_2^2, ...\},$$
$$f^{\langle 3 \rangle} = \{0, 0, 0, f_1^3, 3f_1^2 f_2, ...\},$$
$$f^{\langle 4 \rangle} = \{0, 0, 0, 0, f_1^4, ...\},$$
and then by induction we may show that the first n terms of the sequence $f^{\langle n \rangle}$ are equal to zero. Furthermore, since
$$f_i^{\langle k \rangle} = \sum_{v_1 + \cdots + v_k = i; v_1, ..., v_k \in \mathbf{N}} f_{v_1} f_{v_2} \cdots f_{v_k} = \sum_{l_1 + \cdots + l_k = i; l_1, ..., l_k \in \mathbf{Z}^+} f_{l_1} \cdots f_{l_k},$$
for each $k \in \{0, ..., n\}$, the term $f_n^{\langle k \rangle}$ involves $f_1, ..., f_{n-1}$ only and can be expressed as
$$f_n^{\langle k \rangle} = P(f_1, ..., f_{n-1}), \ n \geq 2, 0 \leq k \leq n,$$
where P is an $(n-1)$-variate polynomial with positive coefficients. Hence the conditions $f_0 = 0$, $f_1 = \mu$ and the iteration formula
$$f_n = F\left(f_n^{\langle 2 \rangle}, ..., f_n^{\langle n \rangle}\right), \ n \geq 2,$$
will define f in a unique manner. For example, if $F(u_2, u_3, ..., u_n) = u_1 + u_2 + \cdots + u_n$, then
$$f_n = \sum_{i=2}^{n} f_n^{\langle i \rangle}, \ n \geq 2,$$
will yield $f_2 = f_2^{\langle 2 \rangle} = f_1^2$, $f_3 = f_3^{\langle 2 \rangle} + f_3^{\langle 3 \rangle} = 2f_1 f_2 + f_1^3 = 2f_1^3 + f_1^3, ...$.

Example 2.12. Let $f = \{f_k\} \in l^{\mathbf{N}}$. If $f_0 = f_1 = 0$, then the first $2n$ terms of the convolution product $f^{\langle n \rangle}$ are equal to zero.

Example 2.13. In case $g = \{g_k\}$ satisfies $g_0 \neq 0$, then $(f/g)^{\langle n \rangle} = f^{\langle n \rangle}/g^{\langle n \rangle}$ for $n \in \mathbf{N}$.

Example 2.14. Let $f, g, p, q \in l^{\mathbf{N}}$ such that $g_0 \neq 0$ and $q_0 \neq 0$. Then
$$\frac{f*p}{g*q} = \frac{f}{g}*\frac{p}{q}$$
since
$$(g*q)*\frac{f}{g}*\frac{p}{q} = f*p.$$

Theorem 2.13. Let $f = \{f_k\}, g = \{g_k\} \in l^{\mathbf{N}}$. Then $\underline{\lambda} \cdot (f*g) = (\underline{\lambda} \cdot f) * (\underline{\lambda} \cdot g)$ for $\lambda \in \mathbf{C}$.

Indeed,
$$\underline{\lambda} \cdot (f*g) = \left\{\sum_{j=0}^{k} \lambda^k f_j g_{k-j}\right\} = \left\{\sum_{j=0}^{k} \lambda^j f_j \lambda^{k-j} g_{k-j}\right\} = (\underline{\lambda} \cdot f) * (\underline{\lambda} \cdot g).$$

Example 2.15. Let $f = \{f_k\} \in l^{\mathbf{N}}$.
$$(\underline{\lambda} \cdot f)^{\langle 2 \rangle} = \left\{\sum_{i=0}^{k} f_i \lambda^i f_{k-i} \lambda^{k-i}\right\} = \left\{\lambda^k \sum_{i=0}^{k} f_i f_{k-i}\right\} = \underline{\lambda} \cdot f^{\langle 2 \rangle}.$$

Similarly, we may show by induction that
$$(\underline{\lambda} \cdot f)^{\langle n \rangle} = \underline{\lambda} \cdot f^{\langle n \rangle}, \; n \in \mathbf{N}.$$

Example 2.16. Note that
$$\sigma * \{f_k\} = \left\{\sum_{i=0}^{k} f_i\right\},$$
$$\sigma * \sigma \equiv \sigma^{\langle 2 \rangle} = \left\{\sum_{i=0}^{k} 1\right\} = \{\lfloor k+1 \rfloor_1\},$$
and in general,
$$\sigma^{\langle n \rangle} = \left\{\frac{\lfloor k+n-1 \rfloor_{n-1}}{(n-1)!}\right\}, \; n \in \mathbf{N}. \tag{2.16}$$

Example 2.17. Note that $\delta = \{1, -1, 0, 0, ...\} = \overline{1} - \hbar$. Also, $\delta^{\langle 2 \rangle} = \{1, -2, 1, 0, ...\}$, and in general,
$$\delta^{\langle n \rangle} = \left\{(-1)^k \frac{\lfloor n \rfloor_k}{k!}\right\}, \; n \in \mathbf{Z}^+.$$

Example 2.18. Let f be a sequence in $l^{\mathbf{N}}$, Δf its first difference, and $\overline{f_0}$ the sequence $\{f_0, 0, 0, ...\}$. Then it is easily checked that

$$\delta * f = \Delta f + \overline{f_0} - \delta * (\Delta f) = \overline{f_0} + \hbar * (\Delta f). \tag{2.17}$$

Example 2.19. As an immediate application of (2.17), let $f = \{2^k\}_{k \in \mathbf{N}}$. Since $\Delta f = f$,

$$f = \delta * f - \overline{1} + \delta * f,$$

from which we obtain

$$\{2^k\} = \frac{\overline{1}}{2\delta - \overline{1}} = \frac{\overline{1}}{\overline{2} * \delta - \overline{1}}.$$

Similarly, the same principle leads to

$$\{\alpha^k\} = \frac{\overline{1}}{\overline{\alpha} * \delta - \overline{\alpha} + \overline{1}}. \tag{2.18}$$

Substituting $\alpha = 1/(1-\beta)$ into the above formula, we obtain

$$\frac{\overline{1}}{\delta - \overline{\beta}} = \left\{\left(\frac{1}{1-\beta}\right)^{k+1}\right\}, \ \beta \neq 1. \tag{2.19}$$

Now in view of Theorem 2.13,

$$\{c^k\} * \{c^k\} = (\underline{c} \cdot \sigma) * (\underline{c} \cdot \sigma) = \underline{c} \cdot \sigma^2 = \{c^k \lfloor k+1 \rfloor_1\},$$

we see that

$$\frac{\overline{1}}{(\delta - \overline{\beta})^{\langle 2 \rangle}} = \{(1-\beta)^{-k-2} \lfloor k+1 \rfloor_1\}. \tag{2.20}$$

By induction, it is not difficult to see that for any scalar $\beta \neq 1$, the following extension of formula (2.16) holds

$$\frac{\overline{1}}{(\delta - \overline{\beta})^{\langle n \rangle}} = \left\{\frac{\lfloor k+n-1 \rfloor_{n-1}}{(n-1)!}(1-\beta)^{-k-n}\right\}, \ n \in \mathbf{Z}^+. \tag{2.21}$$

Recall that $\delta + \hbar = \overline{1}$. By means of this simple relation, some of the previous formulas can also be expressed in terms of the 'translation operator' $\hbar$. For instance, (2.18) can be written as

$$\{\alpha^k\} = \frac{\overline{1}}{\overline{1} - \overline{\alpha} * \hbar}.$$

Similarly, (2.19) and (2.21) are equivalent to

$$\{\gamma^{-k-1}\} = \frac{\overline{1}}{\overline{\gamma} - \hbar}, \ \gamma \neq 0, \tag{2.22}$$

and

$$\frac{\overline{1}}{(\overline{\gamma} - \hbar)^{\langle n \rangle}} = \left\{\frac{\lfloor k+n-1 \rfloor_{n-1}}{(n-1)!}\gamma^{-k-n}\right\}, \ \gamma \neq 0, \ n \in \mathbf{Z}^+, \tag{2.23}$$

or

$$\frac{\overline{1}}{(\overline{1} - \overline{\gamma}\hbar)^{\langle n \rangle}} = \left\{\frac{\lfloor k+n-1 \rfloor_{n-1}}{(n-1)!}\gamma^k\right\}, \ n \in \mathbf{Z}^+. \tag{2.24}$$

2.4.3 Algebraic Derivatives and Integrals

Given a sequence $f = \{f_k\}_{k=0}^{\infty} \in l^{\mathbf{N}}$, we define the *algebraic derivative* of f by
$$Df = \{(k+1)f_{k+1}\}_{k=0}^{\infty}.$$
The higher algebraic derivatives $D^n f$ are defined recursively by $D^n f = D(D^{n-1}f)$. Thus we have
$$D\{f_0, f_1, f_2, ...\} = \{f_1, 2f_2, 3f_3, ...\},$$
and
$$D^n f = \{(k+1)\cdots(k+n)f_{k+n}\}$$
for $n \in \mathbf{Z}^+$. For instance, for any complex number α, $D\overline{\alpha} = \overline{0}$, and we have
$$D^n \sigma = \{(k+1)(k+2)\cdots(k+n)\}$$
for $n \in \mathbf{Z}^+$.

It can easily be verified that for $\alpha, \beta \in C$ and $f, g \in l^{\mathbf{N}}$,
$$D(\alpha f + \beta g) = \alpha Df + \beta Dg,$$
$$D(f * g) = f * Dg + g * Df,$$
$$D(f \cdot g) = (Df) \cdot Eg = (Ef) \cdot Dg$$
and
$$D\left(\frac{f}{g}\right) = \frac{g * Df - f * Dg}{g^{\langle 2 \rangle}},$$
where we recall that f/g is only defined when the zeroth term of g is not 0. For instance, the last equality can be seen from
$$g * x = f \Rightarrow D(g * x) = Df \Rightarrow g * D\left(\frac{f}{g}\right) + \frac{f}{g} * Dg = Df,$$
so that
$$g * g * D\left(\frac{f}{g}\right) + f * Dg = g * Df.$$

Algebraic derivatives of some common sequences can easily be found. More complicated derivatives can be obtained by employing the following list of useful formulae:
$$\hbar * Df = \{kf_k\}, \tag{2.25}$$
$$D\hbar^{\langle n \rangle} = n\hbar^{\langle n-1 \rangle},\ n \in \mathbf{Z}^+, \tag{2.26}$$
$$D\delta^{\langle n \rangle} = -n\delta^{\langle n-1 \rangle},\ n \in \mathbf{Z}^+, \tag{2.27}$$

$$Df^{\langle n\rangle} = D(f^{\langle n-1\rangle} * f) = f^{\langle n-1\rangle} * Df + f * Df^{\langle n-1\rangle}$$
$$= \cdots = \overline{n} * f^{\langle n-1\rangle} * Df, \ n \in \mathbf{Z}^+, \tag{2.28}$$
$$D^n\left(\hbar^{\langle m\rangle} * \{f_k\}\right) = \hbar^{\langle m-n\rangle} * \{\lfloor m+k\rfloor_n f_k\}, \ m \geq n \geq 1, \tag{2.29}$$

and finally
$$\hbar^{\langle m\rangle} * D^n\{f_k\} = \{\lfloor k+n-m\rfloor_n f_{k+n-m}\}, \ n \geq m \geq 0. \tag{2.30}$$

To see that (2.29) holds, let $f = \{f_k\}$, then
$$D\left(\hbar^{\langle m\rangle} * f\right) = \hbar^{\langle m\rangle} * Df + f * D\hbar^{\langle m\rangle}$$
$$= \hbar^{\langle m-1\rangle} * (\hbar * Df + mf)$$
$$= \hbar^{\langle m-1\rangle} * \{(k+m)f_k\}.$$

Similarly,
$$D\left(D\left(\hbar^{\langle m\rangle} * f\right)\right) = \hbar^{\langle m-1\rangle} * D\{(k+m)f_k\} + \{(k+m)f_k\} * ((m-1)\hbar^{\langle m-2\rangle})$$
$$= \hbar^{\langle m-2\rangle} * \{k(k+m)f_k\} + \hbar^{\langle m-2\rangle} * ((m-1)\{(k+m)f_k\})$$
$$= \hbar^{\langle m-2\rangle} * \{(k+m)(k+m-1)f_k\}.$$

The general formula is then obtained by induction.

It is interesting to note that if $D\phi = 0$, then ϕ is a scalar sequence. The algebraic derivatives may also be used to derive identities involving sequences in $l^{\mathbf{N}}$. For instance, the equalities $\hbar * \sigma^{\langle 2\rangle} = \{k\}$ and $D\{k\} = \{(k+1)^2\}$ imply
$$\{(k+1)^2\} = D(\hbar * \sigma^{\langle 2\rangle}) = \hbar * D\sigma^{\langle 2\rangle} + \sigma^{\langle 2\rangle} = 2\hbar * \sigma^{\langle 3\rangle} + \sigma^{\langle 2\rangle},$$

The same principle leads to
$$\{(k+1)^3\} = D\left(\hbar * D\left(\hbar * \sigma^{\langle 2\rangle}\right)\right),$$
$$\{(k+1)^4\} = D\left(\hbar * D\left(\hbar * D(\hbar * \sigma^{\langle 2\rangle})\right)\right),$$

etc., and
$$\{(k+1)^3\} = 6\hbar^{\langle 2\rangle} * \sigma^{\langle 4\rangle} + 6\hbar * \sigma^{\langle 3\rangle} + \sigma^{\langle 2\rangle},$$
$$\{(k+1)^4\} = 24\hbar^{\langle 3\rangle} * \sigma^{\langle 5\rangle} + 36\hbar^{\langle 2\rangle} * \sigma^{\langle 4\rangle} + 14\hbar * \sigma^{\langle 3\rangle} + \sigma^{\langle 2\rangle},$$

etc.

Example 2.20. The equation
$$Da = ra, \ r \in \mathbf{R},$$
can be solved by writing
$$(k+1)a_{k+1} = ra_k, \ k \in \mathbf{N},$$
which yields $a_1 = ra_0$, $a_2 = ra_1/2 = r^2 a_0/2, \ldots,$
$$a = a_0 \left\{\frac{r^k}{k!}\right\}_{k \in \mathbf{N}}.$$

As in calculus, we may define the concept of a primitive of a sequence. Let ϕ be a sequence, if there is a sequence ψ such that $D\psi = \phi$, then ψ is called the primitive of ϕ. In particular, given $\phi = \{\phi_0, \phi_1, \phi_2, ...\}$, the primitive

$$\left\{0, \frac{\phi_0}{1}, \frac{\phi_1}{2}, \frac{\phi_2}{3}, \frac{\phi_3}{4}, ...\right\}$$

is called the *algebraic integral* of ϕ and is denoted by

$$\psi = \int \phi.$$

Hence

$$\int \phi = \hbar * \left\{\frac{\phi_k}{k+1}\right\}_{k \in \mathbf{N}},$$

and clearly, for any $\xi, \zeta \in l^{\mathbf{N}}$ and any $\alpha, \beta \in \mathbf{F}$,

$$\int (\alpha \xi + \beta \zeta) = \alpha \int \xi + \beta \int \zeta.$$

2.4.4 Composition Products

Let $f = \{f_n\}_{n \in \mathbf{N}}$ and $g = \{g_n\}_{n \in \mathbf{N}}$ be sequences in $l^{\mathbf{N}}$. Recall that $g^{\langle 0 \rangle} = \bar{1}$, $g^{\langle 1 \rangle} = g$ and $g^{\langle i \rangle} = g * g^{\langle i-1 \rangle}$ for $i \geq 2$. If no confusion is caused, we will write the n-th term of $g^{\langle i \rangle}$ by $g_n^{\langle i \rangle}$ instead of $(g^{\langle i \rangle})_n$. If

$$\lim_{k \to \infty} \sum_{i=0}^{k} f_i g_n^{\langle i \rangle} = \sum_{i=0}^{\infty} f_i g_n^{\langle i \rangle} < \infty, \; n \in \mathbf{N},$$

then the sequence $\left\{\sum_{i=0}^{\infty} f_i g_n^{\langle i \rangle}\right\}_{n \in \mathbf{N}}$ is called the *composition product* of f and g and denoted by $f \circ g$. For example,

$$f \circ \hbar = f.$$

The products $f \circ f, f \circ f \circ f, ...$, will be denoted by $f^{[2]}, f^{[3]}, ...$, respectively. We also define $f^{[1]} = f$ and $f^{[0]} = \hbar$. The sequence $f^{[p]}$ is called the p-th composition (product) power of f.

Example 2.21. Let $\varpi = \{1/n!\}_{n \in \mathbf{N}}$ and $\bar{c}$ be a scalar sequence. Since $\bar{c}_0^{\langle i \rangle} = c^i$ and $\bar{c}_n^{\langle i \rangle} = 0$ for $i \in \mathbf{N}$ and $n \in \mathbf{Z}^+$, we see that

$$(\varpi \circ \bar{c})_0 = 1 + c + \frac{1}{2!}c^2 + \cdots = e^c$$

and

$$(\varpi \circ \bar{c})_n = 0, \; n \in \mathbf{Z}^+.$$

Thus

$$\varpi \circ \bar{c} = \overline{e^c}.$$

Example 2.22. Let $a = \{a_n\}_{n \in \mathbf{N}}, b = \{b_n\}_{n \in \mathbf{N}} \in l^{\mathbf{N}}$ such that $b_0 = 0$. Then $a \circ b$ is well defined. Indeed, in view of Example 2.11, $b_n^{\langle i \rangle} = 0$ for $i > n$. Thus

$$\sum_{i=0}^{\infty} a_i b_n^{\langle i \rangle} = \sum_{i=0}^{n} a_i b_n^{\langle i \rangle} = \begin{cases} a_0 & n = 0 \\ \sum_{i=1}^{n} a_i b_n^{\langle i \rangle} & n \geq 1 \end{cases}.$$

Example 2.23. Let $g = \{g_n\} \in l^N$ such that $g_0 = 0$. Recall the Heavidside sequence $\mathbf{H}^{(m)}$ defined by $\mathbf{H}_i^{(m)} = 0$ for $i = 0, 1, ..., m-1$, and $\mathbf{H}_i^{(m)} = 1$ for $i \geq m$. Then

$$\left(\mathbf{H}^{(m)} \circ g\right)_n = \sum_{i=0}^{\infty} \mathbf{H}_i^{(m)} g_n^{\langle i \rangle} = \sum_{i=0}^{n} \mathbf{H}_i^{(m)} g_n^{\langle i \rangle}, \ n \in \mathbf{N}.$$

Hence,

$$\left(\mathbf{H}^{(m)} \circ g\right)_n = \begin{cases} \sum_{i=m}^{n} g_n^{\langle i \rangle} & n \geq m \\ 0 & n = 0, 1, ..., m-1 \end{cases}.$$

If we recall the convention that empty sums are equal to 0, then we may write

$$\mathbf{H}^{(m)} \circ g = \left\{\sum_{i=0}^{\infty} \mathbf{H}_i^{(m)} g_n^{\langle i \rangle}\right\} = \left\{\sum_{i=0}^{n} \mathbf{H}_i^{(m)} g_n^{\langle i \rangle}\right\} = \left\{\sum_{i=m}^{n} \mathbf{H}_i^{(m)} g_n^{\langle i \rangle}\right\} = \left\{\sum_{i=m}^{n} g_n^{\langle i \rangle}\right\}.$$

Example 2.24. Let $\varpi = \left\{\frac{1}{n!}\right\}_{n \in \mathbf{N}}$ and $g = \{g_n\}_{n \in \mathbf{N}}$. For fixed $n \in \mathbf{N}$, let $M_n = \max_{0 \leq i \leq n} |g_i|$. Since

$$\left|g_j^{\langle 2 \rangle}\right| = \left|\sum_{i=0}^{j} g_i g_{j-i}\right| \leq \sum_{i=0}^{j} |g_i| |g_{j-i}| \leq M_j^2(j+1), \ j = 0, ..., n,$$

we have

$$\left|g_j^{\langle 3 \rangle}\right| = \left|\sum_{i=0}^{j} g_i^{\langle 2 \rangle} g_{j-i}\right| \leq \sum_{i=0}^{n} \left|g_i^{\langle 2 \rangle}\right| |g_{j-i}| \leq M_j^3(j+1)^2, \ j = 0, ..., n,$$

and by induction,

$$\left|g_j^{\langle k \rangle}\right| \leq M_j^k(j+1)^{k-1}, \ j = 0, ..., n.$$

Thus by the ratio test,

$$|(\varpi \circ g)_n| = \left|\sum_{i=0}^{\infty} f_i g_n^{\langle i \rangle}\right| \leq 1 \left|g_n^{\langle 0 \rangle}\right| + \frac{1}{1!}\left|g_n^{\langle 1 \rangle}\right| + \frac{1}{2!}\left|g_n^{\langle 2 \rangle}\right| + \cdots$$

$$\leq 1 + \frac{1}{1!}M_n + \frac{1}{2!}M_n^2(n+1) + \frac{1}{3!}M_n^3(n+1)^2 + \cdots$$

$$= 1 + \frac{1}{n+1}\left\{e^{M_n(n+1)} - 1\right\}$$

$$< \infty.$$

In other words, $\varpi \circ g$ is a well defined sequence. It is also useful to note that

$$(\varpi \circ g)_0 = 1 + \frac{1}{1!}g_0 + \frac{1}{2!}g_0^2 + \cdots = e^{g_0}.$$

The same principle leads to the following result.

Theorem 2.14. *Let $f = \{f_n\}_{n\in \mathbf{N}} \in l^{\mathbf{N}}$ be a sequence such that $\underline{\lambda} \cdot f$ is absolutely summable for any $\underline{\lambda}$. Then for any $g \in l^{\mathbf{N}}$, $f \circ g$ is well defined.*

Theorem 2.15. *Let $g = \{g_n\}_{n\in \mathbf{N}} \in l^{\mathbf{N}}$ such that $g_0 = 0$ and $g_1 \neq 0$. Then the equation $x \circ g = \hbar$ has a unique solution $x \in l^{\mathbf{N}}$.*

To see the proof, recall that $g_n^{\langle n \rangle} = g_1^n \neq 0$ and $g_n^{\langle m \rangle} = 0$ for $m > n$. The equation $x \circ g = \hbar$ is thus equivalent to the infinite system
$$(x \circ g)_0 = x_0 g_0^{\langle 0 \rangle} = 0,$$
$$(x \circ g)_1 = x_0 g_1^{\langle 0 \rangle} + f_1 g_1^{\langle 1 \rangle} = 1,$$
$$(x \circ g)_2 = x_0 g_2^{\langle 0 \rangle} + x_1 g_2^{\langle 1 \rangle} + x_2 g_2^{\langle 2 \rangle} = 0,$$
$$(x \circ g)_3 = x_0 g_3^{\langle 0 \rangle} + x_1 g_3^{\langle 1 \rangle} + x_2 g_3^{\langle 2 \rangle} + x_3 g_3^{\langle 3 \rangle} = 0,$$
$$... = ...,$$
from which we may obtain $x_0 = 0$, $x_1 = 1/g_1$, $x_2 = -x_1 g_2^{\langle 1 \rangle}/g_1^2, ...$ in a unique manner.

The unique solution in the above result is denoted by $g^{[-1]}$.

For $g \in l^{\mathbf{N}}$, we will call $\varpi \circ g$ the exponential of g. Two reasons for naming it in such a manner are
$$(\varpi \circ f) * (\varpi \circ g) = \varpi \circ (f + g), \qquad (2.31)$$
and
$$D(\varpi \circ f) = (\varpi \circ f) * Df \qquad (2.32)$$
for any $f, g \in l^{\mathbf{N}}$. To show these, we first recall that a sequence $\{f^{(j)}\}_{j \in \mathbf{N}}$ of sequences in $l^{\mathbf{N}}$ is said to converge (pointwise) to $f \in l^{\mathbf{N}}$ if
$$\lim_{j \to \infty} f_k^{(j)} = f_k, \ k \in \mathbf{N}.$$
Clearly, if $\{f^{(j)}\}_{j \in \mathbf{N}}$ and $\{g^{(j)}\}_{j \in \mathbf{N}}$ are two sequences of sequences which converge to f and g respectively, then
$$\lim_{j \to \infty} \left(f^{(j)} + g^{(j)} \right) = f + g,$$
$$\lim_{j \to \infty} \left(f^{(j)} \cdot g^{(j)} \right) = f \cdot g,$$
and
$$\lim_{j \to \infty} f^{(j)} * g^{(j)} = f * g.$$
We may also define the infinite sum of a sequence $\{f^{(j)}\}_{j \in \mathbf{N}}$ of sequences as the limiting sequence of the partial sum sequence $\left\{ \sum_{j=0}^n f^{(j)} \right\}_{n \in \mathbf{N}}$:
$$\sum_{j=0}^{\infty} f^{(j)} = \lim_{n \to \infty} \sum_{j=0}^{n} f^{(j)}.$$

If such a limiting sequence exists, we say that the series $\sum_{j=0}^{\infty} f^{(j)}$ converges. Note that $\sum_{j=0}^{\infty} f^{(j)}$ converges if, and only if,

$$\sum_{j=0}^{\infty} f^{(j)} = \sum_{j=0}^{\infty} \left\{f_n^{(j)}\right\}_{n \in \mathbf{N}} = \left\{\sum_{j=0}^{\infty} f_n^{(j)}\right\}_{n \in \mathbf{N}},$$

that is, the k-th term of the series is obtained by 'adding' all the k-th terms of the individual sequences.

In case the composition product $f \circ g$ of $f, g \in l^{\mathbf{N}}$ is defined, we may now easily see from the previous observation that

$$f \circ g = \left\{\sum_{i=0}^{\infty} f_i g_n^{\langle i \rangle}\right\}_{n \in \mathbf{N}} = \sum_{i=0}^{\infty} f_i g^{\langle i \rangle}.$$

If the infinite sums $\sum_{j=0}^{\infty} f^{(j)}$ and $\sum_{j=0}^{\infty} g^{(j)}$ of two respective sequences $\{f^{(j)}\}_{j \in \mathbf{N}}$ and $\{g^{(j)}\}_{j \in \mathbf{N}}$ of sequences in $l^{\mathbf{N}}$ converge, then it is also easy to see that

$$\sum_{j=0}^{\infty} \left(\alpha f^{(j)} + \beta g^{(j)}\right) = \alpha \sum_{j=0}^{\infty} f^{(j)} + \beta \sum_{j=0}^{\infty} g^{(j)}, \alpha, \beta \in \mathbf{C},$$

$$\sum_{j=0}^{\infty} D f^{(j)} = D \left(\sum_{j=0}^{\infty} f^{(j)}\right),$$

and

$$\sum_{j=0}^{\infty} \int f^{(j)} = \int \left(\sum_{j=0}^{\infty} f^{(j)}\right),$$

and for any $g \in l^{\mathbf{N}}$,

$$\sum_{j=0}^{\infty} f^{(j)} \cdot g = \left(\sum_{j=0}^{\infty} f^{(j)}\right) \cdot g$$

and

$$\sum_{j=0}^{\infty} f^{(j)} * g = \left(\sum_{j=0}^{\infty} f^{(j)}\right) * g.$$

We first show the validity of (2.32):

$$D(\varpi \circ f) = D\left(\sum_{k=0}^{\infty} \frac{1}{k!} f^{\langle k \rangle}\right) = \sum_{k=0}^{\infty} D\left(\frac{1}{k!} f^{\langle k \rangle}\right) = \sum_{k=1}^{\infty} \frac{k f^{\langle k-1 \rangle} * Df}{k!} = (\varpi \circ f) * Df.$$

To show (2.31), we need the following result.

Example 2.25. Consider the equation
$$Dg = \lambda(Dh) * g,$$
where λ is a fixed number different from 0, h is a given sequence in $l^{\mathbf{N}}$ and g is a sequence in $l^{\mathbf{N}}$ to be sought. Then $g = \varpi \circ (\lambda h)$ is such a solution since
$$Dg = D(\varpi \circ (\lambda h)) = (\varpi \circ (\lambda h)) * D(\lambda h) = \lambda(Dh) * g.$$
We assert that any other solution must be a constant multiple of $\varpi \circ (\lambda h)$. To see this, let g be another solution and consider the ratio $g/(\varpi \circ (\lambda h))$ which is defined since $(\varpi \circ (\lambda h))_0$ is not zero by Theorem 2.12. Note that
$$\begin{aligned} D\left(\frac{g}{\varpi \circ (\lambda h)}\right) &= \frac{(\varpi \circ (\lambda h)) * Dg - g * D(\varpi \circ (\lambda h))}{(\varpi \circ (\lambda h))^{\langle 2 \rangle}} \\ &= \frac{(\varpi \circ (\lambda h)) * Dg - g * (\lambda Dh) * (\varpi \circ (\lambda h))}{(\varpi \circ (\lambda h))^{\langle 2 \rangle}} \\ &= \frac{(\varpi \circ (\lambda h)) * (Dg - \lambda(Dh) * g)}{(\varpi \circ (\lambda h))^{\langle 2 \rangle}} \\ &= 0, \end{aligned}$$
thus,
$$\frac{g}{\varpi \circ (\lambda h)} = \overline{\beta},$$
for some scalar sequence $\overline{\beta}$, or,
$$g = \beta(\varpi \circ (\lambda h)).$$

We now show the validity of (2.31). First note that
$$q = (\varpi \circ f) * (\varpi \circ g)$$
satisfies the equation
$$Dq = (D(f+g)) * q,$$
since
$$\begin{aligned} Dq &= (\varpi \circ f) * D(\varpi \circ g) + (D(\varpi \circ f)) * (\varpi \circ g) \\ &= (\varpi \circ f) * \{(Dg) * (\varpi \circ g)\} + \{(Df) * (\varpi \circ f)\} * (\varpi \circ g) \\ &= q * Dg + q * Df \\ &= q * D(f+g). \end{aligned}$$
By the uniqueness in the previous example, we see that
$$(\varpi \circ f) * (\varpi \circ g) = \beta(\varpi \circ (f+g))$$
for some constant β. But since
$$((\varpi \circ f) * (\varpi \circ g))_0 = e^{f_0 + g_0} = (\varpi \circ (f+g))_0,$$
we see that $\beta = 1$.

As an interesting consequence, note that

$$\left(\sum_{k=0}^{\infty}\frac{f^{\langle k\rangle}}{k!}\right) * \left(\sum_{k=0}^{\infty}\frac{g^{\langle k\rangle}}{k!}\right) = (\varpi \circ f) * (\varpi \circ g)$$

$$= \varpi \circ (f+g)$$

$$= \sum_{k=0}^{\infty}\frac{1}{k!}(f+g)^{\langle k\rangle}$$

$$= \sum_{k=0}^{\infty}\frac{1}{k!}\sum_{m=0}^{k}\frac{k!}{m!(m-k)!}(f^{\langle m\rangle} * g^{\langle k-m\rangle})$$

$$= \sum_{k=0}^{\infty}\sum_{m=0}^{k}\left(\frac{f^{\langle m\rangle}}{m!} * \frac{g^{\langle k-m\rangle}}{(m-k)!}\right).$$

As another interesting consequence, we have the following result which will be useful in the calculation of the higher derivatives of composite functions.

Example 2.26. Consider the equation

$$(E\varpi) \cdot (DB) = \lambda\left[(E\varpi) \cdot (Dg)\right] * (\varpi \cdot B), \qquad (2.33)$$

where λ is a fixed number different from 0, g is a given sequence in $l^{\mathbf{N}}$ and B is a sequence in $l^{\mathbf{N}}$ to be sought (recall also that $\{p_k\} \cdot \{q_k\} = \{p_k q_k\}$). Note that (2.33) can be written as

$$D(\varpi \cdot B) = \lambda D(\varpi \cdot g) * (\varpi \cdot B).$$

By our previous Example, we see that

$$\varpi \cdot B = \overline{\beta} * (\varpi \circ (\lambda \varpi \cdot g))$$

for some scalar sequence $\overline{\beta}$. Thus,

$$B = \varpi^{-1} \cdot \left(\overline{\beta} * (\varpi \circ (\lambda \varpi \cdot g))\right).$$

The previous facts are useful in finding the n-th derivative of a composite function. For the sake of convenience, we will use $\mathbf{D}_t^n f(t)$ to denote the n-th derivative $f^{(n)}(t)$ for $n \in \mathbf{N}$.

Theorem 2.16 (Formula of Faà di Bruno). *If $f(t)$ and $g(t)$ are functions for which all the necessary derivatives are defined, then for $n \in \mathbf{Z}^+$,*

$$\mathbf{D}_t^n f(g(t)) = \frac{n!}{k!}\sum_{j_1+\cdots+j_k=n; j_1,\cdots,j_k \in \mathbf{Z}^+} \mathbf{D}_u^k f(u)|_{u=g(t)} \left(\frac{\mathbf{D}_t^{j_1} g(t)}{j_1!}\right) \cdots \left(\frac{\mathbf{D}_t^{j_k} g(t)}{n!}\right).$$

Proof. Let us write $h(t) = f(g(t))$ and

$$h_n = \mathbf{D}_t^n h(t),$$
$$g_n = \mathbf{D}_t^n g(t),$$
$$f_n = \mathbf{D}_u^n f(u)|_{u=g(t)}$$

for $n \in \mathbf{N}$. Then
$$h_1 = \mathbf{D}_t^1 h(t) = \mathbf{D}_u^1 f(u)|_{u=g(t)} \mathbf{D}_t^1 g(t) = f_1 g_1,$$
and similarly
$$h_2 = f_1 g_2 + f_2 g_1^2$$
$$h_3 = f_1 g_3 + f_2 3 g_1 g_2 + f_3 g_1^3.$$
It is easily established by induction that h_n has the form
$$h_n = \sum_{k=1}^{n} f_k l_{n,k}(g_1, ..., g_n) \qquad (2.34)$$
where $l_{n,k}(g_1, ..., g_n)$ does not depend on any of the functions f_j. Now, since we wish only to determine $l_{n,k}(g_1, ..., g_n)$, we are free to choose $f(t)$ arbitrarily. Let us take $f(t) = e^{\lambda t}$ where λ is an arbitrary constant different from 0. Then
$$f_k = \mathbf{D}_u^k f(u)|_{u=g(t)} = \lambda^k e^{\lambda g(t)}, \ k \in \mathbf{N}, \qquad (2.35)$$
and
$$h_n = \mathbf{D}_t^n e^{\lambda g(t)}, \ n \in \mathbf{N}. \qquad (2.36)$$
Substituting (2.35) and (2.36) into (2.34) and multiplying by $e^{-\lambda g(t)}$ gives
$$e^{-\lambda g(t)} \mathbf{D}_t^n e^{\lambda g(t)} = \sum_{k=1}^{n} \lambda^k l_{n,k}(g_1, ..., g_n).$$
If we set $B_n(t) = e^{-\lambda g(t)} \mathbf{D}_t^n e^{\lambda g(t)}$ for $n \in \mathbf{N}$, then $B_0(t) = 1$ and
$$B_n(t) = e^{-\lambda g(t)} \mathbf{D}_t^{n-1} \lambda g_1(t) e^{\lambda g(t)}$$
$$= \lambda e^{-\lambda g(t)} \sum_{k=0}^{n-1} \binom{n-1}{k} g_{k+1}(t) \mathbf{D}_t^{n-k-1} e^{\lambda g(t)}$$
$$= \lambda \sum_{k=0}^{n-1} \binom{n-1}{k} g_{k+1}(t) B_{n-k-1}(t) \qquad (2.37)$$
for $n \in \mathbf{Z}^+$, where we have used Leibniz's formula for the second equality. Now we may think of t as being fixed and define sequences $B = \{B_n\}_{n \in \mathbf{N}}$ and $g = \{g_n\}_{n \in \mathbf{N}}$ where $B_n(t) = B_n$ and $g_n(t) = g_n$ for $n \in \mathbf{N}$.

Equation (2.37) now becomes
$$(E\varpi) \cdot (DB) = \lambda [(E\varpi) \cdot (Dg)] * (\varpi \cdot B) \qquad (2.38)$$
In view of Example 2.26, all its solutions are of the form
$$B = \varpi^{-1} \cdot [\bar{c} * [\varpi \circ (\lambda \varpi \cdot g)]]$$
where c may depend on the fixed t. In order to determine c, we recall that $B_0(t) = 1$ and so
$$1 = B_0(t) = c(\varpi^{-1})_0 [\varpi \circ (\lambda \varpi \cdot g)]_0 = ce^{(\lambda \varpi g)_0} = ce^{\lambda g_0},$$

which implies $c = e^{-\lambda g_0}$. Thus in view of the previous Example,

$$B = \varpi^{-1} \cdot \left[\overline{e^{-\lambda g_0}} * [\varpi \circ (\lambda \varpi \cdot g)]\right]$$
$$= \varpi^{-1} \cdot [(\varpi \circ (\ \overline{\lambda g_0})) * (\varpi \circ (\lambda \varpi \cdot g))] = \varpi^{-1} \cdot \{\varpi \circ [\lambda(\varpi \cdot g - \overline{g_0})]\}.$$

Since $(\varpi g - \overline{g_0})_0 = 0$, by Theorem 2.12 and (2.15),

$$B_n = \frac{n!}{0!}[\lambda(\varpi g - \overline{g_0})]_n^{<0>} + n! \sum_{k=1}^n \frac{\lambda^k}{k!} \sum_{j_1+\cdots+j_k=n;j_1,\ldots,j_k \in \mathbf{Z}^+} \binom{g_{j_1}}{j_1!} \cdots \binom{g_{j_k}}{j_k!}$$

$$= n! \sum_{k=1}^n \frac{\lambda^k}{k!} \sum_{j_1+\cdots+j_k=n;j_1,\ldots,j_k \in \mathbf{Z}^+} \binom{g_{j_1}}{j_1!} \cdots \binom{g_{j_k}}{j_k!}$$

for $n \geq 1$. By equating coefficients of λ^k, where $k \geq 1$, in the two expressions for B_n gives

$$l_{n,k}(g_1, \ldots, g_n) = \frac{n!}{k!} \sum_{j_1+\cdots+j_k=n;j_1,\ldots,j_k \in \mathbf{Z}^+} \binom{g_{j_1}}{j_1!} \cdots \binom{g_{j_k}}{j_k!}, \ n \in \mathbf{Z}^+.$$

This is the desired formula and the proof is complete.

We remark that (see Roman [164])

$$\frac{n!}{k!} \sum_{j_1+\cdots+j_k=n;j_1,\ldots,j_k \in \mathbf{Z}^+} \binom{g_{j_1}}{j_1!} \cdots \binom{g_{j_k}}{j_k!} = \sum \frac{n!}{k_1! \cdots k_n!} f_k \left(\frac{g_1}{1!}\right)^{k_1} \cdots \left(\frac{g_n}{n!}\right)^{k_n},$$

where the last sum is over all $k_1, \ldots, k_n$ for which $k_1 + \cdots + k_n = k$ and $k_1 + 2k_2 + \cdots + nk_n = n$.

Example 2.27. Take

$$g(t) = \frac{1}{1-t} = \sum_{n=0}^\infty t^n, \ t \in (-1, +1)$$

and

$$f(x) = \frac{1}{1 - r(x-1)} = \sum_{n=0}^\infty r^j(x-1)^j, \ |r(x-1)| < 1.$$

Then

$$f(g(t)) = \frac{1-t}{1-(r+1)t} = \frac{1}{1-(r+1)t} - \frac{t}{1-(r+1)t}$$

$$= \sum_{n=0}^\infty (1+r)^n t^n - \sum_{n=0}^\infty (1+r)^n t^{n+1}$$

$$= 1 + \sum_{n=1}^\infty r(1+r)^{n-1} t^n$$

for $|(r+1)t| < 1$. Thus $g^{(j)}(0) = j!$, $f^{(k)}(g(0)) = k!r^k$ and
$$\frac{d^n f(g(0))}{dt} = n!r(1+r)^{n-1}$$
$$= \sum \frac{n!}{k_1!k_2!\cdots k_n!} f(g(0)) \left(\frac{g^{(1)}(0)}{1!}\right)^{k_1} \cdots \left(\frac{g^{(n)}(0)}{n!}\right)^{k_n}$$
$$= \sum \frac{n!}{k_1!k_2!\cdots k_n!} k!r^k$$
where $k = k_1 + \cdots + k_n$ and the sum is taken over all $k_1, ..., k_n$ for which $k_1 + 2k_2 + \cdots + nk_n = n$. Consequently,
$$\sum_{k_1+2k_2+\cdots+nk_n=n; k_1,...,k_n \in \mathbf{N}} \frac{k!}{k_1!k_2!\cdots k_n!} r^k = r(1+r)^{n-1}, \ n \in \mathbf{Z}^+.$$

2.5 Properties of Bivariate Sequences

Let $l^{\mathbf{N}\times\mathbf{N}}$ be the set of all complex bivariate sequences of the form $f = \{f_{ij}\}_{i,j\in\mathbf{N}}$. Such a bivariate sequence f is a function defined on the set of all nonnegative lattice points $\mathbf{N} \times \mathbf{N}$ and it is natural to view a bivariate sequence as an infinite matrix of the form
$$\begin{bmatrix} f_{00} & f_{01} & \cdots \\ f_{10} & f_{11} & \cdots \\ \cdots & \cdots & \cdots \end{bmatrix}.$$
We will also write $\{f_{ij}\}$ instead of $\{f_{ij}\}_{i,j\in\mathbf{N}}$ if no confusion is caused. The number f_{ij} will be called the (i,j)-th component of the bivariate sequence f, while the sequences $\{f_{i0}, f_{i1}, ...\}$ and $\{f_{0j}, f_{1j}, ...\}$ will be called its i-th row and j-th column.

For any complex number α and $f = \{f_{ij}\}, g = \{g_{ij}\} \in l^{\mathbf{N}\times\mathbf{N}}$, we define $-f$, αf, $f \cdot g$ and $f + g$ respectively by $\{-f_{ij}\}, \{\alpha f_{ij}\}, \{f_{ij}g_{ij}\}$ and $\{f_{ij} + g_{ij}\}$ as usual.

There are some common sequences in $l^{\mathbf{N}\times\mathbf{N}}$ which deserve special notations. First of all, let α be a complex number, the sequence whose $(0,0)$-th component is α and others are zero will be denoted by $\overline{\alpha}$ and is called a *scalar bivariate sequence*. In particular, the sequence with all zero components will be denoted by $\overline{0}$. The bivariate sequence whose $(1,0)$-th component is 1 and others are zero will be denoted by $\hbar_x$, while the sequence whose $(0,1)$-th component is 1 and others are zero will be denoted by $\hbar_y$:
$$\hbar_x = \begin{bmatrix} 0 & 0 & 0 & \cdots \\ 1 & 0 & 0 & \cdots \\ 0 & 0 & 0 & \cdots \\ \cdots & \cdots & \cdots & \cdots \end{bmatrix}, \ \hbar_y = \begin{bmatrix} 0 & 1 & 0 & \cdots \\ 0 & 0 & 0 & \cdots \\ 0 & 0 & 0 & \cdots \\ \cdots & \cdots & \cdots & \cdots \end{bmatrix},$$
while the bivariate sequences $\sigma_x, \sigma_y, \delta_x$ and δ_y are defined by
$$\sigma_x = \begin{bmatrix} 1 & 0 & 0 & \cdots \\ 1 & 0 & 0 & \cdots \\ 1 & 0 & 0 & \cdots \\ \cdots & \cdots & \cdots & \cdots \end{bmatrix}, \ \sigma_y = \begin{bmatrix} 1 & 1 & 1 & \cdots \\ 0 & 0 & 0 & \cdots \\ 0 & 0 & 0 & \cdots \\ \cdots & \cdots & \cdots & \cdots \end{bmatrix},$$

$$\delta_x = \begin{bmatrix} 1 & 0 & 0 & \cdots \\ -1 & 0 & 0 & \cdots \\ 0 & 0 & 0 & \cdots \\ \cdots & \cdots & \cdots & \end{bmatrix}, \; \delta_y = \begin{bmatrix} 1 & -1 & 0 & \cdots \\ 0 & 0 & 0 & \cdots \\ 0 & 0 & 0 & \cdots \\ \cdots & \cdots & \cdots & \end{bmatrix}$$

respectively. Note that $\delta_x + \hbar_x = \overline{1}$ and $\delta_y + \hbar_y = \overline{1}$.

The bivariate sequence $\{f_{i+m,j+n}\}_{i,j \in Z}$ will be denoted by $E_x^m E_y^n \{f_{ij}\}$, where $m, n \in \mathbf{N}$. The sequence $E_x^m E_y^n f$ is called a *translated sequence* of f. For the sake of convenience, $E_x^0 E_y^n f$ and $E_x^m E_y^0 f$ are also denoted by $E_y^n f$ and $E_x^m f$ respectively.

For any complex numbers λ and μ, the sequence $\{\lambda^i \mu^j f_{ij}\}$ is called an attenuated bivariate sequence of f and is denoted by $(\lambda, \mu) \cdot f$. It is easily seen that $(0, 0) \cdot f = \overline{f_{00}}$, $(1,1) \cdot f = f$ and $(\lambda, \mu) \cdot \big((\rho, \tau) \cdot f\big) = (\lambda\rho, \mu\tau) \cdot f$.

For any $f = \{f_{ij}\}, g = \{g_{ij}\} \in l^{\mathbf{N} \times \mathbf{N}}$, we define the convolution product $f * g$, by

$$(f * g)_{ij} = \sum_{u=0}^{i} \sum_{v=0}^{j} f_{uv} g_{i-u, j-v}, \; i, j \geq 0.$$

We may evaluate the components of $h = f * g$ in an orderly manner as follows:

$$h_{00} = f_{00} g_{00}; \; h_{10} = f_{10} g_{00} + f_{00} g_{10}; \; h_{01} = f_{01} g_{00} + f_{00} g_{01};$$

$$h_{20} = f_{20} g_{00} + f_{10} g_{10} + f_{00} g_{20}, \; h_{11} = f_{11} g_{00} + f_{01} g_{10} + f_{10} g_{01} + f_{00} g_{11}, \ldots.$$

For the sake of convenience, we will also use the simpler notation fg for the product $f * g$. Note that $f * f$, $f * (f * f)$, ..., will also be written as $f^{\langle 2 \rangle}, f^{\langle 3 \rangle}, \ldots$, respectively.

For example, $\overline{0} * f = \overline{0}$, $\overline{1} * f = f$, $\overline{\alpha} * \overline{\beta} = \overline{\alpha\beta}$, and $\overline{\alpha} * f = (\alpha\overline{1}) * f = \alpha(\overline{1} * f) = \alpha f$. More complicated examples can also be given. First of all, $\hbar_x^{\langle m \rangle}$ (or $\hbar_y^{\langle m \rangle}$) is a bivariate sequence whose $(m, 0)$-th component (respectively $(0, m)$-th component) is 1 and others are zero, while $\hbar_x^{\langle m \rangle} * \hbar_y^{\langle n \rangle}$ is a bivariate sequence whose (m, n)-th component is 1 and others are zero. It is also interesting to note that $\hbar_x^{\langle m \rangle} * \hbar_y^{\langle n \rangle} * \{f_{ij}\} = \{g_{ij}\}$ where

$$g_{ij} = \begin{cases} f_{i-m, j-n} & i \geq m, j \geq n \\ 0 & \text{otherwise} \end{cases}.$$

For instance, the matrix representation of the bivariate sequence $\hbar_x^{\langle 2 \rangle} * \hbar_y^{\langle 1 \rangle} * \{f_{ij}\}$ is

$$\begin{bmatrix} 0 & 0 & 0 & 0 & \cdots \\ 0 & 0 & 0 & 0 & \cdots \\ 0 & f_{00} & f_{01} & f_{02} & \cdots \\ 0 & f_{10} & f_{11} & f_{12} & \cdots \\ \cdots & \cdots & \cdots & \cdots & \end{bmatrix},$$

while $f_{21}\hbar_x^{\langle 2\rangle} * \hbar_y^{\langle 1\rangle}$ is

$$\begin{bmatrix} 0 & 0 & 0 & \cdots \\ 0 & 0 & 0 & \cdots \\ 0 & f_{21} & 0 & \cdots \\ 0 & 0 & 0 & \cdots \\ \cdots & \cdots & \cdots & \cdots \end{bmatrix}.$$

There are several elementary facts related to the convolution product of bivariate sequences. First of all, we may show that for any bivariate sequences $f = \{f_{ij}\}$, $g = \{g_{ij}\}$ and $h = \{h_{ij}\}$, we have $f * g = g * f$ and $f * (g * h) = (f * g) * h$. Indeed, these are due to the fact that the convolution product of sequences of a single integral variable are commutative and associative:

$$\sum_{k=0}^{i} x_k y_{i-k} = \sum_{k=0}^{i} x_{i-k} y_k,$$

and

$$\sum_{k=0}^{m}\left(\sum_{i=0}^{k} x_i y_{k-i}\right) z_{m-k} = \sum_{i=0}^{m}\sum_{k=i}^{m} x_i y_{k-i} z_{m-k} = \sum_{i=0}^{m} x_i \sum_{j=0}^{m-i} y_j z_{m-i-j}.$$

Next, we show that when $f \neq \overline{0}$ and $g \neq \overline{0}$, then $f * g \neq \overline{0}$. Indeed, suppose the components of f and g are ordered by the mapping $\tilde{\Psi}$ defined by (1.3). Then we may assume without loss of generality that

$$f_{00} = f_{10} = f_{01} = \cdots = f_{m+1,n-1} = 0, \ f_{mn} \neq 0,$$

and

$$g_{00} = g_{10} = g_{01} = \cdots = g_{s+1,t-1} = 0, \ g_{st} \neq 0,$$

where $\tilde{\Psi}^{-1}(m,n) \leq \tilde{\Psi}^{-1}(s,t)$. Since when $s + t \geq m + n$,

$$(fg)_{m+s,n+t} = f_{00}g_{m+s,n+t} + \cdots + f_{mn}g_{st} + \cdots + f_{m+s,n+t}g_{00} = f_{mn}g_{st} \neq 0.$$

we see that $f * g \neq 0$.

Theorem 2.17. *Let $f = \{f_{ij}\}$ and $g = \{g_{ij}\}$ be bivariate sequences in $l^{\mathbf{N}\times\mathbf{N}}$. If $g_{00} \neq 0$, then there is a unique bivariate sequence $x = \{x_{ij}\}$ such that $g * x = f$.*

The proof is elementary. We write the component equations of $g * x = f$ in the following orderly manner:

$$g_{00}x_{00} = f_{00},$$
$$g_{00}x_{10} + g_{10}x_{00} = f_{10},$$
$$g_{00}x_{01} + g_{01}x_{00} = f_{01},$$
$$g_{00}x_{20} + g_{10}x_{10} + g_{20}x_{00} = f_{20},$$
$$g_{00}x_{11} + g_{10}x_{01} + g_{01}x_{10} + g_{11}x_{00} = f_{11},$$

and so on, and then obtain $x_{00} = f_{00}/g_{00}$, $x_{10} = (f_{10} - g_{10}x_{00})/g_{00}$, ..., successively in a unique manner.

In case $g = \{g_{ij}\}$ satisfies $g_{00} \neq 0$, the quotient f/g will denote the solution sequence of the equation

$$g * x = f.$$

One important question is how to find the explicit form of a quotient f/g. We remark that although we have mentioned an algorithm to calculate f/g, as we will see below, it may also be found by other means.

Example 2.28. Let $f = \{f_{ij}\} \in l^{\mathbf{N} \times \mathbf{N}}$. If $f_{00} = 0$, then for all $(i,j) \in \{(i,j) \in \mathbf{N}^2 |\ i+j \leq n-1\}$, we have $f_{ij}^{\langle n \rangle} = 0$, where $n \in \mathbf{Z}^+$. Indeed, let $Q_k = \{(i,j) \in \mathbf{N}^2 |\ i+j = k\}$ for $k \in \mathbf{N}$. Assume by induction that $f_{ij}^{\langle k \rangle} = 0$ for $(i,j) \in Q_{k-1}$ where k is a positive integer. Then for $(i,j) \in Q_0 + \cdots + Q_{k-1}$,

$$f_{ij}^{\langle k+1 \rangle} = \sum_{u=0}^{i} \sum_{v=0}^{j} f_{uv}^{\langle k \rangle} f_{i-u,j-v} = \sum_{u=0}^{i} \sum_{v=0}^{j} 0 \cdot f_{i-u,j-v} = 0.$$

For $(i,j) \in Q_k$, let $S = \{0,1,...,i\} \times \{0,1,...,j\}$, then

$$f_{ij}^{\langle k+1 \rangle} = \sum_{(u,v) \in S \setminus \{(i,j)\}} f_{uv}^{\langle k \rangle} f_{i-u,j-v} + f_{ij}^{\langle k \rangle} f_{00} = 0.$$

For instance, when $f_{00} = 0$, the matrix representation of $f^{\langle 4 \rangle}$ is of the form

$$\begin{bmatrix} 0 & 0 & 0 & * & ... \\ 0 & 0 & * & . & ... \\ 0 & * & . & . & ... \\ * & . & . & . & ... \\ . & . & . & . & ... \end{bmatrix}$$

Theorem 2.18. Let $f = \{f_{ij}\}, g = \{g_{ij}\}$ be bivariate sequences in $l^{\mathbf{N} \times \mathbf{N}}$. Then $\underline{(\lambda,\mu)} \cdot (f * g) = \left(\underline{(\lambda,\mu)} \cdot f\right) * \left(\underline{(\lambda,\mu)} \cdot g\right)$ for $\lambda, \mu \in C$.

Indeed,

$$\underline{(\lambda,\mu)} \cdot (f * g) = \left\{ \sum_{u=0}^{i} \sum_{v=0}^{j} \lambda^i \mu^j f_{uv} g_{i-u,j-v} \right\}$$

$$= \left\{ \sum_{u=0}^{i} \sum_{v=0}^{j} \lambda^u \mu^j f_{uv} \lambda^{i-u} \mu^{j-v} g_{i-u,j-v} \right\}$$

$$= \left(\underline{(\lambda,\mu)} \cdot f\right) * \left(\underline{(\lambda,\mu)} \cdot g\right).$$

Theorem 2.19. Let $f = \{f_{ij}\}, g = \{g_{ij}\}$ be bivariate sequences in $l_1^{\mathbf{N} \times \mathbf{N}}$. Then $f * g \in l_1^{\mathbf{N} \times \mathbf{N}}$ and $\sum_{\mathbf{N} \times \mathbf{N}} f * g = \left(\sum_{\mathbf{N} \times \mathbf{N}} f\right) \left(\sum_{\mathbf{N} \times \mathbf{N}} g\right)$.

To see the proof, we first assume that $f, g \geq 0$. Note that

$$\sum_{i=0}^{m}\sum_{j=0}^{n}\left(\sum_{u=0}^{i}\sum_{v=0}^{j} f_{uv}g_{i-u,j-v}\right) \leq \sum_{i=0}^{m}\sum_{j=0}^{n} f_{ij} \sum_{i=0}^{m}\sum_{j=0}^{n} g_{ij} \leq \sum_{\mathbf{N}\times\mathbf{N}} f \sum_{\mathbf{N}\times\mathbf{N}} g.$$

Thus for any $w \in l^{N\times N}$ such that $0 \leq w \leq f * g$ and the support of u is finite, we have

$$\sum_{i=0}^{m}\sum_{j=0}^{n} w_{ij} \leq \sum_{\mathbf{N}\times\mathbf{N}} f \sum_{\mathbf{N}\times\mathbf{N}} g$$

for all sufficiently large m and n. Taking the supremum on the left hand side, we see that $f * g \in l_1^{\mathbf{N}\times\mathbf{N}}$ and $\sum_{\mathbf{N}\times\mathbf{N}} f*g \leq \left(\sum_{\mathbf{N}\times\mathbf{N}} f\right)\left(\sum_{\mathbf{N}\times\mathbf{N}} g\right)$. Next, let $u, v \in l_1^{\mathbf{N}\times\mathbf{N}}$ such that u, v have finite supports and $0 \leq u \leq f$ and $0 \leq v \leq g$. We may assume that the supports $\Phi(u)$ and $\Phi(v)$ are $\{(i,j)|\ i=0,1,...,\alpha; j=0,1,...,\beta\}$ and $\{(i,j)|\ i=0,1,...,\gamma; j=0,1,...,\delta\}$. Let E be the set $\{(i,j)|\ i,j=0,1,...,\alpha\beta\gamma\delta\}$. Since it can easily be checked by listing all the terms of $u*v$ that

$$0 \leq \sum_{\mathbf{N}\times\mathbf{N}} u \sum_{\mathbf{N}\times\mathbf{N}} v \leq \sum_{\mathbf{N}\times\mathbf{N}} f*g.$$

Thus $\left(\sum_{\mathbf{N}\times\mathbf{N}} f\right)\left(\sum_{\mathbf{N}\times\mathbf{N}} g\right) \leq \sum_{\mathbf{N}\times\mathbf{N}} f*g$. For f and g which are not necessary nonnegative, the routine procedure of breaking f and g into real and imaginary parts and/or positive and negative parts will then lead to a proof.

Given a bivariate sequence $f = \{f_{ij}\}$, we denote the sequences $\{(i+1)f_{i+1,j}\}$ and $\{(j+1)f_{i,j+1}\}$ by $D_x f$ and $D_y f$ respectively and call them the (partial) algebraic derivatives of f. The higher algebraic and mixed derivatives are defined recursively. Thus we have

$$D_x \overline{\alpha} = D_y \overline{\alpha} = \overline{0}, \ \alpha \in \mathbf{C},$$

and

$$D_x^m D_y^n f = \{[(i+1)\cdots(i+m)][(j+1)\cdots(j+n)]f_{i+m,j+n}\} = D_y^n D_x^m f$$

for $m, n \in \mathbf{Z}^+$. It is easily verified that for any $\alpha, \beta \in \mathbf{C}$, and $f, g \in l^{\mathbf{N}\times\mathbf{N}}$,

$$D_x(\alpha f + \beta g) = \alpha D_x f + \beta D_x g, \ D_y(\alpha f + \beta g) = \alpha D_y f + \beta D_y g,$$

$$D_x(f * g) = f * D_x g + g * D_x f, \ D_y(f * g) = f * D_y g + g * D_y f,$$

and

$$D_x\left(\frac{f}{g}\right) = \frac{g*D_x f - f*D_x g}{g^{\langle 2 \rangle}}, \ D_y\left(\frac{f}{g}\right) = \frac{g*D_y f - f*D_y g}{g^{\langle 2 \rangle}},$$

where we have assumed that $g_{00} \neq 0$ in the quotient f/g.

Algebraic derivatives of some common operators can easily be found. More complicated derivatives can be obtained by employing the following list of useful formulas:

$$D_x \hbar_y = D_y \hbar_x = \overline{0},$$

$$\hbar_x * D_x\{f_{ij}\} = \{if_{ij}\},$$

$$D_x \hbar_x^{\langle m \rangle} = m\hbar_x^{\langle m-1 \rangle}, \ m \in \mathbf{Z}^+$$

$$D_x \phi^{\langle m \rangle} = D_x(\phi^{\langle m-1 \rangle} \phi)$$
$$= \phi^{\langle m-1 \rangle} * D_x \phi + \phi * D_x \phi^{\langle m-1 \rangle}$$
$$= \cdots$$
$$= m\phi^{\langle m-1 \rangle} * D_x \phi, \ m \in \mathbf{Z}^+,$$

$$D_x^n(\hbar_x^{\langle m \rangle} * \{f_{ij}\}) = \hbar_x^{\langle m-n \rangle} * \{\lfloor m+i \rfloor_n f_{ij}\}, \ m \geq n \geq 1,$$

and finally

$$\hbar_x^{\langle m \rangle} * D_x^n\{f_{ij}\} = \{\lfloor i+n-m \rfloor_n f(i+n-m, j)\}, \ n \geq m \geq 0.$$

Example 2.29. Let us calculate

$$D_x \frac{\bar{1}}{(\bar{1} - 3\hbar_x) * (\bar{1} - 3\hbar_y)}.$$

Since

$$D_x \frac{\bar{1}}{\bar{1} - 3\hbar_y} = \frac{3 D_x \hbar_y}{(\bar{1} - 3\hbar_y)^{\langle 2 \rangle}} = \bar{0},$$

and

$$D_x \frac{\bar{1}}{(\bar{1} - 3\hbar_x)} = \frac{3 D_x \hbar_x}{(\bar{1} - 3\hbar_x)^{\langle 2 \rangle}} = \frac{\bar{3}}{(\bar{1} - 3\hbar_x)^{\langle 2 \rangle}},$$

thus

$$D_x \frac{\bar{1}}{(\bar{1} - 3\hbar_x) * (\bar{1} - 3\hbar_y)} = \frac{\bar{1}}{\bar{1} - 3\hbar_x} * D_x \frac{\bar{1}}{\bar{1} - 3\hbar_y} + \frac{\bar{1}}{\bar{1} - 3\hbar_y} * D_x \frac{\bar{1}}{\bar{1} - 3\hbar_x}$$
$$= \frac{\bar{3}}{(\bar{1} - 3\hbar_y) * (\bar{1} - 3\hbar_x)^{\langle 2 \rangle}}.$$

We conclude this section by remarking that iterated algebraic integrals can be introduced. They are just the primitives of partial algebraic derivatives and thus their properties follow from those of algebraic integrals defined in a previous Section.

2.6 Notes

Most of the material in this Chapter are well known and can be found in standard analysis text books such as Apostol [5], Cheng [28], Krantz and Parks [99], Smith [211], Fichtenholz [62], Balser [13], Kaplan [94], etc. Some of the terminologies used here, however, are slightly different. For instance, instead of the term 'series', we use 'sum'; instead of the term 'sequence of functions', we use 'family of sequences', etc.

We have employed limits of generalized partial sums relative to an ordering for the definition of sums of multiple sequences. This is the usual approach in the theory of several complex variables (see e.g. Krantz [100]). There are other definitions for partial sums of multiple sequences as well (see e.g. Sheffer [169, 170], Wilansky [225]). For instance, the double series

$$\sum f_{ij}$$

is said to converge to s in the sense of Pringsheim if the limit

$$\lim_{m,n\to\infty} \sum_{i=0}^{m} \sum_{j=0}^{n} f_{ij}$$

exists and equals s. It is possible to develope results based on Pringsheim's concept for bivariate power series functions similar to those described in our previous sections. However, as pointed out by Sheffer [169], there are some technical difficulties which have to be circumvented due to the fact that Pringsheim's summability does not imply boundedness of $\{f_{ij}\}$.

Algebraic properties of multiple sequences have also been reported quite extensively. Indeed, some concepts and results in Sections 2.1, 2.4 and 2.5 are taken from Cheng [28]. There are, however, unpublished material in this Chapter. For the sake of convenience, we collect some of the properties of univariate sequences in our Appendix.

There are now active research into functional equations where the unknown functions are sequences. Some of these equations are called recurrence relations, some ordinary or partial difference equations. The former equations are called since their recursive structures are more important, while the latter are called since the concept of rate of changes is more important. In this book, as we shall see, a large number of recurrence relations arising from seeking analytic solutions will be solved. The introduction of algebraic operations and/or limiting operations will enable us to handle the recurrence relations in less cumbersome manners.

The concept of composition product is new. This concept is related to composition of analytic functions. Composition of analytic functions has been studied as abstract mappings, see e.g. Cowen [44].

The formula of Faa di Bruno is well known and is proved in several manners (see e.g. Jordan [90], Roman [164], McKiernan [139]). The one we present in this Chapter is new (and part of the arguments are provided by J. J. Lin). It is based on the idea of Roman [164], but no knowledge of umbral algebra is required. The formula of Faa di Bruno will play important roles in manipulating functional equations with composition of known or unknown analytic functions.

Chapter 3

Power Series Functions

3.1 Univariate Power Series Functions

Let $a = \{a_n\}_{n \in \mathbf{N}} \in l^{\mathbf{N}}$. Let Λ be a subset of $\mathbf{F}$ such that the attenuated sequence $\underline{\lambda} \cdot a$ is summable for each $\lambda \in \Lambda$, that is, such that the limit

$$\lim_{n \to \infty} \sum_{k=0}^{n} (\underline{\lambda} \cdot a)_k = \sum_{k=0}^{\infty} a_k \lambda^k$$

exists for each $\lambda \in \Lambda$, then we may define a function $\widehat{a} : \Lambda \to C$ by

$$\widehat{a}(\lambda) = \sum_{k=0}^{\infty} a_k \lambda^k, \quad \lambda \in \Lambda. \tag{3.1}$$

If Λ is a priori unknown, we will take Λ as the set of all $\lambda \in \mathbf{F}$ such that $\underline{\lambda} \cdot a$ is summable. This function, which is completely determined by a, is called the power series function in λ generated by a, or the *generating function* of a. The function $g(z)$ defined by $g(z) = \widehat{a}(z-c)$ for $z \in c+\Lambda = \{c + z |\ z \in \Lambda\}$ is called the generating function of a about (or with center at) c.

Since properties of power series functions with nonzero centers can easily be deduced from power series functions with center 0, we will therefore concentrate our attention to the latter functions.

Example 3.1. Let f be a complex function defined on a domain Θ of $\mathbf{F}$ which has derivatives of any order at the point $c \in \Theta$. Let

$$a = \left\{ \frac{1}{k!} f^{(k)}(c) \right\}_{k \in \mathbf{N}}.$$

The power series function $\widehat{a}(z - c)$ is called the Taylor series function with center c generated by f.

Theorem 3.1 (Abel's Lemma). *Let $a = \{a_k\}_{k \in \mathbf{N}} \in l^{\mathbf{N}}$. If the attenuated sequence $\underline{\lambda} \cdot a$, where $\lambda \neq 0$, is summable relative to an ordering Ψ for $\mathbf{N}$, then $\underline{\mu} \cdot a$ is absolutely summable for $|\mu| < |\lambda|$. If $\underline{\lambda} \cdot a$ is not summable at $\lambda = \alpha \neq 0$ relative to some ordering for $\mathbf{N}$, then $\underline{\lambda} \cdot a$ is also not summable for all $|\lambda| > |\alpha|$ relative to any ordering for $\mathbf{N}$.*

Indeed, if $\underline{\lambda} \cdot a$ is summable relative to an ordering Ψ, then in view of Theorem 2.5, $\{a_k \lambda^k\}$ is bounded, say by M. Hence for $k \in \mathbf{N}$,

$$\left|a_k \mu^k\right| = \left|a_k \frac{\mu^k \lambda^k}{\lambda^k}\right| = \left|a_k \lambda^k\right| \left|\left(\frac{\mu}{\lambda}\right)^k\right| \leq M \left|\frac{\mu}{\lambda}\right|^k.$$

Thus when $|\mu| < |\lambda|$, the comparison test for series yields

$$\sum_{k=0}^{\infty} |a_k \mu|^k \leq M \sum_{k=0}^{\infty} \left|\frac{\mu}{\lambda}\right|^k < \infty$$

as desired. The second assertion of our Theorem follows from the first.

Suppose we have a sequence $a = \{a_k\}_{k \in \mathbf{N}} \in l^{\mathbf{N}}$. Let Γ be the union of $\{0\}$ and the set of all nonnegative numbers λ such that $\underline{\lambda} \cdot a$ is summable relative to some ordering for $\mathbf{N}$. Since $0 \in \Gamma$, the (extended) real number

$$\rho(a) = \sup \Gamma$$

belongs to $[0, \infty]$. The number $\rho(a)$ is called the radius of convergence of the sequence a or of the power series function $\widehat{a}$ generated by it. Note that by definition, if $\underline{w} \cdot a$ is summable for each w that satisfies $|w| < |c|$, then $\rho(a) \geq |c|$.

Theorem 3.2. *With each* $a = \{a_k\}_{k \in \mathbf{N}} \in l^{\mathbf{N}}$, *there is associated an extended number* $\rho(a) \in [0, \infty]$ *such that* $\underline{\lambda} \cdot a$ *is absolutely summable for* $|\lambda| < \rho(a)$ *and not summable for* $|\lambda| > \rho(a)$ *relative to any ordering for* $\mathbf{N}$. *Furthermore, the family* $\{\underline{\lambda} \cdot a\}$ *is uniformly and absolutely summable for* $|\lambda| \leq r$ *where* $r < \rho(a)$.

The first assertion of the above Theorem follows from Theorem 3.1. The second assertion means the sequence $\{|a_k \lambda^k|\}_{k \in \mathbf{N}}$ is uniformly summable for $|\lambda| < r$, and follows from Weierstrass test for uniform convergence.

In the rest of this section, $a = \{a_k\}$ and $b = \{b_k\}$ are sequences in $l^{\mathbf{N}}$, $\rho(a)$ and $\rho(b)$ respectively are their radii of convergence, and $\widehat{a}(\lambda)$ and $\widehat{b}(\lambda)$ are the corresponding power series generated by them respectively.

Theorem 3.3. *The radius of convergence* $\rho(a)$ *is given by*

$$\frac{1}{\rho(a)} = \limsup_{k \to \infty} |a_k|^{1/k}$$

where $\rho(a) = +\infty$ *if* $\limsup_{k \to \infty} |a_k|^{1/k} = 0$ *and* $\rho(a) = 0$ *if* $\limsup_{k \to \infty} |a_k|^{1/k} = +\infty$.

The above Theorem (due to Cauchy and Hadamard) follows from the root test in elementary analysis.

Example 3.2. Let $a = \{a_k\}_{k \in \mathbf{N}} \in l^{\mathbf{N}}$. Then $\rho(a) = \rho(Da) = \rho\left(\int a\right) = \rho(|a|)$, where Da is the algebraic derivative of a, $\int a$ is the algebraic integral of a, and $|a|$ is the sequence $\{|a_k|\}_{k \in \mathbf{N}}$.

The fact that $\rho(|a|) = \rho(a)$ is clear from the previous result, that $\rho(Da) = \rho(a)$ from

$$\limsup_{n\to\infty} |(Da)_n|^{1/n} = \limsup_{n\to\infty} |na_n|^{1/n} = \limsup_{n\to\infty} |a_n|^{1/n},$$

and that $\rho\left(\int a\right) = \rho(a)$ from

$$\limsup_{n\to\infty} \left|\frac{1}{n+1} a_n\right|^{1/n} = \limsup_{n\to\infty} |a_n|^{1/n}.$$

In case each term a_k in the sequence $a \in l^{\mathbf{N}}$ is not zero, the ratio test for series also yields

$$\liminf_{n\to\infty} \left|\frac{a_n}{a_{n+1}}\right| \leq \rho(a) \leq \limsup_{n\to\infty} \left|\frac{a_n}{a_{n+1}}\right|.$$

Theorem 3.4. *Let $a, b \in l^{\mathbf{N}}$. For any $\alpha, \beta \in \mathbf{F}$,*

$$\rho(\alpha a + \beta b), \rho(a * b) \geq \min\left(\rho(a), \rho(b)\right).$$

Furthermore,

$$\widehat{(\alpha a + \beta b)}(\lambda) = \alpha \widehat{a}(\lambda) + \beta \widehat{b}(\lambda) = \widehat{\alpha a}(\lambda) + \widehat{\beta b}(\lambda)$$

and

$$\widehat{a * b}(\lambda) = \widehat{a}(\lambda)\widehat{b}(\lambda)$$

for $|\lambda| < \min\left(\rho(a), \rho(b)\right).$

To see that $\rho(a * b) \geq \min\left(\rho(a), \rho(b)\right)$, it suffices to show that $\underline{\lambda} \cdot (a * b) = (\underline{\lambda} \cdot a) * (\underline{\lambda} \cdot b)$ is summable for $|\lambda| < \min\left(\rho(a), \rho(b)\right)$. But this is true in view of Theorem 2.13. Furthermore, for $|\mu| < \min\left(\rho(a), \rho(b)\right)$, since $\underline{\mu} \cdot a$ and $\underline{\mu} \cdot b$ are absolutely summable, by Merten's Theorem 2.11 and Theorem 2.13,

$$\widehat{a * b}(\mu) = \sum_{\mathbf{N}} \underline{\mu} \cdot (a * b) = \sum_{\mathbf{N}} (\underline{\mu} \cdot a) * (\underline{\mu} \cdot b) = \left(\sum_{\mathbf{N}} \underline{\mu} \cdot a\right)\left(\sum_{\mathbf{N}} \underline{\mu} \cdot b\right).$$

The other assertions in the above result are proved in similar manners.

As an interesting consequence, we see that

$$\widehat{a^{\langle m \rangle}}(\lambda) = \widehat{a}^m(\lambda), \quad |\lambda| < \rho(a), \tag{3.2}$$

for $m = 2, 3, \ldots$. Recall that $a^{\langle 1 \rangle} = a$, hence (3.2) is also valid for $m = 1$. Furthermore, since $a^{\langle 0 \rangle}$ is defined to be $\bar{1}$, and $(\widehat{a})^0(\lambda) = 1$, we see that (3.2) is valid for $m = 0$.

The following follows from Theorem 2.8.

Theorem 3.5. *The power series function $\widehat{a}(\lambda)$ generated by a is continuous for $|\lambda| < \rho(a)$.*

Example 3.3 (Abel's Limit Theorem). *If a is a real sequence and if $\lambda \cdot a$ is summable at $\lambda = \rho(a) > 0$ (or $\lambda = -\rho(a) < 0$), then $\widehat{a}(\lambda)$, as a function of real variable, is continuous at $\rho(a)$ from the left (respectively continuous at $-\rho(a)$ from the right).*

Proof. Without loss of generality, we will assume that $\rho(a) = 1$ and show that if
$$w = \sum_{i=0}^{\infty} a_i < \infty,$$
then
$$\lim_{x \to 1^-} \widehat{a}(x) = \lim_{x \to 1^-} \sum_{i=0}^{\infty} a_i x^i = \sum_{i=0}^{\infty} a_i.$$

First of all, it can easily be proved from
$$\frac{1 - x^{n+1}}{1 - x} = 1 + x + x^2 + \cdots + x^n$$
that
$$\frac{1}{1-x} = \sum_{i=0}^{\infty} x^i, \quad -1 < x < 1.$$

Since
$$\frac{1}{1-x} \widehat{a}(x) = \sum_{i=0}^{\infty} (\sigma * a)_i x^i, \quad -1 < x < 1,$$
where $\sigma = \{1, 1, 1, \ldots\}$, thus
$$\widehat{a}(x) - w = (1-x) \sum_{i=0}^{\infty} \{(\sigma * a)_i - w\} x^i, \quad -1 < x < 1.$$

By assumption, there exists an integer I such that $i \geq I$ implies
$$|(\sigma * a)_i - w| = \left| \sum_{j=0}^{i} a_j - \sum_{j=0}^{\infty} a_j \right| < \frac{\varepsilon}{2}.$$

Therefore,
$$|\widehat{a}(x) - w| < \left| (1-x) \sum_{i=0}^{I-1} \{(\sigma * a)_i - w\} x^i \right| + \frac{\varepsilon}{2} \left| (1-x) \sum_{i=I}^{\infty} x^i \right|$$
$$\leq (1-x) \cdot I \cdot \max_{0 \leq i \leq I} |(\sigma * a)_i - w| + \frac{\varepsilon}{2}$$

for $0 < x < 1$. If we take x sufficiently close to 1, then the right hand side can be made arbitrary small which is what we need. The proof is complete.

Theorem 3.6 (Representation Theorem). *Let $\widehat{a}(\lambda)$ be the power series function generated by a sequence $a \in l^{\mathbf{N}}$ with $\rho(a) > 0$. Then for each nonzero $\mu \in B(0; \rho(a))$, there exists $\beta > 0$ such that $B(\mu; \beta) \subseteq B(0; \rho(a))$ and*

$$\widehat{a}(\lambda) = \sum_{k=0}^{\infty} \left(\sum_{n=0}^{\infty} C_k^{(n)} a_n \mu^{n-k} \right) (\lambda - \mu)^k < \infty, \ \lambda \in B(\mu; \beta). \tag{3.3}$$

Proof. Choose $\beta > 0$ such that $\beta + |\mu| < \rho(a)$. Then $B(\mu; \beta)$ is contained in $B(0; \rho(a))$, and $\underline{\lambda} \cdot a$ is absolutely summable for each $\lambda \in B(\mu; \beta)$. Since for each $\lambda \in B(\mu; \beta)$,

$$\sum_{n=0}^{\infty} a_n \lambda^n = \sum_{n=0}^{\infty} a_n (\lambda - \mu + \mu)^n = \sum_{n=0}^{\infty} a_n \sum_{k=0}^{n} C_k^{(n)} (\lambda - \mu)^k \mu^{n-k},$$

by Fubini's Theorem 2.4, we may interchange the order of summation and obtain

$$\sum_{n=0}^{\infty} a_n \lambda^n = \sum_{k=0}^{\infty} \sum_{n=k}^{\infty} C_k^{(n)} a_n \mu^{n-k} (\lambda - \mu)^k < \infty.$$

Since $C_k^{(n)} = 0$ for $n < k$, we see that (3.3) holds. The proof is complete.

Theorem 3.7. *Let $a \in l^{\mathbf{N}}$ with $\rho(a) \in (0, \infty]$ and $b = Da$ be the algebraic derivative of a. Then $\rho(b) = \rho(a)$. Furthermore,*

$$\widehat{a}'(\mu) = \widehat{Da}(\mu) = \widehat{b}(\mu)$$

for $|\mu| < \rho(a)$.

Proof. We have already seen that $\rho(b) = \rho(a)$. Let $\mu \in B(0; \rho(a))$. Then by the representation theorem, there exists $\beta > 0$ such that $B(\mu; \beta) \subseteq B(0; \rho(a))$ and for $\lambda \in B(\mu; \beta)$,

$$\frac{\widehat{a}(\lambda) - \widehat{a}(\mu)}{\lambda - \mu} = \frac{1}{\lambda - \mu} \left\{ \sum_{n=0}^{\infty} a_n \mu^n + \sum_{k=1}^{\infty} \sum_{n=k}^{\infty} C_k^{(n)} a_n \mu^{n-k} (\lambda - \mu)^k - \widehat{a}(\mu) \right\}$$

$$= \sum_{n=1}^{\infty} n a_n \mu^{n-1} + \sum_{k=2}^{\infty} \left(\sum_{n=k}^{\infty} C_k^{(n)} a_n \mu^{n-k} \right) (\lambda - \mu)^{k-1}.$$

By taking limits on both sides as $\lambda \to \mu$ and invoking Theorem 3.2, we see that

$$\widehat{a}'(\mu) = \sum_{n=1}^{\infty} n a_n \mu^{n-1} = \widehat{b}(\mu)$$

as required. The proof is complete.

We remark that by repeated application of the above theorem, we see that

$$\widehat{a}^{(m)}(\lambda) = \widehat{D^m a}(\lambda) = \sum_{n=m}^{\infty} \frac{n!}{(n-m)!} a_n \lambda^{n-m}$$

for $|\lambda| < \rho(a)$. If we put $\lambda = 0$ in the above formula, we see that

$$\widehat{a}^{(m)}(0) = m! a_m, \quad m \in \mathbf{Z}^+. \tag{3.4}$$

As an interesting consequence, if

$$\sum_{n=0}^{\infty} a_n(z-c)^n = \sum_{n=0}^{\infty} b_n(z-c)^n < \infty \tag{3.5}$$

for z in a neighborhood of c, then $a_n = b_n$ for $n \in \mathbf{N}$.

Theorem 3.8 (Unique Representation Theorem). *If two power series functions $\sum_{n=0}^{\infty} a_n(z-c)^n$ and $\sum_{n=0}^{\infty} b_n(z-c)^n$ are defined in a neighborhood of c and (3.5) holds, then $a_n = b_n$ for $n \in \mathbf{N}$.*

Theorem 3.9. *Let $a \in l^{\mathbf{N}}$ with $\rho(a) \in (0,\infty]$. Let b be the algebraic integral $\int a$ of a. Then $\rho(b) = \rho(a)$. Furthermore, for $z \in B(0;\rho(a))$,*

$$\int_0^z \widehat{a}(\lambda) d\lambda = \widehat{\int a}(z) = \widehat{b}(z).$$

Indeed, by Theorem 2.9,

$$\int_0^z \widehat{a}(\lambda) d\lambda = \int_0^z \sum_{j=0}^{\infty} a_j \lambda^j d\lambda = \sum_{j=0}^{\infty} \int_0^z a_j \lambda^j d\lambda = \sum_{j=0}^{\infty} \frac{a_j}{j+1} z^{j+1} = \widehat{b}(z)$$

as required.

Recall from Theorem 2.12 that for any sequence $a = \{a_k\} \in l^{\Omega}$ which satisfies $a_0 \neq 0$, there exists a unique $b = \{b_k\} \in l^{\Omega}$ such that

$$a * b = \{1, 0, 0, ...\}.$$

The unique solution b has been denoted by $\overline{1}/a$.

Theorem 3.10 (Inversion Theorem). *Let $a = \{a_k\}_{k \in \mathbf{N}} \in l^{\mathbf{N}}$ with $\rho(a) > 0$ and $a_0 \neq 0$. Let $b = \overline{1}/a$. Then $\rho(b) > 0$. Furthermore,*

$$\widehat{a}(\lambda) \widehat{b}(\lambda) = 1 \tag{3.6}$$

for $|\lambda| < \min(\rho(a), \rho(b))$.

Proof. Since $\lim_{\lambda \to 0} \sum_{n=1}^{\infty} |a_n \lambda^n| = 0$, we may choose $\gamma > 0$ such that

$$\sum_{n=1}^{\infty} |a_n| \gamma^n \leq |a_0|.$$

We assert that

$$|b_n| \leq |a_0|^{-1} \gamma^{-n}, \quad n \in \mathbf{N}. \tag{3.7}$$

Indeed, recall
$$a_0 b_0 = 1,$$
$$a_1 b_0 + a_0 b_1 = 0,$$
$$\ldots = \ldots,$$
$$a_m b_0 + a_{m-1} b_1 + \cdots + a_0 b_m = 0,$$
$$\ldots = \ldots,$$

thus $|b_0| = |a_0|^{-1} \leq |a_0|^{-1} \gamma^{-0}$. Assume by induction that (3.7) holds for $n = 0, \ldots, m-1$, then

$$\begin{aligned}
|a_0 b_m| &\leq |a_1 b_{m-1}| + |a_2 b_{m-2}| + \cdots + |a_m b_0| \\
&\leq |a_1| |a_0|^{-1} \gamma^{-(m-1)} + |a_2| |a_0|^{-1} \gamma^{-(m-2)} + \cdots + |a_m| |a_0|^{-1} \gamma^{-0} \\
&= |a_0|^{-1} \gamma^{-m} \left\{ |a_1| \gamma + |a_2| \gamma^2 + \cdots + |a_m| \gamma^m \right\} \\
&\leq |a_0|^{-1} \gamma^{-m} |a_0| \\
&= \gamma^{-m}
\end{aligned}$$

as desired. Thus for any $|\lambda| < \gamma$,

$$\sum_{n=0}^{\infty} |b_n| |\lambda|^n \leq |a_0|^{-1} \sum_{n=0}^{\infty} \left(\frac{|\lambda|}{\gamma} \right)^n < \infty.$$

Thus $\rho(b) \geq \gamma > 0$. Finally, an application of Theorem 3.4 yields (3.6). The proof is complete.

We remark that in case $a \in l^{\mathbf{N}}$ such that $\rho(a) \in (0, \infty]$, $a_0 \neq 0$ for $z \in B(0; \rho(a))$, then the above result asserts that

$$\frac{1}{\widehat{a}(z)} = \widehat{\overline{1/a}}(z)$$

for $z \in B(0; \min(\rho(a), \rho(\overline{1/a})))$.

Theorem 3.11 (Substitution Theorem). *Let* $a = \{a_k\}_{k \in \mathbf{N}}, b = \{b_k\}_{k \in \mathbf{N}}$ *be sequences in* $l^{\mathbf{N}}$ *with positive* $\rho(a)$ *and* $\rho(b)$ *respectively. For any* $\lambda \in B(0; \rho(b))$ *such that* $\sum_{n=0}^{\infty} |b_n \lambda^n| < \rho(a)$, *we have*

$$\widehat{a}(\widehat{b}(\lambda)) = \widehat{a \circ b}(\lambda) = \sum_{\mathbf{N}} \lambda \cdot (a \circ b) < \infty,$$

where we recall that $a \circ b$ *is the composition product defined by* $(a \circ b)_n = \sum_{i=0}^{\infty} a_i b_n^{\langle i \rangle}$ *for* $n \in \mathbf{N}$.

Proof. Since $|\lambda| < \rho(b)$, we see that $\widehat{b}(\lambda)$ is defined and $\left| \widehat{b}(\lambda) \right| \leq \sum_{n=0}^{\infty} |b_n \lambda^n| < \rho(a)$. Therefore, $\widehat{a}(\widehat{b}(\lambda))$ and $\widehat{|a|}\left(\widehat{|b|}(|\lambda|) \right)$ (where we recall $\widehat{|a|}(\lambda)$ and $\widehat{|b|}(\lambda)$ are the power series functions generated by $|a|$ and $|b|$ respectively) are defined and

$$\widehat{a}(\widehat{b}(\lambda)) = \sum_{n=0}^{\infty} a_n \left(\widehat{b}(\lambda) \right)^n = \sum_{n=0}^{\infty} a_n \left\{ \sum_{k=0}^{\infty} b_k^{\langle n \rangle} \lambda^k \right\}$$

in view of (3.2). By changing the order of summation, we see that the last sum can formally be written as

$$\sum_{n=0}^{\infty}\sum_{k=0}^{\infty} a_k b_n^{\langle k \rangle} \lambda^n$$

or

$$\sum_{n=0}^{\infty} (a \circ b)_n \lambda^n.$$

Since $b^{\langle 0 \rangle} = \bar{1}$ and

$$b_k^{\langle n \rangle} = \sum_{v_1+v_2+\cdots+v_n=k; v_1,\ldots,v_n \in \mathbf{Z}^+} b_{v_1} b_{v_2} \cdots b_{v_n}$$

$$\leq \sum_{v_1+v_2+\cdots+v_n=k; v_1,\ldots,v_n \in \mathbf{Z}^+} |b_{v_1}||b_{v_2}|\cdots|b_{v_n}|$$

for $n \geq 1$, we see that

$$\sum_{n=0}^{\infty}\sum_{k=0}^{\infty} \left| a_n b_k^{\langle n \rangle} \lambda^k \right| \leq \sum_{n=0}^{\infty} |a_n| \left(\sum_{k=0}^{\infty} |b_k \lambda^k| \right)^n = \widehat{|a|}\left(\widehat{|b|}(|\lambda|)\right) < \infty.$$

By Fubini's Theorem 2.4, we see that the change of the order of summation is legal. The proof is complete.

We remark that the series $\sum_{\mathbf{N}} \lambda \cdot (a \circ b)$ is the power series which arises by substituting $w = \widehat{b}(\lambda)$ into $\widehat{a}(w)$ and then formally expand the resulting expression and rearranging terms in increasing powers of λ.

3.2 Univariate Analytic Functions

A function f with domain an open set $\Theta \subseteq \mathbf{F}$ and range $\mathbf{F}$ is said to be analytic at c if there is a sequence $a \in l^{\mathbf{N}}$ and a ball $B(c; \gamma)$ contained in Θ such that

$$f(\lambda) = \sum_{\mathbf{N}} \underline{\lambda - c} \cdot a < \infty, \quad \lambda \in B(c; \gamma).$$

Note that, in view of Abel's Lemma (Theorem 3.1), f is analytic at c if, and only if, there is a sequence $b \in l^{\mathbf{N}}$, an ordering Ψ for $\mathbf{N}$ and a ball $B(c; \delta)$ contained in Θ such that

$$f(\lambda) = \sum_{j=0}^{\infty} b_{\Psi(j)} (\lambda - c)^{\Psi(j)} < \infty, \quad \lambda \in B(c; \delta).$$

For the same reason, f is analytic at c if, and only if, there is a sequence $b \in l^{\mathbf{N}}$ and a ball $B(c; \delta)$ contained in Θ such that

$$f(\lambda) = \sum_{j=0}^{\infty} b_j (\lambda - c)^j < \infty, \quad \lambda \in B(c; \delta).$$

The function f is said to be analytic on or over Θ if it is analytic at each $c \in \Theta$. The set of all analytic functions $f : \Theta \subseteq \mathbf{F} \to \mathbf{F}$ will be denoted by $H(\Theta)$. Analytic functions are plenty as can be seen from the following result.

Theorem 3.12. *Let* $a = \{a_k\}_{k \in \mathbf{N}} \in l^{\mathbf{N}}$ *with positive radius of convergence* $\rho(a)$. *Then the corresponding power series function* $\widehat{a}(\lambda)$ *generated by it belongs to* $H(B(0; \rho(a)))$.

Indeed, $\widehat{a}(\lambda) = \sum_{\mathbf{N}} \underline{\lambda} \cdot a < \infty$ for λ in some $B(0; \delta)$ contained in $B(0; \rho(a))$ and thus $\widehat{a}(\lambda)$ is analytic at 0. For any $\mu \in B(0; \rho(a))$ which is distinct from 0, by Theorem 3.6, there exists $\beta > 0$ such that $B(\mu; \beta) \subseteq B(0; \rho(a))$ and (3.3) holds. Thus $\widehat{a}(\lambda)$ is analytic at μ.

There are several important properties of analytic functions which we shall need in the sequel and follow from the results in the previous section. In the following, Θ, Θ_1 and Θ_2 denote open subsets of $\mathbf{F}$.

Theorem 3.13. *If* $f \in H(\Theta)$, *then its derived function also belongs to* $H(\Theta)$. *If in addition* $\Theta = B(w; \delta)$, *then its primitive function* $g(z) = \int_w^z f(u) du$ *also belongs to* $H(\Theta)$.

If $f = f(z)$ is analytic at c, then its definition asserts that f is the power series function $\widehat{a}$ generated by some sequence $a \in l^{\mathbf{N}}$:

$$f(z) = \sum_{n=0}^{\infty} a_n (z - c)^n.$$

By Theorem 3.7, we see that f has derivatives of any order and hence

$$a_n = \frac{1}{n!} \widehat{a}^{(n)} = \frac{1}{n!} f^{(n)}(c), \ n \in \mathbf{N}.$$

That is

$$f(z) = \sum_{n=0}^{\infty} \frac{1}{n!} f^{(n)}(c)(z - c)^n$$

for z in a neighborhood of c. As a consequence, we have the following result.

Theorem 3.14. *If* $f, g \in H(\Theta)$, *then* $\alpha f + \beta g \in H(\Theta)$ *for any* $\alpha, \beta \in \mathbf{C}$ *and* $f \cdot g \in H(\Theta)$.

Theorem 3.15. *If* $f \in H(\Theta)$ *and* $f(\lambda) \neq 0$ *for* $\lambda \in \Theta$, *then* $1/f \in H(\Theta)$.

The above result follows from the Inversion Theorem 3.10.

Theorem 3.16. *If* f *is analytic at* c *and* g *is analytic at* $f(c)$, *then* $g \circ f$ *is analytic at* $z = c$.

Example 3.4. Polynomials (of one variable) are analytic everywhere, and rational functions $f(z)$ are analytic for all z which are not roots of the denominator.

Example 3.5. The power series
$$1 + z + z^2 + z^2 + \cdots$$
is convergent for each $z \in B(0;1)$ as can be seen by the ratio test. Furthermore, since
$$1 + z + \cdots + z^n = \frac{1 - z^{n+1}}{1 - z}$$
for $z \neq 1$, by taking limits on both sides, we see that the rational function $f(z) = (1-z)^{-1}$ is analytic on $B(0;1)$ and
$$\frac{1}{1-z} = \sum_{n=0}^{\infty} z^n, \ z \in B(0;1).$$
By substituting $w = -z$ into the above equality, we see further that
$$\frac{1}{1+w} = \sum_{n=0}^{\infty} (-1)^n w^n, \ w \in B(0;1).$$

Theorem 3.17 (Unique Continuation Theorem). *Let $f, g \in H(\Theta)$ where Θ is connected. If there is a sequence $\{\mu_i\}_{i \in \mathbf{N}}$ contained in Θ such that $\lim_{i \to \infty} \mu_i = \mu \in \Theta$, $\mu \neq \mu_i$ for $i \in \mathbf{N}$ and $f(\mu_i) = g(\mu_i)$ for $i \in \mathbf{N}$, then $f(\lambda) = g(\lambda)$ for $\lambda \in \Theta$.*

Proof. It suffices to assume that g is the trivial function. We first note that there is some $a \in l^{\mathbf{N}}$ and $\delta > 0$ such that
$$f(\lambda) = \sum_{i=0}^{\infty} a_i (\lambda - \mu), \ \lambda \in B(\mu; \delta) \subseteq \Theta.$$
If $a \neq \overline{0}$, then there is some $I \in \mathbf{N}$ such that $a_0 = a_1 = \cdots = a_{I-1} = 0$ but $a_I \neq 0$ (where the case a_{-1} is taken to be vacuously true). Thus,
$$f(\lambda) = (\lambda - \mu)^I q(\lambda), \ \lambda \in B(\mu; \delta),$$
where
$$q(\lambda) = \left(a_I + \sum_{i=I+1}^{\infty} a_i (\lambda - \mu)^{i-I} \right), \ \lambda \in B(\mu; \delta).$$
Thus, in view of the fact that $f(\mu_k) = 0$ for all large k, we see that
$$0 = (\mu_k - \mu) q(\mu_k)$$
for all large k, so that $q(\mu_k) = 0$ for all large k. By continuity, $q(\mu) = 0$, which is contrary to the fact that $q(\mu) = a_I \neq 0$. We conclude that $f(\lambda) = 0$ for $\lambda \in B(\mu; \delta) \subseteq \Theta$.

Next, let w be any point in Θ and distinct from μ, by connectedness, we may join μ to w by a path h defined on $[0,1]$ and $h(t) \in \Theta$ for $t \in [0,1]$. By what we have just shown, the composite function $f \circ h$ is identically zero on $[0, \alpha]$ for some $\alpha \in (0, 1]$. We assert that $f(w) = f(h(1)) = 0$. To see this, let
$$S = \{ \beta \in (0,1] | \ f(h(t)) = 0 \text{ for } t \in [0, \beta] \}.$$

Since $\alpha \in S$, thus $\beta' = \sup S$ exists and $\beta' \leq 1$. By continuity, we see further that $f(h(\beta')) = 0$. We assert that $\beta' = 1$. Suppose not, by what we have shown above, $f(\lambda)$ is identically zero for $\lambda \in B(h(\beta'), \delta') \subseteq \Theta$ where δ' is some positive number. Thus $\beta' + \delta \in S$ which is contrary to the definition of β'. The proof is complete.

As an immediate corollary, we have the following results.

Theorem 3.18. *If $f \in H(\Theta)$ and $f^{(k)}(c) = 0$ for $k \in \mathbf{N}$ and some $c \in \Theta$, then $f(\lambda) = 0$ for all $\lambda \in \Theta$.*

In particular, if the power series functions $\widehat{a}(\lambda)$ and $\widehat{b}(\lambda)$ generated by $a, b \in l^{\mathbf{N}}$ (are defined and) satisfy $\widehat{a}(\lambda) = \widehat{b}(\lambda)$ in a neighborhood of 0 (or any neighborhood in Θ), then $a = b$. That is, the Uniquenss Representation Theorem 3.8 holds.

As another interesting consequence, let g be an analytic extension of the function $f \in H(\Theta)$, that is, $g \in H(\widetilde{\Theta})$ where $\Theta \subset \widetilde{\Theta}$ and $g(z) = f(z)$ for $z \in \Theta$. If there are two analytic extensions $g_1 \in H(\widetilde{\Theta}_1)$ and $g_2 \in H(\widetilde{\Theta}_2)$ of f, then $g_1(z) = g_2(z)$ for $z \in \widetilde{\Theta}_1 \cap \widetilde{\Theta}_2$. Hence the union $g_1 \cup g_2$ is a well defined function over $\widetilde{\Theta}_1 \cup \widetilde{\Theta}_2$ and belongs to $H\left(\widetilde{\Theta}_1 \cup \widetilde{\Theta}_2\right)$. By similar reasoning, given an analytic function $f \in H(\Theta)$, the following

$$\bigcup \{g : g \text{ is an analytic extension of } f \text{ on an open set containing } \Theta\}$$

is a well defined analytic function. It will be called the analytic continuation of f.

Example 3.6. The Newton binomial expansion formula asserts that

$$(1+x)^\alpha = \sum_{n=0}^{\infty} C_n^{(\alpha)} x^n, \ x \in R, |x| < 1,$$

where we recall that $C_n^{(\alpha)}$ is the extended binomial coefficient defined by $C_0^{(\alpha)} = 1$ for $\alpha \in \mathbf{C}$, and $C_n^{(\alpha)} = \alpha(\alpha-1)\cdots(\alpha-n+1)/n!$ for $n \in \mathbf{Z}^+$ and $\alpha \in \mathbf{C}$. This formula can be proved in several manners. One proof is obtained by observing that $f(x) = (1+x)^\alpha$ for $x \in (-1, 1)$ satisfies $f(0) = 1$ and

$$(1+x)f'(x) = \alpha f(x), \ |x| < 1.$$

Assume that

$$f(x) = \sum_{n=0}^{\infty} a_n x^n$$

is an analytic solution of the above equation. Then $a_0 = f(0) = 1$ and by Theorem 3.18 (or by the Unique Representation Theorem 3.8),

$$(n+1)a_{n+1} + na_n = \alpha a_n, \ n \in \mathbf{N}.$$

The above recurrence is easily solved and

$$a_n = \frac{1}{n!}\alpha(\alpha-1)\cdots(\alpha-n+1), \ n \in \mathbf{Z}^+.$$

The solution is a true solution by checking (by means of the root test) that

$$\sum_{n=0}^{\infty} C_n^{(\alpha)} x^n < \infty$$

for $|x| < 1$.

Example 3.7. By the Newton binomial expansion formula in the previous Example 3.6,

$$(1+x)^{1/2} = \sum_{n=0}^{\infty} C_n^{(1/2)} x^n, \ x \in \mathbf{R}, |x| < 1.$$

Since $(1+x)^{1/2}(1+x)^{1/2} = 1+x$, we see that

$$\left\{C_0^{(1/2)}, C_1^{(1/2)}, C_2^{(1/2)}, ...\right\} * \left\{C_0^{(1/2)}, C_1^{(1/2)}, C_2^{(1/2)}, ...\right\} = \left\{\sum_{k=0}^{n} C_k^{(1/2)} C_{n-k}^{(1/2)}\right\}_{n \in \mathbf{N}}$$
$$= \{1, 1, 0, 0, ...\}.$$

The power series

$$\sum_{n=0}^{\infty} C_n^{(1/2)} z^n,$$

as can be seen from the ratio test, is convergent for complex $z \in B(0; 1)$. According to what we have just shown,

$$\left(\sum_{n=0}^{\infty} C_n^{(1/2)} z^n\right) \left(\sum_{n=0}^{\infty} C_n^{(1/2)} z^n\right) = 1 + z,$$

thus this power series is a square root of the complex number $1 + z$ when $|z| < 1$. We will write

$$\sqrt{1+z} = (1+z)^{1/2} = \sum_{n=0}^{\infty} C_n^{(1/2)} z^n, \ z \in B(0; 1).$$

As a consequence, we see that the function $f(z) = (1+z)^{1/2}$ is analytic over $B(0; 1)$.

Example 3.8. (See pp. 83–87 of [63]) The sine and cosine functions can be introduced as analytic solutions. To see this, we introduce two real power series functions formally defined by

$$C(x) = \sum_{n=0}^{\infty} (-1)^n \frac{x^{2n}}{(2n)!},$$

and

$$S(x) = \sum_{n=0}^{\infty} (-1)^n \frac{x^{2n+1}}{(2n+1)!}.$$

By means of the ratio test, we can easily check that both series converge for every $x \in \mathbf{R}$. Therefore they are infinitely differentiable functions defined on $\mathbf{R}$. We may list several additional properties: (1) by direct term by term multiplications that the two basic formulas

$$C(x+y) = C(x)C(y) - S(x)S(y), \tag{3.8}$$

$$S(x+y) = S(x)C(y) + C(x)S(y), \tag{3.9}$$

are valid for all real x and y; (2) $C(x)$ is a even function, while $S(x)$ is an odd function, that is, $C(-x) = C(x)$ and $S(-x) = -S(x)$, (3) $C(0) = 1$ and $S(0) = 0$, (4) replacing y by $-x$ in (3.8) and invoking the properties just described, we get

$$C^2(x) + S^2(x) = 1, \tag{3.10}$$

(5) by differentiating the functions $C(x)$ and $S(x)$ term by term, we easily find that

$$C'(x) = -S(x), \quad S'(x) = C(x), \tag{3.11}$$

(6) we have $C(2) < 0$ since

$$C(2) = 1 - \frac{2^2}{2!} + \frac{2^4}{4!} - \left(\frac{2^6}{6!} - \frac{2^8}{8!}\right) - \cdots - \left(\frac{2^{2n}}{(2n)!} - \frac{2^{2n+2}}{(2n+2)!}\right) + \cdots,$$

$$1 - \frac{2^2}{2!} + \frac{2^4}{4!} = -\frac{1}{3}$$

and

$$\frac{2^{2n}}{(2n)!} - \frac{2^{2n+2}}{(2n+2)!} = \frac{2^{2n}}{(2n)!}\left[1 - \frac{2 \cdot 2}{(2n+1)(2n+2)}\right] > 0$$

for $n \geq 3$, (7) the function

$$-C'(x) = S(x) = x\left(1 - \frac{x^2}{2 \cdot 3}\right) + \frac{x^5}{5!}\left(1 - \frac{x^2}{6 \cdot 7}\right) + \cdots$$

is obviously positive in $(0, 2]$, so that the derivative $C'(x) = -S(x)$ is negative in $(0, 2)$. Therefore $C(x)$ has exactly one root in $(0, 2)$. Let $\pi'/2$ denote the point at which $C(x)$ vanishes. Thus

$$C\left(\frac{\pi'}{2}\right) = 0, \quad S\left(\frac{\pi'}{2}\right) = 1,$$

where the second formula follows from (3.10), if we note that $S(x)$ is positive in the interval $(0, 2)$. Furthermore, setting first $x = y = \pi'/2$ and then $x = y = \pi'$ in formulas (3.8) and (3.9), we get

$$C(\pi') = -1, \ S(\pi') = 0, \ C(2\pi') = 1, \ S(2\pi') = 0.$$

Hence, holding x fixed in (3.8) and (3.9), we get

$$C(x+\pi') = -C(x), \ S(x+\pi') = -S(x) \tag{3.12}$$

if $y = \pi'$ and
$$C(x + 2\pi') = C(x), \ S(x + 2\pi') = S(x)$$
if $y = 2\pi'$, i.e., the functions $C(x)$ and $S(x)$ are periodic, with period $2\pi'$. Furthermore, $S(x) > 0$ for $0 < x < \pi'/2$. Changing x to $-t$ in the second of the formulas (3.12), we get
$$S(\pi' - t) = -S(-t) = S(t),$$
from which it follows that $S(t) > 0$ for $\pi'/2 \le t < \pi'$. Thus the function $C(t)$, with derivative $-S(t)$, is strictly decreasing over the interval $(0, \pi')$, and hence is one-to-one on $[0, \pi']$. We now show that $C(x) = \cos x$ and $S(x) = \sin x$ for $x \in \mathbf{R}$. To this end, we consider the curve specified by the parametric equations
$$x = C(t), \ y = S(t), \ 0 \le t \le 2\pi'.$$
Since $C^2(t) + S^2(t) = 1$, every point of this curve satisfies the equation $x^2 + y^2 = 1$, and hence lies on the circle of radius 1 with center at the origin. Since $C(t)$ is one-to-one on $[0, \pi']$, there is a one-to-one correspondence between the points of the upper half of our circle and values of the parameter t in the interval $[0, \pi']$. By similar arguments, there is a one-to-one correspondence between the points of the lower half of our circle and values of the parameter t in the interval $[\pi', 2\pi']$. We now calculate the length $s(t)$ of the arc joining the point $(1, 0)$ and the point P on the unit circle with parameter t. By a familiar formula of calculus,
$$s(t) = \int_0^t \sqrt{[C'(t)]^2 + [S'(t)]^2} \, dt = \int_0^t dt = t, \tag{3.13}$$
where (3.10) and (3.11) are used. It follows from (3.13) that t is just the angle subtending our arc, expressed in radians. But then
$$C(\theta) = x = \cos \theta, \ S(\theta) = y = \sin \theta,$$
as desired. By the same token, (3.13) shows that the length of the unit circle is just $2\pi'$, so that our number π' can be identified with the number π.

We recall from Example 1.5 that the exponential function $f(z) = e^z$ is defined by
$$e^z = e^x(\cos y + \mathbf{i} \sin y), \ z = x + \mathbf{i} y.$$
In that example, we have implicitly assumed the existence of the sine and cosine functions. Now that we know the power series expansion of the sine and cosine functions, we may see further that for each $z \in \mathbf{C}$:
$$e^z = \sum_{n=0}^{\infty} \frac{z^n}{n!}, \ z \in \mathbf{C}.$$
and
$$\cos z = \frac{1}{2}\left(e^{\mathbf{i} z} + e^{-\mathbf{i} z}\right) = \sum_{n=0}^{\infty} (-1)^n \frac{z^{2n}}{(2n)!}, \ z \in \mathbf{C},$$

$$\sin z = \frac{1}{2}\left(e^{iz} - e^{-iz}\right) = \sum_{n=0}^{\infty}(-1)^n \frac{z^{2n+1}}{(2n+1)!}, \quad z \in \mathbf{C},$$

$$\sinh z = \frac{1}{2}\left\{e^z - e^{-z}\right\} = \sum_{n=0}^{\infty} \frac{z^{2n+1}}{(2n+1)!}, \quad z \in \mathbf{C},$$

$$\cosh z = \frac{1}{2}\left\{e^z + e^{-z}\right\} = \sum_{n=0}^{\infty} \frac{z^{2n}}{(2n)!}, \quad z \in \mathbf{C}.$$

Example 3.9. Let $\psi = \psi(z)$ be analytic over $B(\xi; \sigma)$ such that $\psi^{(i)}(\xi) = 0$ for $i = 0, 1, 2, ..., m-1$ and $\psi^{(m)}(z)$ is continuous on $\overline{B}(\xi; \sigma)$. Then by Taylor's expansion,

$$\psi(z) = \frac{(z-\xi)^m}{(m-1)!}\int_0^1 (1-t)^{m-1}\psi^{(m)}(\xi + t(z-\xi))dt,$$

so that

$$\sup_{z \in \overline{B}(\xi;\sigma)}|\psi(z)| \leq \frac{\sigma^m}{m!}\sup_{z \in \overline{B}(\xi;\sigma)}\left|\psi^{(m)}(z)\right|.$$

3.3 Bivariate Power Series Functions

Let us first recall the concept of multi-indices and their uses as shorthand notations. Since we will be concerned with bivariate sequences in $l^{\mathbf{N} \times \mathbf{N}}$, we will concentrate on bi-indices. A bi-index is an element of $\mathbf{N}^2$ of the form $v = (v_1, v_2)$. The notations $v!$ and $|v|_1$ stand for $v_1!v_2!$ and $v_1 + v_2$ respectively. Furthermore, for $p = (w, z) \in \mathbf{F}^2$ and $a = \{a_{ij}\}_{i,j \in \mathbf{N}}$,

$$p^v = (w, z)^v = (w, z)^{(v_1, v_2)} = w^{v_1} z^{v_2},$$
$$|p|^v = |(w, z)|^{(v_1, v_2)} = |w|^{v_1} |z|^{v_2},$$
$$p_{\lfloor v \rfloor} = (w, z)_{\lfloor (v_1, v_2) \rfloor} = w_{\lfloor v_1 \rfloor} z_{\lfloor v_2 \rfloor},$$
$$C_v^{(p)} = C_{(v_1, v_2)}^{(w, z)} = \frac{(w, z)_{\lfloor (v_1, v_2) \rfloor}}{(v_1, v_2)!},$$
$$\frac{\partial^{|v|_1}}{\partial p^v} = \frac{\partial^{|(v_1, v_2)|_1}}{\partial (w, z)^{(v_1, v_2)}} = \frac{\partial^{v_1 + v_2}}{\partial w^{v_1} \partial z^{v_2}},$$
$$p \cdot a = (w, z) \cdot a = \{a_{ij}w^i z^j\}.$$

If $u = (u_1, u_2)$ and $v = (v_1, v_2)$ are bi-indices, we will also write $u < v$ if $u_1 < v_1$ and $u_2 < v_2$; and write $u \leq v$ if $u_1 \leq v_1$ and $u_2 \leq v_2$.

Let $a = \{a_{ij}\}_{i,j \in \mathbf{N}} \in l^{\mathbf{N} \times \mathbf{N}}$. Let Λ be the set (or part of the set) of all $p = (w, z) \in \mathbf{F}^2$ such that the attenuated sequence $\{w^i z^j a_{ij}\}$ is absolutely summable. Then we may define a function $\widehat{a} : \Lambda \to C$ by

$$\widehat{a}(p) = \widehat{a}(w, z) = \sum_{v \in \mathbf{N} \times \mathbf{N}} a_v p^v = \sum_{i,j=0}^{\infty} a_{ij} w^i z^j, \ p = (w, z) \in \Lambda.$$

This function, which is completely determined by a, is called a (bivariate) power series function in p generated by a. In practice, for any $q \in \mathbf{F}^2$, the function $g(p)$ defined by $\widehat{a}(p-q)$ is also called a power series function in $p = (w, z)$ about (or with center at) q.

Example 3.10. Let $f = f(w, z)$ be a function defined on a domain Θ of $\mathbf{F}^2$ which has partial derivatives of any order at the point $(c, d) \in \Theta$. Let

$$a = \left\{ \frac{1}{i!j!} \frac{\partial^{i+j} f(c,d)}{\partial w^i \partial z^j} \right\}_{i,j \in \mathbf{N}}.$$

The power series function $\widehat{a}(w-c, z-d)$ is called the Taylor series function with center (c, d) generated by f.

Example 3.11. In view of Example 2.1, the double sequence $\{w^i z^j\}_{i,j \in \mathbf{N}}$ is absolutely summable if $|w| < 1$ and $|z| < 1$. Thus,

$$h(w, z) = \sum_{i=0}^{\infty} \sum_{j=0}^{\infty} w^i z^j, \quad |w| < 1, |z| < 1,$$

is a power series function.

For $p = (w, z) \in \mathbf{F}^2$, we will call the set

$$\mathfrak{S}(p) = \{(\alpha w, \beta z) \in \mathbf{F}^2 |\ |\alpha| < 1, |\beta| < 1\}$$

the silhouette of p.

Theorem 3.19. *Let $a \in l^{\mathbf{N} \times \mathbf{N}}$. If the attenuated sequence $(\lambda, \mu) \cdot a$ is summable relative to an ordering for $\mathbf{N} \times \mathbf{N}$, then the family $\{(w, z) \cdot a\}$ is absolutely and uniformly summable on compact subsets of the silhouette $\mathfrak{S}(\lambda, \mu)$.*

Indeed, suppose $\lambda \neq 0$ and $\mu \neq 0$. Since $\{a_{ij} \lambda^i \mu^j\}$ is bounded, say, by $M > 0$, and since $|w| \leq \rho_1 |\lambda|$ and $|z| \leq \rho_2 |\mu|$ for some $\rho_1, \rho_2 \in (0, 1)$, hence

$$|a_{ij} w^i z^j| \leq M \rho_1^i \rho_2^j, \quad i, j \in \mathbf{N},$$

for any $(w, z) \in \mathfrak{S}(\lambda, \mu)$. Since the double sequence $\{\rho_1^i \rho_2^j\}_{(i,j) \in \mathbf{N} \times \mathbf{N}}$ is absolutely summable, the family $\{a_{ij} w^i z^j\}_{(i,j) \in \mathbf{N} \times \mathbf{N}}$ is absolutely and uniformly summable. The case where $\lambda = 0$ or $\mu = 0$ is similarly proved.

If $(\lambda, \mu) \cdot a$ is absolutely summable, then $\{|a_{ij} \lambda^i \mu^j|\}$ is bounded. Thus, the above arguments also show that for $(w, z) \in \mathfrak{S}(\lambda, \mu)$, there is some $M > 0$ such that

$$|a_{ij}| \leq \frac{M}{|w|^i |z|^j}, \quad i, j \in \mathbf{N}.$$

Let $\mathfrak{B}(a)$ be the set of $(w, z) \in \mathbf{F}^2$ such that $(w, z) \cdot a$ is summable relative to some ordering Ψ (which may depend on (w, z)) for $\mathbf{N} \times \mathbf{N}$. The interior $\mathfrak{D}(a)$ of $\mathfrak{B}(a)$ is called the domain of convergence for the power series function generated by

a. It is known that when $\mathfrak{D}(a)$ is nonempty, it is a complete Reinhart domain with center $(0,0)$ and logarithmically convex subset of $\mathbf{F}^2$. Here a subset $S \subset \mathbf{F}^2$ is said to be logarithmically convex if

$$\{(\ln|w|, \ln|z|)|\ (w,z) \in S, w \neq 0, z \neq 0\}$$

is a convex subset of $\mathbf{R}^2$, and is called complete Reinhart domain with center $(0,0)$ if S is open and connected and $(w,z) \in S$ implies $(\rho_1 w, \rho_2 z) \in S$ for all $\rho_1, \rho_2 \in \mathbf{F}$ satisfying $|\rho_1|, |\rho_2| \leq 1$.

Theorem 3.20. *Suppose $\mathfrak{D}(a)$ is nonempty. Then $\mathfrak{D}(a)$ is a complete Reinhart domain of $\mathbf{F}^2$ with center $(0,0)$. Furthermore, the family $\{\underline{w}_x \cdot \underline{z}_y \cdot a\}$ is absolutely and uniformly summable on each closed dicylinder of the form*

$$\{(z_1, z_2) \in \mathbf{F}^2 |\ |z_1| \leq \rho_1,\ |z_2| \leq \rho_2\}$$

contained in $\mathfrak{D}(a)$.

Proof. Since $\mathfrak{D}(a)$ is nonempty, there exists some point (u,v) in $\mathfrak{D}(a)$ with nontrivial components. By the previous Theorem 3.19, the silhoutte $\mathfrak{S}(u,v)$ is contained in $\mathfrak{D}(a)$. Therefore (u,v) can be joined to $(0,0)$ by a straight line segment completely contained in $\mathfrak{D}(a)$. If $(w,0)$ belongs to $\mathfrak{D}(a)$ where $w \neq 0$, then some ball with center $(w,0)$ and contained in $\mathfrak{D}(a)$ will contain some point (u,v) with nontrivial components. Therefore, we can join $(w,0)$ to (u,v) by a straight line segment completely contained in $\mathfrak{D}(a)$. As a consequence, any two points in $\mathfrak{D}(a)$ can be joined by a broken line completely contained in $\mathfrak{D}(a)$. This shows that $\mathfrak{D}(a)$ is connected. Next note that for any $(w,z) \in \mathfrak{D}(a)$, $(w(1+\varepsilon), z(1+\varepsilon)) \in \mathfrak{D}(a)$ for some positive number ε. Since $(\rho_1 w(1+\varepsilon), \rho_2 z(1+\varepsilon)) \in \mathfrak{D}(a)$ for all $\rho_1, \rho_2 \in \mathbf{F}$ satisfying $|\rho_1|, |\rho_2| < 1$, thus $(\rho_3 w, \rho_4 z) \in \mathfrak{D}(a)$ for all $\rho_3, \rho_4 \in \mathbf{F}$ satisfying $|\rho_3|, |\rho_4| \leq 1$. This shows that $\mathfrak{D}(a)$ is a complete Reinhardt domain. Furthermore, the same reasoning shows that the family $\{\underline{(w,z)} \cdot a\}$ is absolutely and uniformly summable on each closed polycyclinder of the form

$$\{(z_1, z_2) \in \mathbf{F}^2 |\ |z_1| \leq \rho_1,\ |z_2| \leq \rho_2\}$$

contained in $\mathfrak{D}(a)$. The proof is complete.

We remark that when $\mathfrak{D}(a)$ is nonempty, it is also a logarithmically convex subset of $\mathbf{F}^2$. The proof is not difficult and make use of the fact that $a^{1-t}b^t \leq a+b$ whenever $a, b \geq 0$ and $0 \leq t \leq 1$. Since we have no need of this result in the sequel, we refer to pages 181-182 in Kaplan [94] for a proof. It is also remarkable that every logarithmically convex complete Reinhardt domain with center $(0,0)$ is the domain of convergence of the power series function generated by a double sequence.

There are several properties of power series functions generated by double sequences in $l^{\mathbf{N} \times \mathbf{N}}$ which are similar to those for power series functions generated by sequences in $l^{\mathbf{N}}$. In particular, we can show the following.

(i) If $a, b \in l^{\mathbf{N} \times \mathbf{N}}$ such that $\widehat{a}$ and $\widehat{b}$ are defined in a neighborhood of $(0,0)$, then for $\alpha, \beta \in \mathbf{F}^2$, $\alpha \widehat{a} + \beta \widehat{b}$ is also defined in a neighborhood of $(0,0)$ and $\widehat{(\alpha a + \beta b)}(w,z) = \alpha \widehat{a}(w,z) + \beta \widehat{b}(w,z)$ for any (w,z) in this neighborhood.

(ii) If $a, b \in l^{\mathbf{N}\times\mathbf{N}}$ such that $\widehat{a}$ and $\widehat{b}$ are defined in a neighborhood of $(0,0)$, then $\widehat{a*b}$ is also defined in a neighborhood of $(0,0)$ and $\widehat{a*b}(w,z) = \widehat{a}(w,z)\widehat{b}(w,z)$ for any (w,z) in this neighborhood.

(iii) Let $\widehat{a}$ be the power series function generated by a double sequence $a \in l^{\mathbf{N}\times\mathbf{N}}$ and $\mathfrak{D}(a)$ is nonempty. Then for each $(\lambda, \mu) \in \mathfrak{D}(a)$, there exists a ball $B((\lambda, \mu); \gamma) \subseteq \mathfrak{D}(a)$ and

$$\widehat{a}(w,z) = \sum_{\mathbf{N}\times\mathbf{N}} b_{ij}(w-\lambda)^i(z-\mu)^j < \infty, \ (w,z) \in B((\lambda,\mu);\gamma)$$

for some $\{b_{ij}\}_{i,j \in \mathbf{N}} \in l^{\mathbf{N}\times\mathbf{N}}$.

Indeed, as in the proof of Theorem 3.6, we may write

$$\sum_{m=0}^{\infty}\sum_{n=0}^{\infty} a_{mn} w^m z^n = \sum_{m=0}^{\infty}\sum_{n=0}^{\infty} a_{mn}(w-\lambda+\lambda)^m(z-\mu+\mu)^n$$

$$= \sum_{m=0}^{\infty}\sum_{n=0}^{\infty} a_{mn} \sum_{i=0}^{m} C_i^{(m)}(w-\lambda)^i \lambda^{m-i} \sum_{j=0}^{n} C_j^{(n)}(z-\mu)^j \mu^{n-j}$$

$$= \sum_{m=0}^{\infty}\sum_{n=0}^{\infty}\sum_{i=0}^{m}\sum_{j=0}^{n} \left\{ a_{mn} C_i^{(m)} \lambda^{m-i} C_j^{(n)} \mu^{n-j} \right\} (w-\lambda)^i (z-\mu)^j.$$

When $B((\lambda, \mu); \gamma)$ is chosen properly, the last sum, by Fubini's theorem, can be rearranged into the form

$$\sum_{\mathbf{N}\times\mathbf{N}} b_{ij}(w-\lambda)^i(z-\mu)^j = \sum_{i=0}^{\infty}\sum_{j=0}^{\infty}\sum_{m=i}^{\infty}\sum_{n=j}^{\infty} \left\{ a_{mn} C_i^{(m)} \lambda^{m-i} C_j^{(n)} \mu^{n-j} \right\} (w-\lambda)^i(z-\mu)^j.$$

This result may be called the representation theorem in the sequel for bivariate power series functions.

(iv) Let $\widehat{a} = \widehat{a}(w,z)$ be the power series function generated by a double sequence $a \in l^{\mathbf{N}\times\mathbf{N}}$ and $\mathfrak{D}(a)$ is nonempty. Then the domains of convergence of the power series functions generated by the partial algebraic derivatives $D_x a$ and $D_y a$ are equal to $\mathfrak{D}(a)$ and

$$\frac{\partial \widehat{a}}{\partial w}(\lambda,\mu) = \widehat{D_x a}(\lambda,\mu), \ \frac{\partial \widehat{a}}{\partial z}(\lambda,\mu) = \widehat{D_y a}(\lambda,\mu),$$

for $(\lambda, \mu) \in \mathfrak{D}(a)$.

(v) Let $a \in l^{\mathbf{N}\times\mathbf{N}}$ with nonempty $\mathfrak{D}(a)$. If $a_{00} \neq 0$, then $\mathfrak{D}(\overline{1}/a)$ is also nonemty and

$$\widehat{a}(w,z)\widehat{\overline{1}/a}(w,z) = 1$$

in a neighborhood of $(0,0)$.

3.4 Bivariate Analytic Functions

A function f with domain an open set $\Theta \subseteq \mathbf{F}^2$ and range $\mathbf{F}$ is said to be analytic at $q = (\lambda, \mu)$ if there is a sequence $a \in l^{\mathbf{N} \times \mathbf{N}}$ and a ball $B(q; \gamma)$ contained in Θ such that
$$f(w, z) = \sum_{\mathbf{N} \times \mathbf{N}} \overline{(w - \lambda, z - \mu) \cdot a} < \infty, \ (w, z) \in B(q; \gamma).$$
The function f is said to be analytic on Θ if it is analytic at each $q \in \Theta$. The set of all analytic functions $f : \Theta \subseteq \mathbf{F}^2 \to \mathbf{F}$ will be denoted by $H(\Theta)$.

Theorem 3.21. *Let $a \in l^{\mathbf{N} \times \mathbf{N}}$ such that the domain of convergence $\mathfrak{D}(a)$ of the power series $\widehat{a}$ generated by a is nonempty. Then $\widehat{a}$ is analytic on $\mathfrak{D}(a)$.*

Since the family $\left\{\overline{(w, z) \cdot a}\right\}$ is uniformly and absolutely summable on each closed dicylinder of the form
$$\{(z_1, z_2) \in \mathbf{F}^2 | \ |z_1| \leq \rho_1, \ |z_2| \leq \rho_2\}$$
in the domain of convergence $\mathfrak{D}(a)$ of a power series $\widehat{a}$ of two variables generated by a double sequence a, and since every point of $\mathfrak{D}(a)$ can be included in the interior of such a polycylinder, we may conclude that $\widehat{a}$ is continuous everywhere in $\mathfrak{D}(a)$.

There are several other properties of analytic functions which can be proved in manners similar to those for analytic functions in one variables.

Theorem 3.22. *Let U, V be open subsets of $\mathbf{F}^2$. If $f : U \to \mathbf{F}$ and $g : V \to \mathbf{F}$ are analytic, then $f + g$, $f \cdot g$ are analytic on $U \cap V$, and f/g is analytic on $U \cap V \cap \{p \in V | \ g(p) \neq 0\}$.*

Theorem 3.23. *Let $f \in H(\Theta)$. Then f is continuous and has continuous and analytic partial derivatives of all orders in Θ.*

We remark that if $f = f(w, z)$ is analytic at (c, d), then its definition asserts that f is the power series function $\widehat{a}$ generated by some double sequence $a \in l^{\mathbf{N} \times \mathbf{N}}$. By the above results, we see that
$$f(w, z) = \sum_{i=0}^{\infty} \sum_{j=0}^{\infty} \frac{1}{i! j!} \frac{\partial^{i+j} f(c, d)}{\partial w^i \partial z^j} (w - c)^i (z - d)^j, \ i, j \in \mathbf{N},$$
for (w, z) in a neighborhood of (c, d).

Theorem 3.24. *Let $f, g \in H(\Theta)$ where Θ is connected. If there is a sequence $\{(\lambda_i, \mu_i)\}_{i \in \mathbf{N}}$ contained in Θ such that $\lim_{i \to \infty}(\lambda_i, \mu_i) = (\lambda, \mu) \in \Theta$, $(\lambda, \mu) \neq (\lambda_i, \mu_i)$ for $i \in \mathbf{N}$ and $f(\lambda_i, \mu_i) = g(\lambda_i, \mu_i)$ for $i \in \mathbf{N}$, then $f(w, z) = g(w, z)$ for $(w, z) \in \Theta$.*

Theorem 3.25 (Substitution Theorem). *If f_1 and f_2 are analytic on some neighborhood of $(\lambda, \mu) \in \mathbf{F}^2$ and g is analytic on some neighborhood of $(f_1(\lambda, \mu), f_2(\lambda, \mu))$, then the composite function $g(f_1(w, z), f_2(w, z))$ is analytic on a neighborhood of (λ, μ).*

3.5 Multivariate Power Series and Analytic Functions

Multiple sequences, multivariate power series functions and analytic functions are only slightly more complicated than the double sequences, bivariate power series functions and analytic functions. We will therefore be brief in the following discussions.

First of all, κ is positive integer greater than or equal to 2. Elements in $\mathbf{F}^\kappa$ is denoted by $w = (w_1, ..., w_\kappa)$, $z = (z_1, ..., z_\kappa)$, etc. The element $(0, ..., 0)$ in $\mathbf{F}^\kappa$ is denoted by 0 as usual. A multi-index v is an element in $\mathbf{N}^\kappa$ of the form $(v_1, ..., v_\kappa)$. The notations $v!$ and $|v|_1$ stand for $v_1!v_2! \cdots v_\kappa!$ and $v_1 + v_2 + \cdots + v_\kappa$ respectively. The multi-index $e^{(i)}$ is defined by $e_i^{(i)} = 1$ and $e_j^{(i)} = 0$ for $j \neq i$. Multiple sequences are denoted by $f = \{f_v\}_{v \in \mathbf{N}^\kappa}$, $g = \{g_v\}_{v \in \mathbf{N}^\kappa}$, etc. Let $\alpha \in \mathbf{F}$, the multiple sequence whose $(0, ..., 0)$-th component is α and others are zero will be denoted by $\overline{\alpha}$ and is called a *scalar multiple sequence*. For $w \in \mathbf{F}^\kappa$ and $a = \{a_v\}_{v \in \mathbf{N}^\kappa}$,

$$w^v = w_1^{v_1} w_2^{v_2} \cdots w_\kappa^{v_\kappa},$$
$$|w|^v = |w_1|^{v_1} |w_2|^{v_2} \cdots |w_\kappa|^{v_\kappa},$$
$$w_{\lfloor v \rfloor} = (w_1)_{\lfloor v_1 \rfloor} (w_2)_{\lfloor v_2 \rfloor} \cdots (w_\kappa)_{\lfloor v_\kappa \rfloor},$$
$$C_v^{(w)} = \frac{w_{\lfloor v \rfloor}}{v!},$$
$$\frac{\partial^{|v|_1}}{\partial w^v} = \frac{\partial^{v_1}}{\partial w_1^{v_1}} \frac{\partial^{v_2}}{\partial w_2^{v_2}} \cdots \frac{\partial^{v_\kappa}}{\partial w_\kappa^{v_\kappa}} = \frac{\partial^{v_1+v_2+\cdots+v_\kappa}}{\partial w_1^{v_1} \partial w_2^{v_2} \cdots \partial w_\kappa^{v_\kappa}},$$
$$\underline{w} \cdot a = \{a_v w^v\}_{v \in \mathbf{N}^\kappa}.$$

If $u = (u_1, ..., u_\kappa)$ and $v = (v_1, ..., v_\kappa)$ are multi-indices, we will also write $u < v$ if $u_1 < v_1, ..., u_\kappa < v_\kappa$; and write $u \leq v$ if $u_1 \leq v_1, ..., u_\kappa \leq v_\kappa$.

For $a = \{a_v\}_{v \in \mathbf{N}^\kappa}$, we define the partial difference

$$\Delta_i a = a_{v + e^{(i)}} - a_v, \quad i = 1, ..., \kappa,$$

and the mixed partial difference

$$\Delta^u a = (\Delta_1)^{u_1} (\Delta_2)^{u_2} \cdots (\Delta_\kappa)^{u_\kappa} a,$$

where $u = (u_1, ..., u_\kappa)$ is a multi-index. We also define attenuated sequences by

$$\underline{z} \cdot a = \{a_v z^v\}_{v \in \mathbf{N}^\kappa},$$

where $z \in \mathbf{F}^\kappa$.

The convolution product of two multiple sequences $f, g \in l^{\mathbf{N}^\kappa}$ is the multiple sequence $\{h_v\}_{v \in \mathbf{N}^\kappa}$ defined by

$$h_w = \sum_{u+v=w} f_u g_v.$$

The convolution product $f * f$ will be denoted by $f^{\langle 2 \rangle}$. The notation $f^{\langle n \rangle}$ is defined recursively by $f * f^{\langle n-1 \rangle}$ for $n = 2, 3, ...$. For the sake of convenience, $f^{\langle 0 \rangle} = \overline{1}$, and $f^{\langle 1 \rangle} = f$.

By means of arguments similar to those we have seen for bivariate sequences, we can show that if $f, g \in l_1^{\mathbf{N}^\kappa}$, then $f * g \in l_1^{\mathbf{N}^\kappa}$, and $\sum_{\mathbf{N}^\kappa} f * g = (\sum_{\mathbf{N}^\kappa} f)(\sum_{\mathbf{N}^\kappa} g)$. We may also show that if $g_{(0,0,...,0)} \neq 0$, then there exists a unique multiple sequence x such that $x * g = f$. This unique sequence will be denoted by f/g.

The partial algebraic derivatives of a multiple sequence a are defined by

$$D_i a = \{(v_i + 1) f_{v+e^{(i)}}\}_{v \in \mathbf{N}^\kappa}, \quad i = 1, 2, ..., \kappa,$$

and the mixed partial algebraic derivatives by

$$D^u a = D_1^{u_1} D_2^{u_2} \cdots D_\kappa^{u_\kappa} a,$$

where u is a multi-index. Properties of these algebraic derivatives are similar to those for the bivariate sequences.

Let $a = \{a_v\}_{v \in \mathbf{N}^\kappa} \in l^{\mathbf{N}^\kappa}$. Let Λ be the set (or part of the set) of all $\lambda \in \mathbf{F}^\kappa$ such that the attenuated sequence $\underline{\lambda} \cdot a$ is absolutely summable. Then we may define a function $\widehat{a} : \Lambda \to C$ by

$$\widehat{a}(\lambda) = \sum_{v \in \mathbf{N}^\kappa} a_v \lambda^v, \quad \lambda \in \Lambda.$$

This function, which is completely determined by a, is called a (multivariate) power series function in λ generated by a. In practice, for any $\mu \in \mathbf{F}^\kappa$, the function $g(\lambda)$ defined by $\widehat{a}(\lambda - \mu)$ is also called a power series function in λ about (or with center at) μ.

Let $\mathfrak{B}(a)$ be the set of $\lambda \in \mathbf{F}^\kappa$ such that $\underline{\lambda} \cdot a$ is summable relative to some ordering Ψ (which may depend on λ) for $\mathbf{N}^\kappa$. The interior $\mathfrak{D}(a)$ of $\mathfrak{B}(a)$ is called the domain of convergence for the power series function generated by a. It is known that when $\mathfrak{D}(a)$ is nonempty, it is a complete Reinhart domain with center $0 \in \mathbf{F}^\kappa$ and logarithmically convex subset of $\mathbf{F}^\kappa$, where logarithmically convex subsets and complete Reinhart domain are defined in manners similar to those for subsets of $\mathbf{F}^2$.

We can also show the following:

(i) When $\mathfrak{D}(a)$ is nonempty, $\mathfrak{D}(a)$ is a complete Reinhart domain of $\mathbf{F}^2$ with center 0. Furthermore, the family $\{\underline{w} \cdot a\}$ is absolutely and uniformly summable on each closed polycyclinder of the form

$$\{w = (w_1, ..., w_\kappa) \in \mathbf{F}^\kappa | \; |w_1| \leq \rho_1, ..., |w_\kappa| \leq \rho_\kappa\}$$

contained in $\mathfrak{D}(a)$.

(ii) If $a, b \in l^{\mathbf{N}^\kappa}$ such that $\widehat{a}$ and $\widehat{b}$ are defined in a neighborhood of 0, then for $\alpha, \beta \in \mathbf{F}^\kappa$, $\alpha \widehat{a} + \beta \widehat{b}$ is also defined in a neighborhood of 0 and $\widehat{(\alpha a + \beta b)}(z) = \alpha \widehat{a}(z) + \beta \widehat{b}(z)$ for any z in this neighborhood.

(iii) If $a, b \in l^{\mathbf{N}^\kappa}$ such that $\widehat{a}$ and $\widehat{b}$ are defined in a neighborhood of 0, then $\widehat{a * b}$ is also defined in a neighborhood of 0 and $\widehat{a * b}(z) = \widehat{a}(z)\widehat{b}(z)$ for any z in this neighborhood.

(iv) Let $\hat{a}$ be the power series function generated by a double sequence $a \in l^{\mathbf{N}^\kappa}$ and $\mathfrak{D}(a)$ is nonempty. Then for each $\lambda \in \mathfrak{D}(a)$, there exists a ball $B(\lambda;\gamma) \subseteq \mathfrak{D}(a)$ and
$$\hat{a}(w) = \sum_{\mathbf{N}^\kappa} b_v (w-\lambda)^v < \infty, \ w \in B(\lambda;\gamma)$$
for some $\{b_v\}_{v \in \mathbf{N}^\kappa} \in l^{\mathbf{N}^\kappa}$.

(v) Let $\hat{a} = \hat{a}(w)$ be the power series function generated by a double sequence $a \in l^{\mathbf{N}^\kappa}$ and $\mathfrak{D}(a)$ is nonempty. Then the domains of convergence of the power series functions generated by the partial algebraic derivatives $D_1 a, ..., D_\kappa a$ are equal to $\mathfrak{D}(a)$ and
$$\frac{\partial \hat{a}}{\partial w_i}(\mu) = \widehat{D_i a}(\mu), \ i = 1, 2, ..., \kappa,$$
for $\mu \in \mathfrak{D}(a)$.

(vi) Let $a \in l^{\mathbf{N}^\kappa}$ with nonempty $\mathfrak{D}(a)$. If $a_{(0,...,0)} \neq 0$, then $\mathfrak{D}(\overline{1}/a)$ is also nonempty and
$$\hat{a}(w)\widehat{\overline{1}/a}(w) = 1$$
in a neighborhood of 0.

A function f with domain an open set $\Theta \subseteq \mathbf{F}^\kappa$ and range $\mathbf{F}$ is said to be analytic at λ if there is a sequence $a \in l^{\mathbf{N}^\kappa}$ and a ball $B(\lambda;\gamma)$ contained in Θ such that
$$f(w) = \sum_{\mathbf{N}^\kappa} \overline{w-\lambda} \cdot a < \infty, \ w \in B(\lambda;\gamma).$$
The function f is said to be analytic on Θ if it is analytic at each $\lambda \in \Theta$. The set of all analytic functions $f : \Theta \subseteq \mathbf{F}^\kappa \to \mathbf{F}$ will be denoted by $H(\Theta)$.

The following conclusions are similar to those for the bivariate analytic functions.

(a) Let $a \in l^{\mathbf{N}^\kappa}$ such that the domain of convergence $\mathfrak{D}(a)$ of the power series $\hat{a}$ generated by a is nonempty. Then $\hat{a}$ is continuous and analytic on $\mathfrak{D}(a)$.

(b) Let U, V be open subsets of $\mathbf{F}^\kappa$. If $f : U \to F$ and $g : V \to F$ are analytic, then $f+g$, $f \cdot g$ are analytic on $U \cap V$, and f/g is analytic on $U \cap V \cap \{p \in V | \ g(p) \neq 0\}$.

(c) Let $f \in H(\Theta)$. Then f is continuous and has continuous and analytic partial derivatives of all orders in Θ. Further, when Θ is subset of $\mathbf{R}^\kappa$, the indefinite integral of f with respect to any of its independent variable is analytic. Furthermore, if $f = f(w)$ is analytic at λ, then
$$f(w) = \sum_{\mathbf{N}^\kappa} \frac{1}{v!} \frac{\partial^{|v|_1} f(\lambda)}{\partial w^v} (w-\lambda)^v, \ i, j \in \mathbf{N},$$
for w in a neighborhood of λ.

(d) Let $f, g \in H(\Theta)$ where Θ is connected. If there is a sequence $\{\mu_i\}_{i \in \mathbf{N}}$ contained in Θ such that $\lim_{i \to \infty} \mu_i = \mu \in \Theta$, $\mu \neq \mu_i$ for $i \in \mathbf{N}$ and $f(\mu_i) = g(\mu_i)$ for $i \in \mathbf{N}$, then $f(\lambda) = g(\lambda)$ for $\lambda \in \Theta$.

(e) If $f_1, f_2, ..., f_m$ are analytic on some neighborhood of $\lambda \in \mathbf{F}^\kappa$ and g is analytic on some neighborhood of $(f_1(\lambda), f_2(\lambda), ..., f_m(\lambda))$, then the composite function $g(f_1(\lambda), f_2(\lambda), ..., f_m(\lambda))$ is analytic on a neighborhood of λ.

3.6 Matrix Power Series and Analytic Functions

Finite dimensional matrix (and vector) power series functions and matrix analytic functions are ordered tuples of power series functions or analytic functions and they are handled by componentwise manipulations. For instance, a matrix function $F(z) = (f_{ij}(z))$ is said to be analytic at w if each $f_{ij}(z)$ is analytic at w, and it is said be analytic over an open set $\Theta \subseteq \mathbf{F}$ if F is analytic at each $w \in \Theta$. The set of matrix functions are endowed with the usual operations, e.g. let $F(z) = (f_{ij}(z))$ and $G = (g_{ij}(z))$, then

$$(F + G)(z) = (f_{ij}(z) + g_{ij}(z))$$

for z in their common domains. Therefore many of the properties of matrix power series and analytic functions can be inferred from their component functions without too much trouble. For example, the (real) matrix function

$$\begin{bmatrix} \cos t & \sin t \\ e^t & 1/(1-t) \end{bmatrix}$$

is analytic at $t = 0$ since $\cos t$, $\sin t$, e^t and $1/(1-t)$ are analytic at $t = 0$. Furthermore, since

$$\cos t = 1 - \frac{t^2}{2!} + \frac{t^4}{4!} - \frac{t^6}{6!} + \cdots,$$

$$\sin t = t - \frac{t^3}{3!} + \frac{t^5}{5!} - \frac{t^7}{7!} + \cdots,$$

$$e^t = 1 + t + \frac{t^2}{2!} + \frac{t^3}{3!} + \cdots,$$

and

$$\frac{1}{1-t} = 1 + t + t^2 + t^3 + t^4 + \cdots, \quad t \in (-1, 1),$$

we see that our matrix analytic function can be expressed as a matrix power series function of the form

$$\begin{bmatrix} 1 & 0 \\ 1 & 1 \end{bmatrix} + \begin{bmatrix} 0 & 1 \\ 1 & 1 \end{bmatrix} t + \begin{bmatrix} -\frac{1}{2!} & 0 \\ \frac{1}{2!} & 1 \end{bmatrix} t^2 + \cdots$$

for $t \in (-1, 1)$.

Other properties can similarly be obtained whenever they are needed in the sequel. Hence we will not bother with them at this point.

3.7 Majorants

Given a power series function $f(z)$ generated by a sequence $a = \{a_n\}_{n \in \mathbf{N}}$, it is of great interest to estimate the sizes of $f(z)$ or of a. First of all, recall from Theorem 2.5 that if $\lambda \cdot a$ is absolutely summable, then $\{|a_k \lambda^k|\}_{k \in \mathbf{N}}$ is bounded. Thus, for each number $w \in (0, \rho(a))$, there is some $M_w > 0$ such that

$$|a_k| \le \frac{M_w}{|w|^k}, \quad k \in \mathbf{N}. \tag{3.14}$$

The above estimate for the sequence a is called the Cauchy's estimate. Conversely, if there is a number $r \ge \rho(a)$ such that for each $\lambda \in (0, r)$, there exists M_λ such that

$$|a_k| \le \frac{M_\lambda}{\lambda^k}, \quad k \in \mathbf{N},$$

then for any number $\mu \in \mathbf{F}$ that satisfies $|\mu| < \lambda$,

$$\left|a_k \mu^k\right| = \left|a_k \frac{\mu^k \lambda^k}{\lambda^k}\right| = |a_k \lambda^k| \left|\left(\frac{\mu}{\lambda}\right)^k\right| \le M_\lambda \left|\frac{\mu}{\lambda}\right|^k, \quad k \in \mathbf{N}.$$

Thus

$$\sum_{k=0}^{\infty} |a_k \mu|^k \le M_\lambda \sum_{k=0}^{\infty} \left|\frac{\mu}{\lambda}\right|^k < \infty.$$

Since λ is arbitrary, $r \le \rho(a)$ and hence $\rho(a) = r$.

Theorem 3.26 (Cauchy's Estimation). *Suppose $a = \{a_n\}_{n \in \mathbf{N}} \in l^{\mathbf{N}}$ has a positive radius of convergence $\rho(a)$. Then for any $w \in (0, \rho(a))$, there is some $M_w > 0$ such that (3.14) holds. Furthermore, there exists a positive number r such that*

$$|a_n| \le r^{n+1}, \quad n \in \mathbf{N}.$$

Proof. By Cauchy's estimate, for any $\beta \in (0, \rho(a))$, there is $M_\beta > 0$ such that $|a_n z^n| \le |a_n \beta^n| \le M_\beta$ for $n \in \mathbf{N}$ and $|z| \le \beta$. Pick a sufficiently small positive r^{-1} that satisfies $r^{-1} \le \beta$ and the additional property that $M_\beta r^{-1} < 1$. Then $|a_n r^{-n}| \le M_\beta$ so that

$$|a_n| \le M_\beta r^n = M_\beta r^{-1} r^{n+1} < r^{n+1}, \quad n \in \mathbf{N}.$$

The proof is complete.

As a consequence, if

$$f(z) = \sum_{n=0}^{\infty} a_n z^n$$

is analytic over a neighborhood of the origin, then

$$f'(z) = \sum_{n=0}^{\infty} (n+1) a_{n+1} z^n$$

is also analytic over a neighborhood of the origin, so that there exists a positive constant $\tilde{r}$ such that

$$|na_n| < \tilde{r}^n, \ n \in \mathbf{Z}^+.$$

Similarly,

$$f''(z) = \sum_{n=0}^{\infty}(n+2)(n+1)a_{n+2}z^n$$

is analytic on a neighborhood of the origin, thus there is a positive constant β such that

$$|n(n-1)a_n| < \beta^{n-1}, \ n \geq 2.$$

Example 3.12. Let

$$f(z) = sz + \sum_{n=2}^{\infty} a_n z^n$$

be analytic at 0. Then for $|z|$ sufficiently small, there are r and K such that

$$\left|\sum_{n=2}^{\infty} a_n z^n\right| \leq |z|^2 \sum_{n=0}^{\infty} |a_{n+2}| |z|^n \leq |z|^2 \sum_{n=0}^{\infty} r^{n+1} |z|^n = r|z|^2 \frac{1}{1-r|z|} \leq Kr|z|^2.$$

Hence

$$|f(z) - sz| \leq K|z|^{\mu+1}$$

for some $K, \mu > 0$.

Another useful technique for estimating the sizes of a power series function is related to the idea of majorization. Let $a = \{a_k\}_{k \in \mathbf{N}}$, $b = \{b_k\}_{k \in \mathbf{N}} \in l^{\mathbf{N}}$. The sequence a is said to be majorized by b (or b is a majorant of a) if $|a_k| \leq b_k$ for $k \in \mathbf{N}$. In such a case, we will write $a \ll b$. Clearly, if $a \ll b$, then $a \ll \alpha b$ for any real $\alpha \geq 1$, and

$$Da \ll Db$$

as well as

$$\int a \ll \int b.$$

Theorem 3.27. Let $a, b \in l^{\mathbf{N}}$. If a is majorized by b, then $\rho(a) \geq \rho(b)$.

Indeed, this follows from

$$\limsup_{k \to \infty} |a_k|^{1/k} \leq \limsup_{k \to \infty} |a_k|^{1/k}.$$

Alternatively, it suffices to assume that $\rho(b) > 0$ and show that

$$\sum_{k=0}^{\infty} |a_k \lambda^k| < \infty$$

for $|\lambda| < \rho(b)$. But this follows from
$$\sum_{k=0}^{\infty} |a_k \lambda^k| \le \sum_{k=0}^{\infty} b_k \mu^k < \infty$$
where μ is a fixed positive number satisfying $|\lambda| < \mu < \rho(b)$.

We remark that the above Theorem still holds if we only require $|a_k| \le b_k$ for $k \ge K$ for some $K \ge 0$.

It is easy to see that if $a \ll b$, then $a^{\langle 2 \rangle} \ll b^{\langle 2 \rangle}$ since
$$\left| a_k^{\langle 2 \rangle} \right| = \left| \sum_{i=0}^{k} a_{k-i} a_i \right| \le \sum_{i=0}^{k} |a_{k-i}| |a_i| \le \sum_{i=0}^{k} b_{k-i} b_i, \ k \in \mathbf{N}.$$

The same principle leads to the fact that if $a \ll b$ and if $c = \{c_k\}_{k \in \mathbf{N}} \in l^{\mathbf{N}}$ is defined by
$$c_k = P_k\left(a_{i_1}, a_{i_1}, ..., a_{i_j}\right), \ k \in \mathbf{N}, \tag{3.15}$$
where each $P_k(z_1, ..., z_j)$ is a multi-variate polynomial with nonnegative coefficients and j independent variables (where j may depend on k), then c is majorized by the sequence $d = \{d_k\}_{k \in \mathbf{N}}$ defined by
$$d_k = P_k\left(b_{i_1}, b_{i_1}, ..., b_{i_j}\right), \ k \in \mathbf{N}. \tag{3.16}$$
In particular, if $a \ll b$, then $a^{\langle m \rangle} \ll b^{\langle m \rangle}$ for $m \in Z^+$.

Theorem 3.28. Let $a, b \in l^{\mathbf{N}}$, and let $c, d \in l^{\mathbf{N}}$ be defined by (3.15) and (3.16) respectively. If $a \ll b$, then $c \ll d$.

As an immediate application, let $p, q, f, g \in l^{\mathbf{N}}$ such that $p \ll q$ and $f \ll g$. If the composition product $q \circ g$ is defined, then since
$$|(p \circ f)_n| \le \sum_{i=0}^{\infty} |p_i| \left| f_n^{\langle i \rangle} \right| \le \sum_{i=0}^{\infty} q_i g_n^{\langle i \rangle} = (q \circ g)_n,$$
we see that
$$p \circ f \ll q \circ g. \tag{3.17}$$

Let $\widehat{a}(\lambda)$ and $\widehat{b}(\lambda)$ be the power series functions generated by the $l^{\mathbf{N}}$ sequences a and b respectively. If a is majorized by b, then we say that $\widehat{a}$ is majorized by $\widehat{b}$ on their common domain Ω of definition (which contains $B(0; \beta(b))$ by Theorem 3.27). In such a case, we write $\widehat{a} \ll \widehat{b}$, or $\widehat{a}(\lambda) \ll \widehat{b}(\lambda)$ for $\lambda \in \Omega$.

In view of the properties of majorizing sequences, we see that
$$\widehat{a}(\lambda) \ll \widehat{b}(\lambda), \widehat{b}(\lambda) \ll \widehat{c}(\lambda) \Rightarrow \widehat{a}(\lambda) \ll \widehat{c}(\lambda),$$
$$\widehat{a}(\lambda) \ll \widehat{b}(\lambda) \Rightarrow \widehat{a}^m(\lambda) \ll \widehat{b}^m(\lambda), \ m \in \mathbf{N},$$
$$\widehat{a}(\lambda) \ll \widehat{b}(\lambda) \Rightarrow \widehat{a}'(\lambda) \ll \widehat{b}'(\lambda),$$

$$\widehat{a}(\lambda) \ll \widehat{b}(\lambda) \Rightarrow \int_0^z \widehat{a}(\lambda)d\lambda \ll \int_0^z \widehat{b}(\lambda)d\lambda,$$

and

$$\widehat{a}(\lambda) \ll \widehat{b}(\lambda), \widehat{c}(\lambda) \ll \widehat{d}(\lambda) \Rightarrow \widehat{a}(\widehat{c}(\lambda)) \ll \widehat{b}(\widehat{d}(\lambda)),$$

where the last assertion, in view of Theorem 3.11, needs the additional assumption that $\sum_{n=0}^\infty |d_n \lambda^n| < \rho(b)$ for $|\lambda| < \rho(d)$, $\sum_{n=0}^\infty |c_n \lambda^n| < \rho(a)$ for $|\lambda| < \rho(c)$ (and conditions that guarantee $(a \circ c)_n, (b \circ d)_n < \infty$).

Example 3.13. Let $a \in l^\mathbf{N}$ such that $\{a_k \lambda_0^k\}_{k \in \mathbf{N}}$ is relatively summable, where $\lambda_0 \neq 0$. Then in view of Theorem 2.5, $\{|a_k \lambda_0^k|\}$ is bounded, say by M. Hence

$$|a_k| \leq \frac{M}{|\lambda_0|^k}, \quad k \in \mathbf{N},$$

so that

$$a \ll \left\{M/|\lambda_0|^k\right\} \ll \{M/r^k\}, \quad 0 < r \leq |\lambda_0|.$$

Note that

$$\sum_{k=0}^\infty \frac{M}{r^k} z^k = M \frac{1}{1 - z/r}, \quad |z| < r \in (0, |\lambda_0|].$$

Thus there is some positive M such that

$$\widehat{a}(z) \ll \frac{M}{1 - z/r}, \quad |z| < r \in (0, |\lambda_0|].$$

We have defined majorant series functions generated by univariate sequences. Similar definitions hold for bivariate or multivariate sequences. For the sake of simplicity, we quickly go through the corresponding facts for multiple sequences. Let $a = \{a_v\}_{v \in \mathbf{N}^\kappa}$ and $b = \{b_v\}_{v \in \mathbf{N}^\kappa}$ be multiple sequences in $l^{\mathbf{N}^\kappa}$. The sequence a is said to be majorized by b if $|a_n| < b_v$ for $v \in \mathbf{N}^\kappa$. In such a case, we will write $a \ll b$. If $a \ll b$ and if $c = \{c_v\}_{v \in \mathbf{N}^\kappa} \in l^{\mathbf{N}^\kappa}$ is defined by

$$c_v = P_v\left(a_{u^{(1)}}, a_{u^{(2)}}, ..., a_{u^{(j)}}\right), \quad v, u^{(1)}, ..., u^{(j)} \in \mathbf{N}^\kappa, \tag{3.18}$$

where each P_v is a multivariate polynomial with nonnegative coefficients and j independent variables (where j may depend on the multi-index v), then c is majorized by the sequence $d = \{d_v\}_{v \in \mathbf{N}^\kappa}$ defined by

$$d_v = P_v\left(b_{u^{(1)}}, ..., b_{u^{(j)}}\right), \quad v, u^{(1)}, ..., u^{(j)} \in \mathbf{N}^\kappa. \tag{3.19}$$

Theorem 3.29. Let $a, b, c, d \in l^{\mathbf{N}^\kappa}$ where c_v and d_v are defined by (3.18) and (3.19) respectively. If $a \ll b$, then $c \ll d$.

If $a \ll b$, then we say that the power series functions $\widehat{a}(w)$ generated by a is majorized by the power series function $\widehat{b}(w)$ generated by b on their common domain Ω of convergence. In such a case, we write $\widehat{a} \ll \widehat{b}$, or $\widehat{a}(w) \ll \widehat{b}(w)$ for $\lambda \in \Omega$.

Example 3.14. By Example 2.2, if $w, z \in \mathbf{F}$ and $|w| + |z| < 1$, then

$$\sum_{k=0}^{\infty} \sum_{i+j=k; i,j \geq 0} \frac{k!}{i!j!} w^i z^j = \sum_{k=0}^{\infty} (w+z)^k = \frac{1}{1-w-z}.$$

Since the power series

$$\sum_{i,j=0}^{\infty} w^i z^j$$

is generated by the double sequence $a = \{a_{ij}\}$ where $a_{ij} = 1$ for all $i, j \in \mathbf{N}$, and since the coefficient of the term $w^i z^j$ in the power series function $1/(1-w-z)$ is greater than or equal to 1, we see from Example 2.1 that

$$\frac{1}{(1-w)(1-z)} = \sum_{i=0}^{\infty} w^i \sum_{j=0}^{\infty} z^j = \sum_{i,j=0}^{\infty} w^i z^j \ll \frac{1}{1-w-z}.$$

Example 3.15. If $\{a_v \lambda^v\}_{v \in \mathbf{N}^\kappa}$ is relatively summable for some $\lambda = (\lambda_1, ..., \lambda_\kappa)$ with positive components, then $\{|a_v \lambda^v|\}$ is bounded, say, by M. Thus

$$|a_v| \leq \frac{M}{|\lambda|^v}, \quad v \in \mathbf{N}^\kappa,$$

so that $a \ll \{M/|\lambda|^v\}_{v \in \mathbf{N}^\kappa}$. Thus for each $z = (z_1, ..., z_\kappa) \in \mathbf{F}^\kappa$ which satisfies $0 < |z_1| \leq \lambda_1, ..., 0 < |z_\kappa| \leq \lambda_\kappa$, we have $a \ll \{M/|z|^v\}$. Note that for $|w_1| < |z_1|, ..., |w_\kappa| < |z_\kappa|$, by Example 2.1,

$$\sum_{v \in \mathbf{N}^\kappa} \frac{M}{|z|^v} w^v = \sum_{(v_1, ..., v_\kappa) \in \mathbf{N}^\kappa} M \left(\frac{w_1}{|z_1|} \right)^{v_1} \cdots \left(\frac{w_\kappa}{|z_\kappa|} \right)^{v_\kappa}$$

$$= M \left\{ \sum_{i=0}^{\infty} \left(\frac{w_1}{|z_1|} \right)^i \right\} \cdots \left\{ \sum_{i=0}^{\infty} \left(\frac{w_\kappa}{|z_\kappa|} \right)^i \right\}$$

$$= \frac{M}{(1 - w_1/|z_1|) \cdots (1 - w_\kappa/|z_\kappa|)}.$$

As a consequence, if $\{a_v \lambda^v\}$ is relatively summable at λ with positive components $\lambda_1, ..., \lambda_\kappa$, then there is some positive number M such that for $0 < |z_1| \leq \lambda_1, ..., 0 < |z_\kappa| \leq \lambda_\kappa$,

$$\widehat{a}(w) \ll \frac{M}{(1 - w_1/|z_1|) \cdots (1 - w_\kappa/|z_\kappa|)}, \quad |w_1| < |z_1|, ..., |w_\kappa| < |z_\kappa|.$$

Note that if we set $\rho = \min\{|z_1|, ..., |z_\kappa|\}$, then for $w \in \mathbf{F}^\kappa$ which satisfies $|w_1| + |w_2| + \cdots + |w_\kappa| < \rho$, by the same reasoning used in Example 3.14, we have

$$\widehat{a}(w) \ll \frac{M}{\left(1 - \frac{w_1}{|z_1|}\right)\cdots\left(1 - \frac{w_\kappa}{|z_\kappa|}\right)} = M \sum_{(v_1,...,v_\kappa)\in\mathbf{N}^\kappa} \left(\frac{w_1}{|z_1|}\right)^{v_1} \cdots \left(\frac{w_\kappa}{|z_\kappa|}\right)^{v_\kappa}$$

$$\ll M \sum_{k=0}^{\infty} \sum_{|v|_1=k; v\in\mathbf{N}^\kappa} \frac{k!}{v!} \left(\frac{w_1}{|z_1|}\right)^{v_1} \cdots \left(\frac{w_\kappa}{|z_\kappa|}\right)^{v_\kappa} \ll M \sum_{k=0}^{\infty} \frac{1}{\rho^k} \sum_{|v|_1=k; v\in\mathbf{N}^\kappa} \frac{k!}{v!} w^v$$

$$= M \sum_{k=0}^{\infty} \left(\frac{w_1 + w_2 + \cdots + w_\kappa}{\rho}\right)^k = \frac{M}{1 - \frac{w_1+w_2+\cdots+w_\kappa}{\rho}}.$$

Theorem 3.30. *Let f, F be analytic functions from $B(0; \rho) \subset \mathbf{F}^\kappa$ into $B(0; \mu) \subset \mathbf{F}^\tau$ such that $f(0) = F(0) = 0$. Let g, G be analytic functions from $B(0; \mu) \subset \mathbf{F}^\tau$ into $\mathbf{F}$. If $f(z) \ll F(z)$ for $z \in B(0; \rho)$ and $g(w) \ll G(w)$ for $w \in B(0; \mu)$, then $g(f(z)) \ll G(F(z))$ for $z \in B(0; \rho)$.*

3.8 Siegel's Lemma

An important result in the method of majorants is due to Siegel [203]. Before going into the details, let us make the following definition.

Definition 3.1. A complex number α is called a Siegel number if $|\alpha| = 1$, $\alpha^n \neq 1$ for $n \in \mathbf{Z}^+$ and

$$\log |\alpha^n - 1|^{-1} \leq T \log n, \quad n = 2, 3, ... \tag{3.20}$$

for some positive constant T.

By writing α as $e^{2\pi i \omega}$, the condition (3.20) may be expressed in the form

$$\left|\omega - \frac{m}{n}\right| > \lambda n^{-\mu},$$

for arbitrary $m, n \in \mathbf{Z}^+$, where λ and μ denote positive numbers depending upon ω. It is then easily seen that (3.20) holds for all points of the unit circle $|\alpha| = 1$ with the exception of a set of 'measure' 0.

Before stating an important result due to Siegel [203], it is convenient to introduce the following notation. Let $d = \{d_k\} \in l^\mathbf{N}$ which satisfies $d_0 = 0$. Then we set $\Upsilon_n(d)$ to be

$$\max\{d_{v_1} d_{v_2} \cdots d_{v_t} | v_1 + v_2 + \cdots + v_t = n, 0 < v_1 \leq v_2 \leq \cdots \leq v_t, 2 \leq t \leq n\}.$$

For instance,

$$\Upsilon_2(d) = \max\{d_1^2\},$$
$$\Upsilon_3(d) = \max\{d_1 d_2, d_1^3\},$$
$$\Upsilon_4(d) = \max\{d_1 d_3, d_2^2, d_1^2 d_2, d_1^4\},$$

etc. In view of the definition of $\Upsilon_n(d)$, it is easily seen that it depends only on the terms $d_1, d_2, ..., d_{n-1}$.

Next recall from Example 2.11 that for a sequence $f \in l^{\mathbf{N}}$ that satisfies $f_0 = 0$, we have
$$\left\{f_2^{\langle 2 \rangle}\right\} = \{f_1^2\},$$
$$\left\{f_3^{\langle 2 \rangle}, f_3^{\langle 3 \rangle}\right\} = \{2f_1 f_2, f_1^2\},$$
$$\left\{f_4^{\langle 2 \rangle}, f_4^{\langle 3 \rangle}, f_4^{\langle 4 \rangle}\right\} = \{2f_1 f_3 + f_2^2, 3f_1^2 f_2, f_1^4\},$$

and
$$f_n^{\langle t \rangle} = \sum_{v_1 + \cdots + v_t = n; v_1, ..., v_t \in \mathbf{Z}^+} f_{v_1} f_{v_2} \cdots f_{v_t}.$$

Thus it is also easily realized that $f_{v_1} f_{v_2} \cdots f_{v_t}$, where $v_1 + v_2 + \cdots + v_t = n, 0 < v_1 \leq v_2 \cdots \leq v_t$, is one of the terms in $f_n^{\langle t \rangle}$. Consequently, if $q, c \in l^{\mathbf{N}}$ satisfy $q_0 = c_0 = 0$, $c_k \geq 0$, $q_k \geq 0$ for $k \geq 1$, then
$$(c \cdot q)_n^{\langle t \rangle} \leq \Upsilon_n(c) q_n^{\langle t \rangle}, \ 2 \leq t \leq n. \tag{3.21}$$

Theorem 3.31 (Siegel's Lemma). *Let α be a Siegel number. Then there is a positive number δ such that $|\alpha^n - 1|^{-1} < (2n)^\delta$ for $n \in \mathbf{Z}^+$. Furthermore, the sequence $\{d_n\}_{n=0}^\infty$ defined by $d_0 = 0$, $d_1 = 1$ and*
$$d_n = \frac{1}{|\alpha^{n-1} - 1|} \Upsilon_n(d), \ 2 \leq t \leq n, \ n \geq 2, \tag{3.22}$$

will satisfy
$$d_n \leq \left(2^{5\delta+1}\right)^{n-1} n^{-2\delta}, \ n \in \mathbf{Z}^+.$$

The proof is rather long and we will therefore require several assertions.

First of all, we show that if $x_1, ..., x_r$ and $y_1, ..., y_s$ are positive integers, where $r \geq 0, s \geq 2$, such that
$$\sum_{p=1}^r x_p + \sum_{q=1}^s y_q = k,$$
$$\sum_{q=1}^s y_q > \frac{k}{2},$$

and
$$y_q \leq \frac{k}{2}, \ q = 1, ..., s,$$

then
$$\prod_{p=1}^r x_p \prod_{q=1}^s y_q^2 \geq k^3 8^{1-t}, \ t = r + s. \tag{3.23}$$

To see this, denote the left-hand side of (3.23) by L. Consider first the case $k < 2t - 2$. Then

$$k^{-3} L \geq k^{-3} > (2t - 2)^{-3}. \tag{3.24}$$

Assume now $k \geq 2t - 2$ and let $g = [k/2]$, $\eta = r + \sum_{q=1}^{s} y_q$. Then $t \leq g + 1 \leq g + 1 + r \leq \eta \leq k$ and

$$\sum_{p=1}^{r} x_p = k - \eta + r,$$

so that

$$\prod_{p=1}^{r} x_p \geq k - \eta + 1,$$

and

$$\prod_{q=1}^{s} y_q \geq \begin{cases} \eta - t + 1, & \eta \leq g - 1 + t \\ (\eta - g - t + 2) g, & \eta \geq g - 1 + t \end{cases}.$$

In the interval $g + 1 \leq \eta \leq g - 1 + t$,

$$(k - \eta + 1)(\eta - t + 1)^2 \geq \min\left\{(k - g)(g - t + 2)^2, (k - g - t + 2) g^2\right\};$$

in the interval $g - 1 + t \leq \eta \leq k$,

$$(k - \eta + 1)(\eta - g - t + 2)^2 g^2 \geq (k - g - t + 2) g^2;$$

and in the interval $0 \leq \zeta \leq g$,

$$(k - g)(g - \zeta)^2 - (k - g - \zeta) g^2 = \{(k - g)\zeta - (2k - 3g) g\}\zeta \leq g (2g - k) \zeta \leq 0;$$

consequently

$$L \geq (k - g)(g - t + 2)^2$$

and

$$k^{-3} L \geq \frac{k - g}{k} \left(\frac{g - t + 2}{k}\right)^2 \geq \frac{1}{2}(2t - 2)^{-2} \geq (2t - 2)^{-3}. \tag{3.25}$$

Now

$$t - 1 \leq 2^{t-2}, \; t \geq 2,$$

and our assertion follows from (3.24) and (3.25).

Next, let

$$\varepsilon_n = \frac{1}{|\alpha^n - 1|}, \; n \in \mathbf{Z}^+.$$

In view of (3.20), there is a positive number δ such that

$$\varepsilon_n < (2n)^\delta, \; n \in \mathbf{Z}^+.$$

Let $N_1 = 2^{2\delta+1}$ and $N_2 = 8^\delta N_1 = 2^{5\delta+1}$. We assert that if $m_1, ..., m_r$ are positive integers, where $r \geq 0$, such that $m_0 > m_1 > \cdots > m_r > 0$, then

$$\prod_{l=0}^{r} \varepsilon_{m_l} < N_1^{r+1} \left\{ m_0 \prod_{l=1}^{r} (m_{l-1} - m_l) \right\}. \qquad (3.26)$$

Indeed, the assertion is true in the case $r = 0$. Assume $r \geq 1$. Note that

$$a_1^q \left(a_1^{p-q} - 1 \right) = (a_1^p - 1) - (a_1^q - 1), \; 0 < q < p,$$

hence

$$\varepsilon_{p-q}^{-1} \leq \varepsilon_p^{-1} + \varepsilon_q^{-1}$$

and

$$\min(\varepsilon_p, \varepsilon_q) \leq 2\varepsilon_{p-q} < 2^{\delta+1} (p-q)^\delta.$$

Let $\min\{\varepsilon_{m_1}, ...\varepsilon_{m_r}\} = \varepsilon_{m_h}$. Then

$$\varepsilon_{m_h} < 2^{\delta+1} \min\left\{ (m_{h-1} - m_h)^\delta, (m_h - m_{h+1})^\delta \right\}, \qquad (3.27)$$

provided we define $m_{-1} = \infty$ and $m_{r+1} = -\infty$. Assume by induction that (3.26) holds for $r-1$ instead of r, we have

$$\varepsilon_{m_h}^{-1} \prod_{l=0}^{r} \varepsilon_{m_l} < N_1^r \left\{ \frac{m_0 (m_{h-1} - m_{h+1})}{(m_{h-1} - m_h)(m_h - m_{h+1})} \prod_{l=1}^{r} (m_{l-1} - m_l) \right\}. \qquad (3.28)$$

Since

$$\frac{m_{h-1} - m_{h+1}}{(m_{h-1} - m_h)(m_h - m_{h+1})} = \frac{1}{m_{h-1} - m_h} + \frac{1}{m_h - m_{h+1}}$$

$$\leq \frac{2}{\min(m_{h-1} - m_h, m_h - m_{h+1})},$$

the inequality (3.26) follows from (3.27) and (3.28).

We now turn to the proof of Siegel's Lemma (Theorem 3.31). Let the positive sequence $\{d_n\}_{n=1}^\infty$ be defined by $d_1 = 1$ and (3.22). We assert that

$$d_k \leq k^{-2\delta} N_2^{k-1}, \; k \in \mathbf{Z}^+. \qquad (3.29)$$

Our assertion is true in the case $k = 1$. Assume $k \geq 2$. The numbers $\alpha_k = k^{-2\delta} N_2^{k-1}$ satisfy

$$\frac{\alpha_k \alpha_l}{\alpha_{k+l}} = \left(k^{-l} + l^{-l} \right)^{2\delta} N_2^{-1} \leq 2^{2\delta} N_2^{-1} < 1,$$

for $k, l \geq 1$, and consequently

$$d_{j_1} d_{j_2} \cdots d_{j_f} \leq j^{-2\delta} N_2^{j-1}, \; 1 \leq j_1 + \cdots + j_f = j < k; \; f \geq 1. \qquad (3.30)$$

By (3.22), there exists a decomposition

$$d_k = \varepsilon_{k-1} d_{g_1} d_{g_2} \cdots d_{g_\alpha}, \; g_1 + \cdots + g_\alpha = k > g_1 \geq \cdots \geq g_\alpha \geq 1.$$

In the case $g_1 > k/2$, we use this formula with g_1 instead of k and find a decomposition
$$d_{g_1} = \varepsilon_{g_1-1} d_{h_1} d_{h_2} \cdots d_{h_\beta}, \quad h_1 + \cdots + h_\beta = g_1 > h_1 \geq \cdots \geq h_\beta \geq 1;$$
if also $h_1 > k/2$, we decompose again
$$d_{h_1} = \varepsilon_{h_1-1} d_{i_1} d_{i_2} \cdots d_{i_\gamma}, \quad i_1 + \cdots + i_\gamma = h_1 > i_1 \geq \cdots \geq i_\gamma \geq 1;$$
and so on. Writing $k_0 = k$, $k_1 = g_1$, $k_2 = h_1$, ..., we obtain in this manner the formula
$$d_k = \prod_{p=0}^{r} \left(\varepsilon_{k_{p-1}} \Delta_p \right)$$
with $k = k_0 > k_1 > \cdots > k_r > k/2$, where Δ_p denotes for $p = 0, ..., r$ a certain product $d_{j_1} \cdots d_{j_f}$ and
$$j_1 + \cdots + j_f = \begin{cases} k_p - k_{p+1}, & p = 0, ..., r-1 \\ k_r, & p = r \end{cases},$$
all subscripts $j_1, ..., j_f$ being $\leq k/2$. The number f depends upon p; let $f = s$ for $p = r$. Using (3.29) for the s single factors of Δ_r and applying (3.30) for the estimation of Δ_p, $p = 0, ..., r-1$, we find the inequality
$$\prod_{p=0}^{r} \Delta_p \leq N_2^{k-r-s} \left\{ \prod_{q=1}^{s} j_q \prod_{p=1}^{r} (k_{p-1} - k_p) \right\}^{-2\delta},$$
where $1 \leq j_q \leq k/2$ for $q = 1, ..., s$ and $j_1 + \cdots + j_s = k_r$. In view of (3.26),
$$\prod_{p=0}^{r} \varepsilon_{k_p-1} < N_1^{r+1} \left\{ k \prod_{p=1}^{r} (k_{p-1} - k_p) \right\}^{\delta},$$
and consequently
$$d_k < N_1^{r-1} N_2^{k-t} \left(k^{-1} \prod_{p=1}^{r} x_p \prod_{q=1}^{s} y_q^2 \right)^{-\delta}$$
with $t = r + s$, $x_p = k_{p-1} - k_p$, $y_q = j_q$. In view of (3.23),
$$N_2^{1-k} k^{2\delta} \delta_k < N_1^{r+1} N_2^{1-t} 8^{\delta(t-1)} \leq \left(\frac{8^\delta N_1}{N_2} \right)^{t-1} = 1,$$
and (3.29) is proved.

Siegel's Lemma can be used to obtain majorants. For instance, we have the following result.

Theorem 3.32. *Let α be a Siegel number. Let $\{u_n\}_{n \in \mathbf{N}}$ be a complex sequence defined by $u_0 = 0$, $u_1 = \mu > 0$ and*
$$u_n = \frac{M}{|\alpha^{n-1} - 1|} \sum_{i=2}^{n} a_i u_n^{\langle i \rangle}, \quad n \geq 2,$$

where $M > 0$ and $a_i > 0$ for $i \geq 2$. Let $\{v_n\}_{n \in \mathbf{N}}$ be a complex sequence defined by $v_0 = 0$, $v_1 = \eta \geq \mu$ and

$$v_n = M \sum_{i=2}^{n} b_i v_n^{\langle i \rangle}, \ n \geq 2,$$

where $a_i \leq b_i$ for $i \geq 2$. Then there is $\delta > 0$ such that

$$u_n \leq v_n \left(2^{5\delta+1}\right)^{n-1} n^{-2\delta}, \ n \geq 2.$$

Proof. It is not difficult to see by induction that $u_k, v_k > 0$ for $k \geq 1$. Let $d = \{d_n\}_{n=0}^{\infty}$ be defined by $d_0 = 0$, $d_1 = 1$ and (3.22). Note that $u_1 \leq v_1 = d_1 v_1$. Assume by induction that $u_k \leq d_k v_k$ for $k = 1, 2, ..., n-1$ where $n \geq 2$. Then by (3.21) and Siegel's Lemma, there is $\delta > 0$ such that

$$u_n = \frac{M}{|\alpha^{n-1} - 1|} \sum_{i=2}^{n} a_i u_n^{\langle i \rangle} \leq \frac{M}{|\alpha^{n-1} - 1|} \sum_{i=2}^{n} b_i (d \cdot v)_n^{\langle i \rangle}$$

$$\leq \frac{1}{|\alpha^{n-1} - 1|} \Upsilon_n(d) M \sum_{i=2}^{n} b_i v_n^{\langle i \rangle}$$

$$= d_n v_n$$

$$\leq v_n \left(2^{5\delta+1}\right)^{n-1} n^{-2\delta}.$$

The proof is complete.

We remark that under the assumptions in Theorem 3.32 and the additional assumption that there is some $r > 0$ such that $v_n \leq r^n$ for $n \in \mathbf{N}$, then we may conclude that u has a positive radius of convergence as can be seen from

$$\limsup_{n \to \infty} u_n^{1/n} \leq \limsup_{n \to \infty} r \left(2^{5\delta+1}\right)^{(n-1)/n} n^{-2\delta/n} = r 2^{5\delta+1}.$$

3.9 Notes

Most of the material in this Chapter are well known and can be found in standard analysis text books such as Balser [13], Hille [78], Krantz and Parks [99], Krantz [100], Smith [211], Sneddon [212], Valiron [216].

Example 3.8 is adopted from Fichtenholz (see pp. 83-87 of [63]).

Siegel's Lemma and its proof can be found in Siegel [203]. It is also asserted in [203] that 'almost all' complex numbers on the unit circle are Siegel numbers.

Chapter 4

Functional Equations without Differentiation

4.1 Introduction

Roughly, a functional equation is a mathematical relation involving at least one unknown function and one or more known functions. Such equations and their extensions arise in many mathematical models. For instance, the defining relation for the unit circle is given by

$$x^2 + y^2 = 1, \ x, y \in \mathbf{R}.$$

Intuitively, it is clear that the upper part of the unit circle can be described by a function $y = y(x)$. Therefore, we have a functional equation of the form

$$x^2 + y^2(x) - 1 = 0.$$

The unknown function $y = y(x)$ is said to be 'implicit' in the above relation.

More generally, given a relation of the form

$$F(x, y) = 0,$$

where F is some given function of two variables, it is desirable to 'extract' a function $y = f(x)$ from such a relation so that

$$F(x, f(x)) = 0$$

for x in some appropriate domain.

In this Chapter, we will assume that the derivatives of the unknown functions are not involved in the implicit relations.

Example 4.1. Consider the functional equation

$$F(x, f(x)) = f^2(x) - f(x) - 1 = 0. \tag{4.1}$$

Suppose we are interested in finding a solution $f(x)$ which renders the equation (4.1) into an identity on some interval. Although there may be many possible types of solutions, an analytic solution is a good candidate. We therefore assume that $f(x)$ is an analytic solution defined over a neighborhood of 0 and is generated by the sequence $a = \{a_k\}_{k \in \mathbf{N}}$, that is,

$$f(x) = \widehat{a}(x) = \sum_{n=0}^{\infty} a_n x^n.$$

Then substituting it into (4.1), we obtain $\widehat{a^{\langle 2\rangle}}(x) - \widehat{a}(x) - 1 = 0$ for x in a neighborhood of 0. By the Unique Representation Theorem 3.8, we see that $a^{\langle 2\rangle} - a - \overline{1} = 0$, that is,

$$a_0^2 - a_0 - 1 = 0,$$
$$2a_0 a_1 - a_1 = 0,$$
$$2a_0 a_2 + a_1^2 - a_2 = 0,$$
$$2a_0 a_3 + 2a_1 a_2 - a_3 = 0,$$

and

$$a_k^{\langle 2\rangle} - a_k = 0$$

for $k \geq 4$. Thus $a_0 = (1 \pm \sqrt{5})/2$. Since $a_0 \neq 0$, we see from $2a_0 a_1 - a_1 = 0$ that $a_1 = 0$, and by induction, $a_k = 0$ for $k \in \mathbf{Z}^+$. This shows that any analytic solution of (4.1) is necessarily of the form $f(x) = (1 + \sqrt{5})/2$ or $f(x) = (1 - \sqrt{5})/2$. On the other hand, we may easily show by direct verification that either $f(x) = (1 + \sqrt{5})/2$ or $f(x) = (1 - \sqrt{5})/2$ is a solution of (4.1). Since both functions are analytic on $\mathbf{F}$, we have found all the analytic solutions of (4.1) on $\mathbf{F}$.

Example 4.2. Let us consider the functional equation

$$f^2(z) - \frac{\beta^2}{\beta + M} f(z) + \frac{M\beta^2}{\beta + M} \frac{z}{\alpha - z} = 0.$$

where $\alpha, \beta, M > 0$. Solving the quadratic equation

$$w^2 - \frac{\beta^2}{\beta + M} w + \frac{M\beta^2}{\beta + M} \frac{z}{\alpha - z} = 0, \qquad (4.2)$$

we see that

$$w = \frac{\beta^2}{2(\beta + M)} \left\{ 1 \pm \sqrt{1 - \frac{4M(\beta + M)}{\beta^2} \frac{z}{\alpha - z}} \right\}.$$

Setting

$$\beta_1 = \alpha \left(\frac{\beta}{\beta + 2M} \right)^2,$$

we see that

$$w = \frac{\beta^2}{2(\beta + M)} \left\{ 1 \pm \left(1 - \frac{z}{\beta_1}\right)^{1/2} \left(1 - \frac{z}{\alpha}\right)^{-1/2} \right\}.$$

Thus we are led to the solutions

$$f_+(z) = \frac{\beta^2}{2(\beta + M)} \left\{ 1 + \left(1 - \frac{z}{\beta_1}\right)^{1/2} \left(1 - \frac{z}{\alpha}\right)^{-1/2} \right\}$$

and

$$f_-(z) = \frac{\beta^2}{2(\beta + M)} \left\{ 1 - \left(1 - \frac{z}{\beta_1}\right)^{1/2} \left(1 - \frac{z}{\alpha}\right)^{-1/2} \right\}.$$

Recall from Example 3.7 that the functions $\left(1 - \frac{z}{\beta_1}\right)^{1/2}$ and $\left(1 - \frac{z}{\alpha}\right)^{1/2}$ are analytic over $B(0; \beta_1)$ and $B(0; \alpha)$ respectively. Thus $f_+(z)$ and $f_-(z)$ are solutions of (4.2) which are analytic over a neighborhood of the origin.

Example 4.3. To see another simple example, let us consider
$$f(2x) = 2f^2(x) - 1. \tag{4.3}$$
Assume that $f(x) = \hat{a}(x)$ is an analytic solution of (4.3) defined over a neighborhood of 0 and is generated by the sequence $a = \{a_k\}_{k \in \mathbf{N}}$, that is,
$$\hat{a}(x) = \sum_{n=0}^{\infty} a_n x^n,$$
then substituting it into (4.3), we obtain $2 \cdot a = 2a^{\langle 2 \rangle} - \bar{1}$, that is,
$$a_0 = 2a_0^2 - 1,$$
$$2a_1 = 4a_0 a_1,$$
$$2^2 a_2 = 2\left(a_1^2 + 2a_0 a_2\right),$$
$$2^3 a_3 = 2\left(2a_0 a_3 + 2a_1 a_2\right),$$
and
$$2^k a_k = 2a_k^{\langle 2 \rangle}$$
for $k \geq 4$. Thus, $a_0 = 1$ or $a_0 = -1/2$, and $a_1 = 0$. By induction, we may easily see that
$$a_{2k-1} = 0, \; k \in \mathbf{Z}^+.$$
Furthermore, if $a_0 = -1/2$, then $a_{2k} = 0$ for $k \in \mathbf{Z}^+$; while if $a_0 = 1$, then a_2 may be arbitrary, say, $a_2 = \alpha$, and
$$a_{2k} = \frac{(2\alpha)^k}{(2k)!}, \; k \in \mathbf{Z}^+.$$
Thus
$$\hat{a}(x) = 1 + \sum_{k=1}^{\infty} \frac{(2\alpha)^k}{(2k)!} x^{2k}$$
or
$$\hat{a}(x) = -\frac{1}{2}.$$
In both cases, the power series function $\hat{a}$ are analytic on $\mathbf{F}$ (by applying the ratio test to the former case). We remark that one particular case occurs when $\alpha < 0$, since letting $b^2 = -2\alpha$, we have
$$\hat{a}(x) = \sum_{k=0}^{\infty} (-1)^k \frac{(bx)^{2k}}{(2k)!} = \cos bx,$$
and another particular case occurs when $\alpha > 0$, since letting $d^2 = 2\alpha$, we have
$$\hat{a}(x) = \sum_{k=0}^{\infty} \frac{(dx)^{2k}}{(2k)!} = \cosh dx.$$

4.2 Analytic Implicit Function Theorem

In the previous section, the functional relations contain enough information to yield recurrence relations which define explicit sequences for generating the desired analytic solutions. In general, recurrence relations may only yield implicitly defined sequences, and alternate technique has to be used to show that the power series functions generated by these sequences are analytic.

As an example, let F be a function of two (real or complex) variables and consider the functional equation

$$F(x, f(x)) = 0. \tag{4.4}$$

Consider the possibility of determining a solution $f(x)$ such that substituting $f(x)$ into (4.4) renders it into an identity. The function F is in general not specified. However, if we assume that $F = F(x,y)$ is analytic at the point (x_0, y_0) and

$$F(x_0, y_0) = C_{00} = 0, \quad F'_y(x_0, y_0) \neq 0, \tag{4.5}$$

we may show that there is a solution $y = f(x)$ of (4.4) which is analytic at x_0. To see this, we assume without loss of generality that $x_0 = y_0 = 0$. Then in view of (4.4),

$$0 = F(x, y) = \sum_{j=0}^{\infty} \sum_{i=0}^{\infty} C_{ij} x^i y^j$$

in a neighborhood of $(0,0)$. Since $F(0,0) = 0$ and $F_y(0,0) = C_{01} \neq 0$, we may divide the above equation by C_{01} to obtain

$$y = c_{10}x + c_{20}x^2 + c_{11}xy + c_{02}y^2 + c_{30}x^3 + c_{21}x^2y + c_{12}xy^2 + c_{03}y^3 + \cdots, \tag{4.6}$$

where $c_{ij} = -C_{ij}/C_{01}$. Assuming an analytic solution $y = \widehat{a}(x)$ of (4.4) in the form

$$y = \sum_{k=0}^{\infty} a_k x^k \tag{4.7}$$

in a sufficiently small neighborhood of 0, we obtain from the assumption that $y = 0$ for $x = 0$ that $a_0 = 0$ and from (4.6) that

$$a_1 x + a_2 x^2 + a_3 x^3 + \cdots = c_{10}x + c_{20}x^2 + c_{11}x \left(a_1 x + a_2 x^2 + a_3 x^3 + \cdots \right)$$
$$+ c_{02} \left(a_1 x + a_2 x^2 + a_3 x^3 + \cdots \right)^2 + c_{30} x^3$$
$$+ c_{21} x^2 \left(a_1 x + a_2 x^2 + a_3 x^3 + \cdots \right)$$
$$+ c_{12} x \left(a_1 x + a_2 x^2 + a_3 x^3 + \cdots \right)^2$$
$$+ c_{03} \left(a_1 x + a_2 x^2 + a_3 x^3 + \cdots \right)^3 + \cdots$$

By Theorems 3.16 and 3.11 (which show that substituting one power series into another and combining coefficients of like powers of x is legitimate), and equating coefficients of like powers of x, we arrive at the system of equations

$$a_1 = c_{10},$$
$$a_2 = c_{20} + c_{11}a_1 + c_{02}a_1^2,$$
$$a_3 = c_{11}a_2 + 2c_{02}a_1 a_2 + c_{30} + c_{21}a_1 + c_{12}a_1^2 + c_{03}a_1^3,$$
$$\cdots = \cdots \tag{4.8}$$

Although the explicit forms of $a_1, a_2, \ldots$ are not given, their calculations involve only with additions and multiplications, therefore, it is not difficult to see by induction that the coefficients $a_2, a_3, \ldots$ are of the form $a_2 = P_2(a_1, c_{11}, c_{20})$, $a_3 = P_3(a_1, a_2, c_{11}, c_{02}, c_{30}, c_{21}, c_{12}, c_{03}), \ldots, a_k = P_k(a_1, a_2, \ldots, a_{k-1}, c_{11}, \ldots, c_{0k})$ where $P_2, P_3, \ldots$ are uniquely determined polynomials with positive coefficients. These show the uniqueness of the solution $\widehat{a}(x)$.

To see that (4.7) is indeed an analytic solution of (4.4), we only need to show that it converges in a neighborhood of 0. This is accomplished by finding a majorant function for $\widehat{a}(x)$.

To this end, suppose there is a double sequence $\{d_{ij}\}_{i,j \in \mathbf{N}}$ such that $d_{00} = 0$ and $d_{ij} > 0$ for $(i,j) \neq (0,0)$ and satisfies

$$|c_{ij}| \leq d_{ij}, \ i, j \in \mathbf{N}.$$

Then letting the sequence $\{b_k\}_{k \in \mathbf{N}}$ be defined by $b_0 = 0$, $b_1 = d_{10}$ and $b_k = P_k(b_1, b_2, \ldots, b_{k-1}, d_{11}, \ldots, d_{0k})$ for $k \geq 2$, in view of the similarity between the sequence a and b, it is easily seen that the coefficients $b_1, b_2, \ldots$ are all positive and satisfy

$$|a_i| \leq b_i, \ i \in \mathbf{Z}^+,$$

i.e., a is majorized by b; furthermore, its corresponding power series function

$$Y = \widehat{b}(x) = b_1 x + b_2 x^2 + b_3 x^3 + \cdots$$

satisfies

$$Y = d_{10}x + d_{20}x^2 + d_{11}xY + d_{02}Y^2 + d_{30}x^3 + d_{21}x^2 Y + d_{12}xY^2 + d_{03}Y^3 + \cdots. \quad (4.9)$$

Such a double sequence $\{d_{ij}\}$ exists. Indeed, since $F(x,y)$ is analytic in a neighborhood of $(0,0)$, there exist positive numbers α and β such that the double series

$$|c_{10}|\alpha + |c_{20}|\alpha^2 + |c_{11}|\alpha\beta + |c_{02}|\beta^2 + \cdots$$

converges. Then

$$|c_{ij}|\alpha^i \beta^j \leq M, \ i \in \mathbf{Z}^+; j \in \mathbf{N},$$

for some positive constant M so that we may set

$$d_{ij} = \frac{M}{\alpha^i \beta^j}, \ i \in \mathbf{Z}^+; j \in \mathbf{N}.$$

To complete our investigation, we need to show that $Y = \widehat{b}(x)$ is analytic at 0. To this end, note from (4.9) that

$$Y = \frac{M}{\alpha}x + \frac{M}{\alpha^2}x^2 + \frac{M}{\alpha\beta}xY + \frac{M}{\beta^2}Y^2 + \cdots = \frac{M}{\left(1 - \frac{x}{\alpha}\right)\left(1 - \frac{Y}{\beta}\right)} - M - \frac{M}{\beta}Y,$$

or,

$$Y^2 - \frac{\beta^2}{\beta + M}Y + \frac{M\beta^2}{\beta + M}\frac{x}{\alpha - x} = 0.$$

In view of Example 4.2, $Y = Y_+$ or $Y = Y_-$ where

$$Y_\pm = \frac{\beta^2}{2(\beta + M)} \left\{ 1 \pm \left(1 - \frac{x}{\beta_1}\right)^{1/2} \left(1 - \frac{x}{\alpha}\right)^{-1/2} \right\},$$

and

$$\beta_1 = \alpha \left(\frac{\beta}{\beta + 2M}\right)^2,$$

The correct one is Y_- since $Y = 0$ when $x = 0$. Since the function $Y = Y_- = \widehat{b}(x)$ is analytic at 0, hence (4.7) is also analytic over a neighborhood of 0. The proof is complete.

We summarize the above discussions as follows.

Theorem 4.1 (Analytic Implicit Function Theorem). *Suppose $F = F(x, y)$ is analytic at the point (x_0, y_0), $F(x_0, y_0) = 0$ and $F'_y(x_0, y_0) \neq 0$. Then there exists a unique function*

$$f(x) = y_0 - \frac{F'_x(x_0, y_0)}{F'_y(x_0, y_0)}(x - x_0) + \sum_{k=2}^{\infty} a_k (x - x_0)^k$$

which is analytic on a neighborhood of x_0 and satisfies $F(x, f(x)) = 0$ for x near x_0.

We remark that the assumption that $F_y(x_0, y_0) \neq 0$ in (4.5) is a sufficient but not a necessary condition.

Example 4.4. Consider the analytic function

$$F(x, y) = ((1-x)y - x)(x^2 + y^2) = (1-x)y^3 - xy^2 + x^2(1-x)y - x^3$$

for $x, y \in \mathbf{R}$. It is easily checked that $F(0, 0) = 0$ and $F'_y(0, 0) = 0$. But

$$y = f(x) = \frac{x}{1-x}, \quad |x| < 1,$$

is an analytic solution of the equation $F(x, y) = 0$.

We remark further that an immediate consequence of the above implicit function theorem is the analytic inverse function theorem.

Theorem 4.2 (Analytic Inverse Function Theorem). *If $y = g(x)$ is an analytic function of x at x_0 such that $g'(x_0) \neq 0$, then the inverse function g^{-1} exists and is analytic at $y_0 = g(x_0)$.*

Indeed, let $g(x) = \sum_{k=0}^{\infty} a_k (x - x_0)^k$, and let $\{c_{ij}\}_{i,j \in \mathbf{N}}$ be defined by $c_{ij} = a_j$ if $i = 0$ and $c_{ij} = 0$ if $i > 0$. Then the bivariate function

$$F(u, v) = u - \sum_{j=0}^{\infty} \sum_{i=0}^{\infty} c_{ij} u^i (v - x_0)^j = u - \sum_{j=0}^{\infty} a_j (v - x_0)^j$$

is analytic at $(u, v) = (y_0, x_0)$ and
$$F'_v(y_0, x_0) = -a_1 = g'(x_0) \neq 0.$$
By our implicit function theorem, there exists a function $v = f(u)$ defined in a neighborhood Ω of the point $u = y_0$ such that
$$0 = F(u, f(u)) = u - \sum_{j=0}^{\infty} a_j \left(f(u) - x_0\right)^j = u - g(f(u)).$$
That is, $f(u) = g^{-1}(u)$ for $u \in \Omega$.

Example 4.5. Recall from Example 1.6 that the inverse function of exp is the function $\log_0$ defined on $\{z \in C |\ \Im(z) \in (-\pi, \pi]\}$. Furthermore, as can be verified directly, the exponential function of a complex variable maps the domain $\{z \in \mathbf{C} |\ \Im(z) \in (-\pi, \pi)\}$ one-to-one onto the domain $\mathbf{C}\backslash(-\infty, 0]$. Since $\exp'(w) = \exp w \neq 0$ and since exp is analytic at each $w \in \{z \in \mathbf{C} |\ \Im(z) \in (-\pi, \pi)\}$, its inverse function $\log_0$ is analytic on $\mathbf{C}\backslash(-\infty, 0]$. Furthermore, since
$$z \log'_0(z) = \exp\left(\log_0(z)\right) \log'_0(z)$$
$$= \exp'\left(\log_0(z)\right) \log'_0(z)$$
$$= \frac{d}{dz} \exp\left(\log_0(z)\right) = \frac{d}{dz} z = 1,$$
we see that
$$\log'_0(z) = \frac{1}{z}, \quad z \in \mathbf{C}\backslash\{z \in \mathbf{C} : \Re(z) \leq 0\}.$$
Thus
$$\log'_0(1 + z) = \frac{1}{1 + z} = \sum_{n=0}^{\infty} (-1)^n z^n, \quad z \in B(0; 1),$$
and
$$\log_0(1 + z) = \int_0^z \log'_0(1 + w) dw = \int_0^z \sum_{n=0}^{\infty} (-1)^n w^n dw = \sum_{n=0}^{\infty} (-1)^n \frac{z^{n+1}}{n+1}$$
for $z \in B(0; 1)$.

Example 4.6. Recall from Example 1.7 that for $w \in \mathbf{C}$ and $z \in \mathbf{C}\backslash(-\infty, 0]$, $z^w = e^{w \log_0(z)}$. Hence the power function $f : \mathbf{C}\backslash(-\infty, 0] \to \mathbf{C}$ defined by
$$f(z) = z^w$$
is analytic on $\mathbf{C}\backslash(-\infty, 0]$. Furthermore,
$$\frac{d}{dz} z^w = w z^{w-1}.$$
The Newton binomial expansion formula in Example 3.6 takes the form
$$(1 + x)^\alpha = \sum_{n=0}^{\infty} C_n^{(\alpha)} x^n$$

for real α and x that satisfies $|x| < 1$. If $z \in \mathbf{C}$ and $|z| < 1$, then for any $w \in \mathbf{C}$, in view of the Substitution Theorem 3.11,

$$(1+z)^w = e^{w \log_0(1+z)} = \exp\left\{ w \sum_{n=0}^{\infty} (-1)^n \frac{z^{n+1}}{n+1} \right\}$$

$$= 1 + wz + \frac{w(w-1)}{2!} z^2 + \cdots .$$

To find the coefficients of z^n, we argue as follows: Clearly this coefficient is some polynomial $Q_n(w)$ of degree n in w. Since there is no term involving z^n in the above expansion if $w = 0, 1, 2, ..., n-1$, the polynomial must vanish at the indicated points. But then,

$$Q_n(w) = dw(w-1)\cdots(w-n-1)$$

where d is a constant. For $w = n$, the coefficient of z^n is just 1, and hence $Q_n(n) = 1$. It follows that

$$d = \frac{1}{n!},$$

so that $Q_n(w) = C_n^{(w)}$. Thus we have

$$(1+z)^w = \sum_{n=0}^{\infty} C_n^{(w)} z^n, \ |z| < 1.$$

Example 4.7. An alternate derivation of the Newton Binomial Expansion Theorem can be seen by observing that $f(z) = (1+z)^w$ satisfies $f(0) = 1$ and

$$(1+z)f'(z) = wf(z), \ |z| < 1.$$

Assume that

$$f(z) = \sum_{n=0}^{\infty} a_n z^n.$$

Then $a_0 = f(0) = 1$ and by the Unique Representation Theorem 3.8,

$$(n+1)a_{n+1} + na_n = wa_n, \ n \in \mathbf{N}.$$

The above recurrence is easily solved and

$$a_n = \frac{1}{n!} w(w-1)\cdots(w-n+1), \ n \in \mathbf{Z}^+.$$

4.3 Polynomial and Rational Functional Equations

Given a polynomial $P(z)$ of the form

$$P(z) = z^m + a_{m-1} z^{m-1} + \cdots + a_1 z + a_0,$$

then formally 'replacing' each i-th power by the i-th power G^i of a function G will result in a polynomial in G, which is naturally denoted by $P(G)$. Given a polynomial $P(G)$ in the function $G = G(z)$ and a function $h = h(z)$, the equation

$$P(G)(z) = h(z)$$

is called a polynomial functional equation. For instance,

$$G^m(z) + a_{m-1}G^{m-1}(z) + \cdots + a_1 G(z) + a_0 = h(z)$$

is such an equation. A rational functional equation is similarly defined. For instance,

$$G(z) - \alpha z - \beta \frac{G(z)}{1 - G(z)} = H(z)$$

is a rational functional equation.

In this section, we find analytic solutions to several polynomial and rational functional equations. As our first example, consider the polynomial equation

$$G^2(z) - \frac{1}{\mu}G(z) + \frac{\alpha}{\mu}z = 0, \qquad (4.10)$$

where $\alpha \neq 0$ and $\mu \neq 0$. Solving the quadratic equation

$$w^2 - \frac{1}{\mu}w + \frac{\alpha}{\mu}z = 0,$$

we see that

$$w = \frac{1}{2\mu}\left\{1 \pm \sqrt{1 - 4\alpha\mu z}\right\}.$$

Thus we may conclude that either

$$G_+(z) = \frac{1}{2\mu}\left\{1 + \sqrt{1 - 4\alpha\mu z}\right\}$$

or

$$G_-(z) = \frac{1}{2\mu}\left\{1 - \sqrt{1 - 4\alpha\mu z}\right\}$$

are formal solutions of our polynomial equation. Furthermore, since $(1 - 4\alpha\mu z)^{1/2}$ is analytic on $B(0; 1/|4\alpha\mu|)$ (see Example (3.21)), we see that both are solutions which are analytic on $B(0; 1/|4\alpha\mu|)$. They are the only solutions that are analytic near 0. Indeed, if $G(z)$ is defined at $z = 0$, then

$$0 = G^2(0) - \frac{1}{\mu}G(0) + 0 = G(0)\left(G(0) - \frac{1}{\mu}\right),$$

so that $G(0) = 0$ or $G(0) = 1/\mu$. But then the additional condition $G(0) = 0$ will lead us to $G(z) = G_-(z)$, while the additional condition $G(0) = 1/\mu$ to the $G(z) = G_+(z)$.

There is another approach to solving (4.10). We may treat $G(z)$ as an implicit function to be sought in the relation

$$F(z, G) \equiv G^2 - \frac{1}{\mu}G + \frac{\alpha}{\mu}z = 0.$$

Since $F(z,w)$ is analytic for each (z,w) in a neighborhood of $(0,0)$, $F(0,0) = 0$ and $F'_w(0,0) = -1/\mu \neq 0$, by the Analytic Implicit Function Theorem 4.1, we see that the implicit relation $F(z,G) = 0$ has a solution

$$G(z) = \sum_{n=0}^{\infty} g_n z^n$$

which is analytic on a neighborhood of the origin, and the sequence $g = \{g_n\}_{n \in \mathbf{N}}$ satisfies $g_0 = 0$ as well as $g_1 = -F'_z(0,0)/F'_w(0,0) = \alpha$. The other terms of g can also be determined. To see this, we substitute $G(z) = \sum_{n=0}^{\infty} g_n z^n$ into our equation (4.10), then

$$0 = \sum_{n=0}^{\infty} g_n^{\langle 2 \rangle} z^n - \frac{1}{\mu} \sum_{n=0}^{\infty} g_n z^n + \frac{\alpha}{\mu} z = \sum_{n=2}^{\infty} g_n^{\langle 2 \rangle} z^n - \frac{1}{\mu} \sum_{n=1}^{\infty} g_n z^n + \frac{\alpha}{\mu} z.$$

By the Unique Representation Theorem 3.8, we see that

$$g_n = \mu g_n^{\langle 2 \rangle}, \ n \geq 2.$$

Or equivalently, by Theorem 3.4, we may view (4.10) as

$$\widehat{g^{\langle 2 \rangle}}(z) - \frac{1}{\mu} \widehat{g}(z) + \frac{\alpha}{\mu} \hbar = 0,$$

which leads us to

$$g_1^{\langle 2 \rangle} - \frac{1}{\mu} g_1 + \frac{\alpha}{\mu} = -\frac{1}{\mu} g_1 + \frac{\alpha}{\mu} = 0,$$

and

$$g_n^{\langle 2 \rangle} - \frac{1}{\mu} g_n = 0, \ n \geq 2,$$

yielding the same conclusion about the sequence g.

The solution just determined satisfies $G(0) = 0$, and hence it must be equal to $G_-(z)$ found above. Therefore,

$$\frac{1}{2\mu} \left\{ 1 - \sqrt{1 - 4\alpha\mu z} \right\} = \sum_{n=0}^{\infty} g_n z^n,$$

where $g = \{g_n\}_{n \in \mathbf{N}}$ is determined by $g_0 = 0$, $g_1 = \alpha$ and $g_n = \mu g_n^{\langle 2 \rangle}$ for $n \geq 2$ (cf. Example 2.11).

We can also find the solution $G_+(z)$ by the Analytic Implicit Function Theorem 4.1 since $F(0, 1/\mu) = 0$ and $F'_w(0, 1/\mu) = 1/\mu \neq 0$. The ideas are not much different and hence the details are skipped.

Similar techniques will lead us to analytic solutions of several polynomial and rational functional equations as follows.

Example 4.8. Let $\mu \neq 0$ and $K \in \mathbf{Z}^+$. Then the equation

$$G(z) - \sum_{i=1}^{K} \alpha_i z^i - \frac{1}{\mu} \frac{G^{K+1}(z)}{1 - G(z)} = 0$$

has a solution $G(z)$ which is analytic at each z in a neighborhood of the origin and $G(z)$ can be defined in a recursive manner. To this end, let

$$F(z,w) \equiv w - \sum_{i=1}^{K} \alpha_i z^i - \frac{1}{\mu} \frac{w^{K+1}}{1-w}.$$

Since $F(z,w)$ is analytic for each (z,w) in a neighborhood of $(0,0)$, $F(0,0) = 0$ and $F'_w(0,0) = 1$, by the Analytic Implicit Function Theorem 4.1, we see that the implicit relation $F(z,w) = 0$ has an analytic solution $w = G(z)$ which satisfies $G(0) = 0$ and $G'(0) = -F'_z(0,0)/F'_w(0,0) = \alpha_1$, and is defined on a disk with the origin as the center and with a positive radius. This solution can be determined in a recursive manner. To see this, let

$$G(z) = \sum_{k=0}^{\infty} g_k z^k.$$

Then $g_0 = G(0) = 0$ and $g_1 = G'(0) = \alpha_1$. Furthermore, by substituting $w = G(z)$ into $F(z,w) = 0$, we see that

$$G(z) - \sum_{i=1}^{K} \alpha_i z^i = \frac{1}{\mu} G^{K+1}(z) \sum_{n=0}^{\infty} G^n(z)$$

$$= \frac{1}{\mu} \sum_{n=K+1}^{\infty} G^n(z)$$

$$= \frac{1}{\mu} \sum_{n=K+1}^{\infty} \left\{ \sum_{k=K+1}^{n} g_n^{\langle k \rangle} \right\} z^n.$$

By comparing coefficients, we see that $g_i = \alpha_i$ for $i = 1, ..., K$ and

$$g_n = \frac{1}{\mu} \sum_{k=K+1}^{n} g_n^{\langle k \rangle}, \ n \geq K+1.$$

The proof is complete.

We remark that the sequence g in the above result can be obtained in the following manner as well. First note that

$$w - \sum_{i=1}^{K} \alpha_i z^i - \frac{1}{\mu} \frac{w^{K+1}}{1-w} = w - \sum_{i=1}^{K} \alpha_i z^i - \frac{1}{\mu} \sum_{i=K+1}^{\infty} w^i = w - \sum_{i=1}^{K} \alpha_i z^i - \frac{1}{\mu} \widehat{H^{(K+1)}}(w),$$

where $\mathbf{H}^{(m)}$ is the Heaviside sequence defined before by $\mathbf{H}_n^{(m)} = 1$ for $n \geq m$ and $\mathbf{H}_n^{(m)} = 0$ for $0 \leq n \leq m-1$. Since $G(z) = \widehat{g}(z)$ where $g \in l^{\mathbf{N}}$, thus substituting $w = \widehat{g}(z)$ into $F(z,w) = 0$ then leads us to the equation

$$g - \sum_{i=1}^{K} \alpha_i \hbar^{\langle i \rangle} - \frac{1}{\mu} \mathbf{H}^{(K+1)} \circ g = 0.$$

Since
$$\mathbf{H}^{(K+1)} \circ g = \left\{ \sum_{i=K+1}^{n} g_n^{\langle i \rangle} \right\}$$
by Example 2.23, we see that
$$\widehat{g}(z) - \sum_{i=1}^{K} \alpha_i z^i = \frac{1}{\mu} \sum_{n=K+1}^{\infty} \sum_{i=K+1}^{n} g_n^{\langle i \rangle} z^n$$
as desired.

Example 4.9. The equation
$$G(z) - \alpha z - \beta \frac{G^2(z)}{1 - G(z)} - \gamma \frac{z}{1-z} G(z) = 0$$
has a solution $G(z)$ which is analytic at each z in a neighborhood of the origin and $G(z)$ can be defined in a recursive manner. To see this, let
$$F(z, w) \equiv w - \alpha z - \beta \frac{w^2}{1 - w} - \gamma \frac{z}{1-z} w.$$
Since $F(z, w)$ is analytic for each (z, w) in a neighborhood of $(0,0)$, $F(0,0) = 0$ and $F'_w(0,0) = 1 \neq 0$, by the Analytic Implicit Function Theorem 4.1, we see that the implicit relation $F(z,w) = 0$ has an analytic solution $w = G(z)$ which satisfies $G(0) = 0$ and $G'(0) = -F'_z(0,0)/F'_w(0,0) = \alpha$, and is defined on a disk with the origin as the center and with a positive radius. Let $G(z) = \widehat{g}(z)$ where $g \in l^\mathbf{N}$. Then $g_0 = G(0) = 0$ and $g_1 = G'(0) = \alpha$. Furthermore, by writing
$$F(z,w) = w - \alpha z - \beta \widehat{\mathbf{H}^{(2)}}(w) - \gamma w \widehat{\mathbf{H}^{(1)}}(z),$$
then substituting $w = \widehat{g}(z)$ into it leads us to
$$g = \alpha \hbar + \beta \mathbf{H}^{(2)} \circ g + \gamma g * \mathbf{H}^{(1)}.$$
By Example 2.23 and the fact that
$$g * \mathbf{H}^{(1)} = \left\{ \sum_{i=0}^{n} g_i \mathbf{H}_{n-i}^{(1)} \right\} = \left\{ \sum_{i=1}^{n-1} g_i \right\},$$
we may then see that
$$g_n = \beta \sum_{k=2}^{n} g_n^{\langle k \rangle} + \sum_{k=1}^{n-1} g_k, \quad n \geq 2.$$

Example 4.10. The equation
$$G(z) - \alpha z - \beta z^2 - \gamma z G(z) - \delta \frac{z G^2(z)}{1 - G(z)} - \xi \frac{z^2 G(z)}{1 - z} = 0$$
has a solution $G(z)$ which is analytic at each z in a neighborhood of the origin and $G(z)$ can be defined in a recursive manner. To see this, let
$$F(z, w) = w - \alpha z - \beta z^2 - \gamma z w - \delta \frac{z w^2}{1 - w} - \xi \frac{z^2 w}{1 - z}.$$

Since $F(z,w)$ is analytic for each (z,w) in a neighborhood of $(0,0)$, $F(0,0)=0$ and $F'_w(0,0)=1\neq 0$, by the Analytic Implicit Function Theorem 4.1, we see that the implicit relation $F(z,w)=0$ has an analytic solution $w=G(z)$ which satisfies $G(0)=0$ and $G'(0)=-F'_z(0,0)/F'_w(0,0)=\alpha$, and is defined on a disk with the origin as the center and with a positive radius. Let $G(z)=\widehat{g}(z)$ where $g\in l^{\mathbf{N}}$. Then $g_0=G(0)=0$ and $g_1=G'(0)=\alpha$. Furthermore, by writing

$$F(z,w) = w - \alpha z - \beta z^2 - \gamma zw - \delta z\widehat{\mathbf{H}^{(2)}}(w) - \xi w\widehat{\mathbf{H}^{(2)}}(z),$$

then substituting $w=\widehat{g}(z)$ into it leads us to

$$g = \alpha\hbar + \beta\hbar^{\langle 2\rangle} + \gamma\hbar * g - \delta\hbar * \left(\mathbf{H}^{(2)}\circ g\right) + \xi\left(g * \mathbf{H}^{(2)}\right).$$

By equating the corresponding terms on both sides, we see that

$$g_2 = \alpha\gamma + \beta$$

and

$$g_n = \gamma g_{n-1} + \delta\sum_{k=2}^{n-1} g_{n-1}^{\langle k\rangle} + \xi\sum_{k=1}^{n-2} g_k, \quad n\geq 3.$$

Example 4.11. The equation

$$G(z) - \alpha z - z\frac{\beta G(z)}{1-\beta G(z)} - zQ(z) = 0$$

where

$$Q(z) = \widehat{q}(z), \quad q = \{q_k\} \in l^{\mathbf{N}},$$

is analytic at each z in a neighborhood of the origin, has a solution $G(z)$ which is analytic at each z in a neighborhood of the origin and $G(z)$ can be defined in a recursive manner. To see this, let

$$F(z,w) = w - \alpha z - z\frac{\beta w}{1-\beta w} - zQ(z).$$

Since $F(z,w)$ is analytic for each (z,w) in a neighborhood of $(0,0)$, $F(0,0)=0$ and $F'_w(0,0)=1$, the Analytic Implicit Function Theorem 4.1 asserts that the implicit relation $F(z,w)=0$ has a unique solution $w=G(z)$ which satisfies $G(0)=0$ and $G'(0)=-F'_z(0,0)/F'_w(0,0)=\alpha+q_0$, and is defined on a disk with the origin as the center and with a positive radius. Let $G(z)=\widehat{g}(z)$ where $g\in l^{\mathbf{N}}$. Then $g_0=0$ and $g_1=G'(0)=\alpha+q_0$. Furthermore, by writing

$$F(z,w) = w - \alpha z - z\widehat{\mathbf{H}^{(1)}}(\beta w) - zQ(z)$$

and substituting $w=\widehat{g}(z)$ into it, we see that

$$g = \alpha\hbar + \hbar * \left(\mathbf{H}^{(1)}\circ(\beta g)\right) - \hbar * q.$$

By equating the corresponding terms on both sides, we see that

$$g_n = \sum_{k=1}^{n-1} \beta^k g_{n-1}^{\langle k\rangle} + q_{n-1}, \quad n\geq 2.$$

Example 4.12. The equation

$$G(z) - \alpha z - \beta z^2 - z^2 \frac{\delta G(z)}{1 - \delta G(z)} - z^2 Q(z) = 0,$$

where

$$Q(z) = \widehat{q}(z), \quad q = \{q_k\} \in l^{\mathbf{N}},$$

is analytic at each z in a neighborhood of the origin, has a solution $G(z)$ which is analytic at each z in a neighborhood of the origin and $G(z)$ can be defined in a recursive manner. To see this, let

$$F(z, w) = w - \alpha z - \beta z^2 - z^2 \frac{\delta w}{1 - \delta w} - z^2 Q(z).$$

Since $F(z, w)$ is analytic for each (z, w) in a neighborhood of $(0,0)$, $F(0,0) = 0$ and $F'_w(0,0) = 1$, the Analytic Implicit Function Theorem 4.1 asserts that the implicit relation $F(z, w) = 0$ has a unique solution $w = G(z)$ which satisfies $G(0) = 0$ and $G'(0) = -F'_z(0,0)/F'_w(0,0) = \alpha + q_0$, and is defined on a disk with the origin as the center and with a positive radius. $G(z) = \widehat{g}(z)$ where $g \in l^{\mathbf{N}}$. Then $g_0 = 0$ and $g_1 = G'(0) = \alpha + q_0$. Furthermore, by writing

$$F(z, w) = w - \alpha z - \beta z^2 - z^2 \widehat{\mathbf{H}^{(1)}}(\delta w) - z^2 Q(z)$$

and substituting $w = \widehat{g}(z)$ into it, we see that

$$g = \alpha \hbar + \beta \hbar^{\langle 2 \rangle} + \hbar^{\langle 2 \rangle} * \left(\mathbf{H}^{(1)} \circ (\delta g) \right) + \hbar^{\langle 2 \rangle} * q$$

By equating corresponding terms on both sides, we see that

$$g_2 = \beta$$

and

$$g_n = \sum_{k=1}^{n-2} \delta^k g_{n-2}^{\langle k \rangle} + h_{n-2}, \quad n \geq 3.$$

Example 4.13. Let $\mu \neq 0$. The equation

$$G(z) - \alpha z - \beta \frac{\delta G^2(z)}{1 - \delta G(z)} - \frac{\gamma z^2}{\mu^2 - \mu z} = 0$$

has a solution $G(z)$ which is analytic at each z in a neighborhood of the origin and $G(z)$ can be defined in a recursive manner. To see this, let

$$F(z, w) \equiv w - \alpha z - \beta \frac{\delta w^2}{1 - \delta w} - \frac{\gamma z^2}{\mu^2 - \mu z}. \tag{4.11}$$

Since $F(z, w)$ is analytic for each (z, w) in a neighborhood of $(0,0)$, $F(0,0) = 0$ and

$$\frac{\partial F}{\partial w} = 1 - \beta \left\{ \frac{2\delta w}{1 - \delta w} + \frac{\delta^2 w^2}{(1 - \delta w)^2} \right\}$$

which is not zero at $(0,0)$, by the Analytic Implicit Function Theorem 4.1, we see that the implicit relation $F(z,w) = 0$ has an analytic solution $w = G(z)$ which satisfies $G(0) = 0$ and $G'(0) = -F'_z(0,0)/F'_w(0,0) = \alpha$, and is defined on a disk with the origin as the center and with a positive radius. $G(z) = \widehat{g}(z)$ where $g \in l^{\mathbf{N}}$. Then $g_0 = G(0) = 0$ and $g_1 = G'(0) = \alpha$. By writing

$$F(z,w) = w - \alpha z - \beta w \widehat{\mathbf{H}^{(1)}}(\delta w) - \frac{\gamma}{\mu^2} z^2 \widehat{\mathbf{H}^{(0)}}(z/\mu)$$

and substituting $w = \widehat{g}(z)$ into it, we see that

$$g = \alpha \hbar + \beta g * \left(\mathbf{H}^{(1)} \circ (\delta g) \right) + \frac{\gamma}{\mu^2} \hbar^{(2)} * \left(\mu^{-1} \cdot \mathbf{H}^{(0)} \right)$$

By equating the corresponding terms on both sides, we see that

$$g_n = \beta \sum_{i=0}^{n} g_{n-i} \left\{ \sum_{k=1}^{i} \delta^k g_i^{\langle k \rangle} \right\} + \frac{\gamma}{\mu^n}, \ n \geq 2.$$

We remark that in case $\delta = 1$, the function in (4.11) reduces to

$$F(z,w) \equiv w - \alpha z - \frac{\beta w^2}{1-w} - \frac{\gamma z^2}{\mu^2 - \mu z}.$$

The corresponding analytic solution

$$G(z) = \sum_{n=0}^{\infty} g_n z^n.$$

can then be defined recursively by $g_0 = 0$, $g_1 = \alpha$ and

$$g_n = \beta \sum_{k=2}^{n} g_n^{\langle k \rangle} + \frac{\gamma}{\mu^n}$$

for $n \geq 2$.

Example 4.14. Let $\mu \neq 0$ and $M \in \mathbf{Z}^+$. The equation

$$G^2(z) - \frac{1}{\mu} G(z) + \frac{\alpha_1}{\mu} z + \sum_{n=2}^{M} \left(\frac{\alpha_n}{\mu} - \sum_{k=1}^{n-1} \alpha_k \alpha_{n-k} \right) z^n = 0$$

has a solution $G(z)$ which is analytic at each z in a neighborhood of the origin and $G(z)$ can be defined in a recursive manner. To see this, let

$$F(z,w) = w^2 - \frac{1}{\mu} w + \frac{\alpha_1}{\mu} z + \sum_{n=2}^{M} \left(\frac{\alpha_n}{\mu} - \sum_{k=1}^{n-1} \alpha_k \alpha_{n-k} \right) z^n, \ M \in \mathbf{Z}^+.$$

Since $F(z,w)$ is analytic for each (z,w) in a neighborhood of $(0,0)$, $F(0,0) = 0$ and

$$\frac{\partial}{\partial w} F(0,0) = -\frac{1}{\mu} \neq 0,$$

the Analytic Implicit Function Theorem 4.1 asserts that there is a function $G(z)$ which is a solution of $F(z,G(z)) = 0$ and is analytic at each z in a neighborhood of the origin. To determine $G(z) = \hat{g}(z)$, we write $F(z,w) = 0$ as

$$\mu w^2 - w + \alpha_1 + \sum_{n=2}^{M}\left(\alpha_n - \mu\sum_{k=1}^{n-1}\alpha_k\alpha_{n-k}\right)z^n = 0,$$

which leads us to

$$g = \mu g^{\langle 2\rangle} + \alpha_1 \hbar + \sum_{n=2}^{M}\left(\alpha_n - \mu\sum_{k=1}^{n-1}\alpha_k\alpha_{n-k}\right)\hbar^{\langle n\rangle}.$$

By comparing coefficients on both sides, we see that $g_0 = 0$,

$$g_n = \alpha_n, \quad n = 1, ..., M,$$

and

$$g_{n+1} = \mu\sum_{k=0}^{n-1} g_{k+1}g_{n-k} = \mu g_{n+1}^{\langle 2\rangle}, \quad n \geq M.$$

We remark that $F(z, G(z)) = 0$ in the above Example can be verified directly as follows:

$$G^2(z) = \sum_{n=2}^{\infty}\left(\sum_{k=1}^{n-1} g_k g_{n-k}\right)z^n$$

$$= \sum_{n=2}^{M}\left(\sum_{k=1}^{n-1}\alpha_k\alpha_{n-k}\right)z^n + \sum_{n=M+1}^{\infty}\left(\sum_{k=1}^{n-1} g_k g_{n-k}\right)z^n$$

$$= \sum_{n=2}^{M}\left(\sum_{k=1}^{n-1}\alpha_k\alpha_{n-k}\right)z^n + \frac{1}{\mu}\sum_{n=M}^{\infty} g_{n+1}z^{n+1}$$

$$= \sum_{n=2}^{M}\left(\sum_{k=1}^{n-1}\alpha_k\alpha_{n-k}\right)z^n + \frac{1}{\mu}G(z) - \frac{1}{\mu}\sum_{n=1}^{M}\alpha_n z^n.$$

We remark further that the case $M = 1$ has been discussed in depth at the beginning of this section.

Example 4.15. Let $\mu \neq 0$. The equation

$$G^3(z) - 2\alpha G^2(z) - \left(\frac{1}{\mu} - \alpha^2\right)G(z) + \frac{1}{\mu}(\beta z + \alpha) = 0$$

has a solution $G(z)$ which is analytic at each z in a neighborhood of the origin and $G(z)$ can be defined in a recursive manner. To see this, let

$$F(z,w) = w^3 - 2\alpha w^2 - \left(\frac{1}{\mu} - \alpha^2\right)w + \frac{1}{\mu}(\beta z + \alpha).$$

Since $F(z,w)$ is analytic for each (z,w) in a neighborhood of $(0,\alpha)$, $F(0,\alpha) = 0$ and $F'_w(0,\alpha) = -1/\mu \neq 0$, the Analytic Implicit Function Theorem 4.1 asserts there exists a function $G(z)$ which is a solution of $F(z, G(z)) = 0$ and is analytic at each z in a neighborhood of the origin. To determine $G(z) = \hat{g}(z)$, we write $F(z, w) = 0$ as

$$\mu w (w - \alpha)^2 - w + \beta z + \alpha = 0,$$

which leads us to

$$\mu g * (g - \overline{\alpha})^{\langle 2 \rangle} - g + \beta \hbar + \overline{\alpha} = 0.$$

By comparing coefficients on both sides, we see that $g_0 = \alpha$, $g_1 = \beta$ and

$$g_{n+2} = \mu \sum_{k=0}^{n} \sum_{j=0}^{n-k} g_{k+1} g_{j+1} g_{n-k-j} = \mu \left(g * (g - \overline{\alpha})^{\langle 2 \rangle} \right)_{n+2}, \quad n \in \mathbf{N}.$$

We remark that $F(z, G(z)) = 0$ in the above Example can be verified directly as follows:

$$G^2(z) = \left(g_0 + \sum_{n=0}^{\infty} g_{n+1} z^{n+1} \right) \left(\sum_{n=0}^{\infty} g_n z^n \right) = g_0 \sum_{n=0}^{\infty} g_n z^n + \sum_{n=0}^{\infty} \left(\sum_{k=0}^{n} g_{k+1} g_{n-k} \right) z^{n+1}$$

and

$$G^3(z) = \left(g_0 + \sum_{n=0}^{\infty} g_{n+1} z^{n+1} \right) \left(g_0 \sum_{n=0}^{\infty} g_n z^n + \sum_{n=0}^{\infty} \left(\sum_{k=0}^{n} g_{k+1} g_{n-k} \right) z^{n+1} \right)$$

$$= g_0^2 \sum_{n=0}^{\infty} g_n z^n + 2 g_0 \sum_{n=0}^{\infty} \left(\sum_{k=0}^{n} g_{k+1} g_{n-k} \right) z^{n+1}$$

$$+ \sum_{n=0}^{\infty} \left(\sum_{k=0}^{n} \sum_{j=0}^{n-k} g_{k+1} g_{j+1} g_{n-k-j} \right) z^{n+2}$$

$$= g_0^2 G(z) + 2 g_0 \left(G^2(z) - g_0 G(z) \right) + \frac{1}{\mu} \sum_{n=0}^{\infty} g_{n+2} z^{n+2}$$

$$= g_0^2 G(z) + 2 g_0 \left(G^2(z) - g_0 G(z) \right) + \frac{1}{\mu} \left(G(z) - g_0 - g_1 z \right)$$

$$= 2 g_0^2 G^2(z) + \left(\frac{1}{\mu} - g_0^2 \right) G(z) - \frac{1}{\mu} (g_1 z + g_0)$$

$$= 2 \alpha G^2(z) + \left(\frac{1}{\mu} - \alpha^2 \right) G(z) - \frac{1}{\mu} (\beta z + \alpha),$$

thus

$$G^3(z) - 2\alpha G^2(z) + \left(\frac{1}{\mu} - \alpha^2 \right) G(z) - \frac{1}{\mu} (\beta z + \alpha) = 0$$

as desired.

4.4 Linear Equations

We have considered some specific implicit relations and their analytic solutions. Now we will consider more general functional equations of the form

$$F(z, \phi(z), \phi(f_1(z)), ..., \phi(f_m(z))) = 0. \tag{4.12}$$

In general, z belongs to a subset of $\mathbf{F}$, while the functions $F, f_1, ..., f_m$ are assumed known. We will be interested in finding analytic solutions ϕ or to derive existence theorems for these equations. When the function F in (4.12) is linear, we obtain the following 'nonhomogeneous' linear functional equation

$$\sum_{i=0}^{m} a_i(z)\phi(f_i(z)) + h(z) = 0. \tag{4.13}$$

A fair amount of investigations have been carried out for this equation. We select a few simple ones to illustrate how power series solutions are found.

4.4.1 Equation I

A simple case of equation (4.13) is [112]

$$\phi(z) = \sum_{i=1}^{m} a_i \phi(\lambda_i z) + G(z), \quad z \in \mathbf{C}. \tag{4.14}$$

Theorem 4.3. *Suppose* $|\lambda_1|, |\lambda_2|, ..., |\lambda_m| < 1$ *and* $a_1 \lambda_1^k + a_2 \lambda_2^k + \cdots + a_m \lambda_m^k \neq 1$ *for all* $k \in \mathbf{N}$. *Suppose further that*

$$G(z) = \widehat{g}(z) = \sum_{k=0}^{\infty} g_k z^k$$

is analytic over a neighborhood of the origin. Then equation (4.14) has a solution which is analytic over $B(0; \rho(g))$ *where* $\rho(g)$ *is the radius of convergence of the sequence* $g = \{g_n\}_{n \in \mathbf{N}}$.

Proof. Assume that

$$\phi(z) = \widehat{b}(z) = \sum_{k=0}^{\infty} b_k z^k$$

is an analytic solution of (4.14) on a neighborhood of 0. Then inserting ϕ into (4.14) and employing the Unique Representation Theorem 3.8, we see that

$$b = g + \sum_{i=1}^{m} a_i \underline{\lambda_i} \cdot b.$$

Hence

$$b_k = a_1 b_k \lambda_1^k + a_2 b_k \lambda_2^k + \cdots + a_m b_k \lambda_m^k + g_k, \quad k \in \mathbf{N},$$

or
$$b_k = \frac{g_k}{1 - (a_1\lambda_1^k + a_2\lambda_2^k + \cdots + a_m\lambda_m^k)}, \quad k \in \mathbf{N}.$$
This proves the uniqueness of the analytic solution $\phi(z)$. To see that $\phi(z)$ is indeed an analytic solution, we only need to show that it converges in a neighborhood of 0. This is accomplished by finding a majorant function for $\phi(z)$. To this end, recall from Cauchy's Estimation (Theorem 3.26) that for each $r \in (0, \rho(g))$, there is some $M_r > 0$ such that
$$|g_k| \le \frac{M_r}{r^k}, \quad k \in \mathbf{N}.$$
Next, since $|\lambda_1|, ..., |\lambda_m| < 1$, there is some positive integer T such that
$$|a_1||\lambda_1|^k + |a_2||\lambda_2|^k + \cdots + |a_m||\lambda_m|^k < 1, \quad k > T.$$
Let the sequence $c^{(r)} = \{c_k^{(r)}\}_{k \in \mathbf{N}}$ be defined by
$$c_k^{(r)} = \frac{M_r}{r^k} \frac{1}{1 - \left(|a_1||\lambda_1|^k + |a_2||\lambda_2|^k + \cdots + |a_m||\lambda_m|^k\right)}, \quad k > T,$$
and
$$c_k \ge |b_k|, \quad k = 0, 1, ..., T.$$
Then $b \ll c^{(r)}$ since
$$\left| \frac{g_k}{1 - (a_1\lambda_1^k + a_2\lambda_2^k + \cdots + a_m\lambda_m^k)} \right|$$
$$\le \frac{M_r}{r^k} \frac{1}{1 - \left(|a_1||\lambda_1|^k + |a_2||\lambda_2|^k + \cdots + |a_m||\lambda_m|^k\right)}$$
for $k > T$. Since
$$\lim_{k \to \infty} \frac{c_{k+1}^{(r)}}{c_k^{(r)}} = \frac{1}{r},$$
we see from the ratio test that
$$r = \rho\left(c^{(r)}\right) \le \rho(b).$$
Since r is an arbitrary number satisfying $0 < r < \rho(g)$, we see further that $\rho(g) \le \rho(b)$. Finally, since $\phi(z)$ is majorized by $\sum_{k=0}^{\infty} c^{(r)} z^k$, we see that $\phi(z)$ defined by (5.2) is an analytic solution on $B(0; \rho(g))$. The proof is complete.

Example 4.16. As an example, consider the functional equation
$$\phi(z) = 2\phi\left(\frac{z}{2}\right) + 3\phi\left(\frac{z}{3}\right) + \frac{\sin z}{z},$$
where we take
$$\frac{\sin z}{z} = \sum_{k=0}^{\infty} \frac{(-1)^k z^{2k}}{(2k+1)!}, \quad z \in \mathbf{C}.$$
By means of our Theorem 4.3, we see that it has the unique analytic solution
$$\phi(z) = \sum_{k=0}^{\infty} \frac{(-1)^k z^{2k}}{(2k+1)!\left(1 - 2^{-2k+1} - 3^{-2k+1}\right)}, \quad z \in \mathbf{C}.$$

4.4.2 Equation II

Next we consider another simple case of (4.13) [112]:

$$\phi(z) = P(z)\phi(\alpha z) + Q(z), \quad z \in \mathbf{C}. \tag{4.15}$$

Theorem 4.4. *Suppose $|\alpha| < 1$ and $P(0)\alpha^k \neq 1$ for $k \in \mathbf{N}$. Suppose further that $P(z) = \widehat{p}(z)$ and $Q(z) = \widehat{q}(z)$ are power series functions generated respectively by $p = \{p_k\}_{k \in \mathbf{N}}$ and $q = \{q_k\}_{k \in \mathbf{N}}$ in $l^{\mathbf{N}}$ with positive $\rho(p)$ and $\rho(q)$. Then (4.15) has a solution which is analytic over $B(0, \rho)$ where $\rho = \min\{\rho(p), \rho(q)\}$.*

Proof. Assume that

$$\phi(z) = \widehat{b}(z) = \sum_{k=0}^{\infty} b_k z^k$$

is an analytic solution of (4.15) in a neighborhood of 0. Then inserting ϕ into (4.15) and employing the Unique Representation Theorem 3.8, we see that

$$b = p * (\underline{\alpha} \cdot b) + q,$$

that is,

$$b_0 = p_0 + q_0,$$

and

$$b_k\left(1 - p_0 \alpha^k\right) = p_1 \alpha^{k-1} b_{k-1} + \cdots + p_k \alpha^0 b_0 + q_k, \quad k \in Z^+.$$

Since $P(0)\alpha^k = p_0 \alpha^k \neq 1$ for $k \in \mathbf{N}$, we can easily show by induction that $\{b_k\}_{k \in \mathbf{N}}$ is uniquely defined. This proves the uniqueness of the analytic solution $\phi(z)$. To see that $\widehat{b}(z)$ is indeed an analytic solution, we only need to show that it converges in a neighborhood of 0. This is accomplished by finding majorant functions for $\phi(z)$. To do this, first observe that the convergence of the series $\widehat{p}(z)$ and $\widehat{q}(z)$ implies, by Cauchy's Estimation 3.26, that for each $r \in (0, \rho)$ where $\rho = \min\{\rho(p), \rho(q)\}$, there is some $M_r > 0$ such that

$$|p_k|, |q_k| \le \frac{M_r}{r^k}, \quad k \in \mathbf{N}.$$

Next, since $|\alpha| < 1$, there is some positive integer T such that $M_r |\alpha|^k < 1$ for $k > T$. Let the sequence $c^{(r)} = \left\{c_k^{(r)}\right\}_{k \in \mathbf{N}}$ be defined by

$$c_k^{(r)} = \frac{1}{1 - M_r |\alpha|^k} \left\{ \left(\frac{M_r}{r} |\alpha|^{k-1} c_{k-1}^{(r)} + \cdots + \frac{M_r}{r^k} |\alpha|^0 c_0^{(r)} \right) + \frac{M_r}{r^k} \right\}$$

for $k > T$, and

$$c_k^{(r)} \ge |b_k|, \quad k = 0, 1, ..., T.$$

Note that

$$|b_{T+1}| \le \frac{1}{1-|p_0||\alpha|^{T+1}}\left\{\left(|p_1||\alpha|^T|b_T|+\cdots+|p_{T+1}||\alpha|^0|b_0|\right)+|q_{T+1}|\right\}$$
$$\le \frac{1}{1-M_r|\alpha|^{T+1}}\left\{\left(\frac{M_r}{r}|\alpha|^T c_T^{(r)}+\cdots+\frac{M_r}{r^{T+1}}|\alpha|^0 c_0^{(r)}\right)+\frac{M_r}{r^{T+1}}\right\}$$
$$= c_{T+1},$$

and by induction, it is easy to show that

$$|b_k| \le c_k^{(r)}, \ k \ge T+2.$$

Thus b is majorized by c. Furthermore, since

$$r\left(1-M_r|\alpha|^{k+1}\right)c_{k+1}^{(r)} = M_r|\alpha|^k c_n^{(r)} + \left\{\left(\frac{M_r}{r}|\alpha|^{k-1}+\cdots+\frac{M_r}{r^k}c_0^{(r)}\right)+\frac{M_r}{r^k}\right\}$$
$$= M|\alpha|^k c_k^{(r)} + \left(1-M_r|\alpha|^k c_k^{(r)}\right)c_k^{(r)}$$
$$= c_k^{(r)},$$

we see that

$$\lim_{k\to\infty}\frac{c_{k+1}^{(r)}}{c_k^{(r)}} = \lim_{k\to\infty}\frac{1}{r\left(1-M_r|\alpha|^{k+1}\right)} = \frac{1}{r}.$$

Thus $r = \rho\left(c^{(r)}\right) \le \rho(b)$. Since r is an arbitrary number satisfying $0 < r < \rho$, we see further that $\rho \le \rho(c)$. Finally, since $\phi(z)$ is majorized by $\widehat{c}(z)$, we see that $\phi(z)$ is an analytic solution on $B(0;\rho)$. The proof is complete.

Example 4.17. Consider the functional equation

$$\phi(z) = (1-\alpha z)\phi(\alpha z)$$

subject to the condition $\phi(0) = 1$, where $|\alpha| < 1$. By means of Theorem 4.4, we see that it has the analytic solution

$$\phi(z) = 1 + \sum_{k=1}^{\infty}\frac{\alpha^{k(k+1)/2}z^k}{(\alpha-1)(\alpha^2-1)\cdots(\alpha^k-1)}, \ z \in \mathbf{C}.$$

4.4.3 Equation III

As our third simple case of (4.13), consider the equation [145]

$$\phi(\alpha z) = P(z)\phi(z) + Q(z), \ z \in \mathbf{C}. \quad (4.16)$$

Theorem 4.5. *Suppose $0 < |\alpha| < 1$. Suppose that $P(z) = \widehat{p}(z)$ and $Q(z) = \widehat{q}(z)$ are power series functions generated respectively by $p = \{p_k\}_{k\in\mathbf{N}}$ and $q = \{q_k\}_{k\in\mathbf{N}}$ in $l^{\mathbf{N}}$ with positive $\rho(p)$ and positive $\rho(q)$. If $P(z) \ne 0$ for $z \in B(0;\rho(p))$ and $\alpha^k \ne p_0$ for $k \in \mathbf{N}$, then (4.16) has a solution which is analytic over a neighborhood of 0.*

Proof. Assume

$$\phi(z) = \widehat{c}(z) = \sum_{k=0}^{\infty} c_k z^k$$

is an analytic solution of (4.16) over some ball $B(0;\rho)$. Then substituting ϕ into (4.16) and invoking the Unique Representation Theorem 3.8, we see that

$$\underline{\alpha} \cdot c = p * c + q,$$

or

$$(\alpha^n - p_0)c_n = p_1 c_{n-1} + \cdots + p_n c_0 + q_n, \ n \in \mathbf{N}.$$

If $\alpha^n \ne p_0$ for $n \in \mathbf{N}$, then c can be uniquely determined by induction. This shows that an analytic solution of (4.16) is unique. To show that $\widehat{c}(z)$ is analytic on some ball $B(0;\rho)$, we may try to calculate $\rho(c)$ directly. This turns out to be difficult. We therefore proceed in a different manner. First observe that if $\phi_0(z)$ is a solution of the equation

$$\phi_0(\alpha z) = P(z)\phi_0(z) + Q(z) + P(z)\bar{\phi}(z) - \bar{\phi}(\alpha z), \tag{4.17}$$

where

$$\bar{\phi}(z) = \sum_{i=0}^{m-1} c_i z^i, \ m \in \mathbf{Z}^+,$$

then $\phi_0(z) + \bar{\phi}(z)$ is a solution of (4.16). Now pick $m \in \mathbf{Z}^+$ such that

$$|\alpha|^m < |p_0|, \tag{4.18}$$

and let

$$\tilde{Q}(z) = Q(z) + P(z)\bar{\phi}(z) - \bar{\phi}(\alpha z). \tag{4.19}$$

Then

$$\phi_0(z) = -\sum_{n=0}^{\infty} \tilde{Q}(\alpha^n z) \left(\prod_{j=0}^{n} P(\alpha^j z)\right)^{-1}$$

is a formal solution of (4.17) since

$$P(z)\phi_0(z) + \tilde{Q}(z) = P(z) \left\{ -\frac{\tilde{Q}(z)}{P(z)} - \sum_{n=1}^{\infty} \tilde{Q}(\alpha^n z) \left(\prod_{j=0}^{n} P(\alpha^j z)\right)^{-1} \right\} + \tilde{Q}(z)$$

$$= -\sum_{n=1}^{\infty} \tilde{Q}(\alpha^n z) \left(\prod_{j=0}^{n} P(\alpha^j z)\right)^{-1}$$

$$= -\sum_{n=0}^{\infty} \tilde{Q}(\alpha^{n+1} z) \left(\prod_{j=0}^{n} P(\alpha^{j+1} z)\right)^{-1}$$

$$= \phi_0(\alpha z).$$

It suffices to show that ϕ_0 just defined is analytic in some $B(0;\rho)$. To see this, note that $\tilde{Q}(z)$ is analytic in $B(0;r)$ where $r = \min\{\rho(p), \rho(q)\}$. If we write $\tilde{Q}(z) = \hat{h}(z)$ where $h = \{h_k\}_{k \in \mathbf{N}} \in l^{\mathbf{N}}$, then we may easily check that $h_0 = h_1 = \cdots = h_{m-1} = 0$ in view of (4.19). Thus there is $B > 0$ such that $|\tilde{Q}(z)| \leq B|z|^m$ for $|z|$ sufficiently small. Next, we note that $\tilde{P}(z) = 1/P(z)$ is analytic in some ball $B(0;s)$ and $\tilde{P}(0) = p_0^{-1}$. Thus by continuity, there is $A > 1$ such that $A|\alpha|^m |p_0|^{-1} < 1$ and $|\tilde{P}(z)| \leq A|p_0^{-1}|$ for $|z|$ sufficiently small. We may now conclude that

$$|\tilde{Q}(z)| \leq B|z|^m, |\tilde{P}(z)| \leq A|p_0|^{-1}, A|\alpha|^m |p_0|^{-1} < 1, z \in \overline{B}(0;\tau)$$

where τ is some small positive number in $(0,r)$.

Now let $\overline{B}(0;\xi)$ be any closed ball inside $\overline{B}(0;\tau)$. Then there exists some positive integer T such that $\alpha^k z \in \bar{B}(0;\xi)$ for $k > T$ and $z \in \bar{B}(0;\tau)$. Thus for $z \in \bar{B}(0;\xi)$,

$$\left|\tilde{Q}(\alpha^n z)\prod_{j=0}^{n}\tilde{P}(\alpha^j z)\right| \leq \{B|\alpha^m|^n |z|^m\}\left\{\max_{z \in \bar{B}(0;\xi)}\left|\prod_{j=0}^{T}\tilde{P}(\alpha^j z)\right|\right\}\left\{A^{n-T}|p_0|^{-(n-T)}\right\}$$

$$\leq \left\{\max_{z \in \bar{B}(0;\xi)}\left|\prod_{j=0}^{T}\tilde{P}(\alpha^j z)\right|\right\}\left\{B|\alpha|^{mT}|z|^m \left(A|p_0|^{-1}|\alpha|^m\right)^{n-T}\right\}.$$

Since $A|\alpha|^m |p_0|^{-1} < 1$, the series $\sum_{n=0}^{\infty}\left(A|\alpha|^m |p_0|^{-1}\right)^{n-T}$ converges. This shows, in view of the Weierstrass Test (Theorem 2.7), that $\sum_{n=0}^{\infty}\tilde{Q}(\alpha^n z)\left(\prod_{j=0}^{n}P(\alpha^j z)\right)^{-1}$ is analytic on $B(0;\xi)$. The proof is complete.

We remark that if the condition that $\alpha^k \neq p_0$ for $k \in \mathbf{N}$ in the above result is replaced by the alternate condition

$$\alpha^k = p_0, \ p_1 c_{n-1} + \cdots + p_n c_0 + q_n = 0, \ n \in \mathbf{N}.$$

Then the sequence c can be chosen in an arbitrary manner. The uniqueness now does not hold. However, the rest of the proof goes through so that we may now conclude that (4.16) has a solution which is analytic on a neighborhood of 0.

4.4.4 Equation IV

We next consider linear equations of the form [145]

$$\phi(z) = g(z)\phi(f(z)) + h(z), \quad (4.20)$$

where ϕ is the unknown function, and f, g and h are assumed to be given. In case $f(z) = \alpha z$, equation (4.20) reduces to (4.15).

Assume $f: B(0;r) \to B(0;r)$ and $g, h: B(0;r) \to \mathbf{C}$ are analytic over $B(0;r)$ and $f(0) = 0$. Then for any analytic function $\phi: B(0;r) \to \mathbf{C}$, we have

$$\frac{d}{dz}[g(z)\phi(f(z))] = g(z)\phi'(f(z))f'(z) + g'(z)\phi(f(z)),$$

and in general, by the Formula of Faa di Bruno (Theorem 2.16), we have

$$\frac{d^k}{dz^k}[g(z)\phi(f(z))] = g(z)(f'(z))^k \phi^{(k)}(f(z)) + \sum_{i=0}^{k-1} P_{ki}(z)\phi^{(i)}(f(z)) \quad (4.21)$$

for $k \in \mathbf{Z}^+$, where each P_{ki} is analytic in U. Therefore, if we look for an analytic solution of (4.20), then $c_k = \phi^{(k)}(0)$ satisfies

$$c_k = g(0)(f'(0))^k c_k + \sum_{i=0}^{k-1} P_{ki}(0)c_i + h^{(k)}(0) \quad (4.22)$$

for each $k \in \mathbf{N}$.

Theorem 4.6. *Suppose $f : B(0;\sigma) \to B(0;\sigma)$ and $g, h : B(0;\sigma) \to \mathbf{C}$ are analytic over $B(0;\sigma)$ and $f(0) = 0$. Suppose further that $|f'(0)| < 1$. Then given a sequence $\{c_k\}_{k \in \mathbf{N}}$ which satisfies (4.22) for each $k \in \mathbf{N}$, there is $B(0;\delta)$ and a solution $\phi(z)$ of (4.20) which is analytic over $B(0;\delta)$ and satisfies*

$$\phi^{(k)}(0) = c_k, \ k \in \mathbf{N}.$$

Proof. The formal power series function

$$\phi(z) = \sum_{i=0}^{\infty} c_i \frac{z^i}{i!},$$

in view of (4.22), is easily seen to be a formal solution of (4.20). We need to show that it is analytic at 0. To see this, choose a positive integer r, a real number $\theta \in (0,1)$ and $B(0;\delta)$ strictly inside $B(0;r)$, and

$$|f'(z)|^r |g(z)| < \theta, \ z \in \overline{B}(0;\delta).$$

Let $\phi(z) = Q(z) + \psi(z)$, where

$$Q(z) = \sum_{i=0}^{r} c_i \frac{z^i}{i!}, \quad (4.23)$$

and $\psi : B(0;\delta) \to \mathbf{C}$ is analytic over $B(0;\delta)$ and satisfies

$$\psi^{(k)}(0) = 0, \ k = 0, ..., r.$$

Then

$$\psi(z) = g(z)\psi(f(z)) + H(z), \quad (4.24)$$

where

$$H(z) = h(z) - Q(z) + g(z)Q(f(z)) \quad (4.25)$$

is analytic in $B(0;\sigma)$ and $H^{(k)}(0) = 0$ for $k = 0, ..., r$.

Let Ξ be the set of all analytic functions ψ on $B(0;\delta)$ such that $\psi^{(k)}(0) = 0$ for $k = 0, ..., r$ and $\psi^{(r)}$ is continuous on $\overline{B}(0;\delta)$. When endowed with the usual linear structure and the norm

$$\|\psi\| = \sup_{z \in \overline{B}(0;\delta)} \left|\psi^{(r)}(z)\right|,$$

it is easy to show that Ξ is a Banach space. Define a mapping $T : \Psi \to \Psi$ by
$$(T\psi)(z) = g(z)\psi(f(z)) + H(z), \quad z \in \overline{B}(0;\delta).$$
Then for any $\psi_1, \psi_2 \in \Xi$, in view of (4.21), we have
$$\frac{d^r}{dz^r}\left(g(z)\left(\psi_1(f(z)) - \psi_2(f(z))\right)\right) = g(z)\left(f'(z)\right)^r \left(\psi_1^{(r)}(f(z)) - \psi_2^{(r)}(f(z))\right)$$
$$+ \sum_{i=0}^{r-1} P_{ki}(z)\left(\psi_1^{(i)}(f(z)) - \psi_2^{(i)}(f(z))\right).$$

By Example 3.9,
$$\sup_{w\in\overline{B}(0;\delta)}\left|\psi_1^{(i)}(w) - \psi_2^{(i)}(w)\right| \le \delta^{r-i}\|\psi_1 - \psi_2\|, \quad i = 0, 1, ..., r-1.$$

Thus,
$$\|T\psi_1 - T\psi_2\| \le \left\{\theta + \sum_{i=0}^{r-1} \delta^{r-i}\sup_{z\in\overline{B}(0;\delta)}|P_{ki}(z)|\right\}\|\psi_1 - \psi_2\|.$$

If we choose δ so small that
$$\theta + \sum_{i=0}^{r-1} \delta^{r-i}\sup_{z\in\overline{B}(0;\delta)}|P_{ki}(z)| < 1,$$
then T is a contraction mapping. Thus (4.24) has a unique solution $\psi \in \Psi$. This in turn implies (4.20) has a unique analytic solution $\phi : U_0 \to C$ of the form $\phi(z) = \sum_{i=0}^{\infty} c_i z^i/i!$ which satisfies $\phi^{(k)}(0) = c_k$ for $k \in \mathbf{N}$. The proof is complete.

We also remark that the condition $f(0) = 0$ asserts that 0 is a fixed point of f.

4.4.5 Equation V

Next we consider equations of the form
$$\phi(f(z)) = g(z)\phi(z) + h(z), \tag{4.26}$$
where ϕ is the unknown function, and f, g and h are assumed to be given function analytic at 0. In case $f(z) = \alpha z$, equation (4.26) reduces to (4.16).

First we consider the situation
$$f(0) = f'(0) = g(0) = 0. \tag{4.27}$$

To avoid trivial cases, we assume that f and g are nontrivial. There exist integers p, q, r such that
$$f(z) = z^p F(z),$$
$$g(z) = z^q G(z),$$
and
$$h(z) = z^r H(z),$$

where $F(0) \neq 0$, $G(0) \neq 0$ and $H(0) \neq 0$ unless $h = 0$. By (4.27) we have $p \geq 2$, $q \geq 1$ and $r \geq 0$ (we take $r = \infty$ if $h = 0$). We exclude from further considerations the trivial solutions of equation (4.26) (occurring iff $h = 0$). Thus we seek analytic solutions of (4.26) in the form

$$\varphi(z) = z^s \Phi(z), \quad \Phi(0) \neq 0, \quad s \in \mathbf{N}.$$

By substituting φ into (4.26), we see that

$$z^{ps} (F(z))^s \Phi(f(z)) = z^{q+s} G(z) \Phi(z) + z^r H(z), \qquad (4.28)$$

and thus by comparing the orders of the zeros at $z = 0$ we arrive at the four possibilities (i) $ps = q + s = r$, (ii) $ps = q + s < r$, (iii) $ps > q + s = r$, and (iv) $ps = r < q + s$. (Note that the case $h = 0$ comes under (ii).)

We need to find nonnegative integers s satisfying one of the relations (i)-(iv). This leads to the following conditions for p, q, r:

$$p - 1 \text{ divides } q \text{ and } pq = (p-1)r \quad (s = q/(p-1)); \qquad (4.29)$$

$$p - 1 \text{ divides } q \text{ and } pq < (p-1)r \quad (s = q/(p-1)); \qquad (4.30)$$

$$p(r - q) > r \quad (s = r - q); \qquad (4.31)$$

$$p \text{ divides } r \text{ and } pq > (p-1)r \quad (s = r/p). \qquad (4.32)$$

In each of the above cases, there is a unique $s \in \mathbf{N}$ (given in the parentheses) satisfying the corresponding cases (i)-(iv). Conditions (4.29)-(4.32) exclude each other, with the only exception that (4.30) is a special case of (4.31) (since $pq < (p-1)r \Leftrightarrow p(r-q) > r$). Of course, the s from (4.30) is smaller than that from (4.31), when calculated for the same p, q, r. Moreover, each of conditions (4.29)-(4.31) implies $q < r$.

The first thing to be determined now is the value $c_0 = \Phi(0)$ of an analytic solution Φ of (4.28) in each of the cases (4.29)-(4.32). Whenever it exists, all the remaining $c_k = \Phi^{(k)}(0)$ for $k \geq 1$, as we shall see, can be uniquely determined by applying Theorem 4.6 to equation (4.28).

Let us examine cases (4.29)-(4.31). In case (4.29), equation (4.28) becomes

$$\Phi(z) = \frac{(F(z))^s}{G(z)} \Phi(f(z)) - \frac{H(z)}{G(z)}, \quad s = \frac{q}{p-1}.$$

Since $H(0) \neq 0$, c_0 exists if, and only if, $(F(0))^s \neq G(0)$.

In case (4.30), we obtain from (4.28) that

$$\Phi(z) = \frac{(F(z))^s}{G(z)} \Phi(f(z)) - \frac{z^{r-ps} H(z)}{G(z)}, \quad s = \frac{q}{p-1}. \qquad (4.33)$$

Now, if $(F(0))^s = G(0)$, then $c_0 = \tau$ may be taken as arbitrary. Then equation (4.26) has a unique one-parameter family of analytic solution $\varphi_\tau(z) = z^s \Phi_\tau(z)$, where $s = q/(p-1)$. For $\tau = 0$ we obtain the solution φ_0 which at the origin has a zero of an order higher than $q/(p-1)$. If $h \neq 0$, this order must be $r - q$ (Case (4.31)), while if $h = 0$, then $\varphi_0 = 0$.

In case (4.31), let us exclude (4.30). Now (4.28) may be written as

$$\Phi(z) = \frac{(F(z))^s}{G(z)} z^{ps-r} \Phi(f(z)) - \frac{H(z)}{G(z)}, \quad s = r - q.$$

Then $c_0 = \Phi(0) = -H(0)/G(0)$ always exists.

The observations obtained so far can be summarized as follows.

Theorem 4.7. *Let the functions $f(z) = z^p F(z)$, $g(z) = z^q G(z)$ and $h(z) = z^r H(z)$ be analytic at 0 and $F(0) \neq 0$, $G(0) \neq 0$ as well as $H(0) \neq 0$ except when $h = 0$. Further assume that $p \geq 2$, $q \geq 1$ and $r \geq pq/(p-1)$. Then (i) when (4.29) and $(F(0))^{q/(p-1)} \neq G(0)$ hold, equation (4.26) has a unique analytic solution φ defined over a neighborhood of 0; (ii) when (4.30) and $(F(0))^{q/(p-1)} = G(0)$ hold, equation (4.26) has a unique one-parameter family of analytic solutions φ_t defined over a neighborhood of 0; (iii) when (4.31) and $q/(p-1) \notin \mathbf{N}$ hold, equation (4.26) has a unique analytic solution φ defined over a neighborhood of 0; (iv) for other cases covered by (4.29), (4.30) and (4.31), equation (4.26) does not have any solutions which are analytic at 0.*

Now we turn to the case (4.32). We need some extra work before we can apply Theorem 4.6. Put $s = r/p$ and let m be the smallest integer fulfilling

$$m \geq q/(p-1). \tag{4.34}$$

Of course, $m > s$ by (4.32). Suppose that equation (4.26) has an analytic solution φ. Since the order of zero of φ at the origin must be s, we can write

$$\varphi(z) = P(z) + \varphi^*(z), \tag{4.35}$$

where

$$P(z) = d_s z^s + \cdots + d_{m-1} z^{m-1}, \tag{4.36}$$

and $\varphi^*(z) = z^S \Phi^*(z)$, $S \geq m$, $\Phi^*(0) \neq 0$. It follows from (4.26) that

$$\varphi^*(f(z)) = g(z)\varphi^*(z) + h^*(z), \tag{4.37}$$

where

$$h^*(z) = h(z) - P(f(z)) + g(z)P(z). \tag{4.38}$$

Write $h^*(z) = z^R H^*(z)$, where $H^*(0) \neq 0$ unless $h^* = 0$. By (4.34) we have $Sp \geq q + S$, whence $R \geq q + S \geq q + m$.

We conclude from the above remarks that equation (4.26) cannot have analytic solutions unless there exists a polynomial (4.36) such that function (4.38) has at the origin a zero of an order at least $m + q$. If such a polynomial does exist, it is uniquely determined by (4.38). Indeed, we have by (4.38) and (4.36) that

$$h^*(z) = h(z) - \sum_{i=s}^{m-1} d_i z^{ip} (F(z))^i + \sum_{i=s}^{m-1} d_i z^{i+q} G(z). \tag{4.39}$$

The integer m being the smallest one to fulfil (4.34), we have $ip < i + q$ for $i < m$. Thus the coefficient of z^{ip} on the right-hand side of (4.39) has the form $d_i \left(F\left(0\right)\right)^i + A_i$, where A_i depends only on f, g, h and on d_j for $j < i$. All the coefficients of z^k in (4.39) must vanish whenever $k < m + q$. If $k = ip$ for $i = s, ..., m - 1$, then $d_i = A_i \left(F\left(0\right)\right)^{-i}$ and the $d_s, ..., d_{m-1}$ in (4.36) are uniquely determined (if they exist). The existence of P depends on whether the coefficients of the remaining z^k in (4.39), i.e. of those where k, not an integral multiple of p and $k < m + q$, vanish for the d_i just determined.

Suppose the polynomial P exists, then equation (4.26) has analytic solutions if and only if equation (4.37) has, and these solutions are linked by formula (4.35). As we have $R \geq m + q > q$ for h^* given by (4.38). Theorem 4.6 applies to equation (4.37), yielding the following result.

Theorem 4.8. *Let the hypotheses of Theorem 4.7 be satisfied, except that now $r < pq/\left(p - 1\right)$. Assume that there exists a polynomial (4.36) such that function (4.38) has at the origin a zero of an order $R \geq q + m$, where m is the smallest integer fulfilling (4.34). Then (i) when $q/\left(p-1\right) \in \mathbf{N}$, $pq = R\left(p-1\right)$ and $\left(F\left(0\right)\right)^{q/(p-1)} \neq G\left(0\right)$ hold, equation (4.26) has a unique analytic solution φ over a neighborhood of the origin; (ii) when $q/\left(p-1\right) \in \mathbf{N}$, $pq < R\left(p-1\right)$ and $\left(F\left(0\right)\right)^{q/(p-1)} = G\left(0\right)$ hold, equation (4.26) has a unique one-parameter family of analytic solutions φ_t over neighborhood of the origin; (iii) when $q/\left(p-1\right) \notin \mathbf{N}$, and $pq < R\left(p-1\right)$ hold, equation (4.26) has a unique analytic solution φ on a neighborhood of the origin; and (iv) when remaining cases hold, equation (4.26) does not have any analytic solutions near the origin.*

4.4.6 Schröder and Poincaré Equations

Schröder equation is

$$\phi(f(z)) = s\phi(z), \tag{4.40}$$

where $z \in \mathbf{F}$, s is a given number in $\mathbf{F}$ different from 0 or 1, and f is a given function (which will be taken to be analytic at 0 and satisfies $f(0) = 0$). This equation can be regarded as a special case of (4.20) (by taking $g(z) = 1/s$). Since formally,

$$\phi(f(f(z))) = s\phi(f(z)) = s^2\phi(z),$$

and

$$\phi(f^{[n]}(z)) = s^n\phi(z), \ n \in \mathbf{Z}^+,$$

where we recall that $f^{[n]}$ denotes the n-th iterate of f. It appeared for the first time in Schröder [173] in connection with the problem of continuous iteration. After the proof of a fundamental existence and uniqueness of analytic solutions by Koenigs [96, 97], it has been studied by many authors.

Note that if $f(0) = 0$ and if ϕ is a differentiable function that satisfies (4.40), then the number $\phi'(0) = \eta$ satisfies
$$f'(0)\eta = s\eta.$$
If $\eta \neq 0$, then $f'(0) = s$ is necessarily true; while if $\eta = 0$, $f'(0)$ and s can be arbitrary. For this reason, we will consider the existence of analytic solutions of the Schröder equation under two different assumptions on the given number $s = f'(0)$.

We first assume $0 < |s| < 1$. We have the following theorem of Koenigs [96, 97].

Theorem 4.9. *Suppose f is analytic at 0 and $f(0) = 0$, $f'(0) = s$ where $0 < |s| < 1$. Then (4.40) has an analytic solution ϕ defined over a neighborhood of the origin and satisfies $\phi'(0) = \eta$.*

Indeed, if ϕ is an analytic solution of (4.40), then
$$\phi(0) = \phi(f(0)) = s\phi(0)$$
so that $\phi(0) = 0$. Next, by (4.40),
$$\phi'(f(z))f'(z) = s\phi'(z),$$
so that
$$\phi'(0)s = s\phi'(0).$$
Thus $\phi'(0)$ can be any given number η. Furthermore, by (4.22), we see that $c_k = \phi^{(k)}(0)$ satisfies
$$c_k = \frac{1}{s}(f'(0))^k c_k + \sum_{i=0}^{k-1} P_{ki}(0)c_i = s^{k-1}c_k + \sum_{i=0}^{k-1} P_{ki}(0)c_i$$
for $k \geq 2$. Hence
$$c_k = \frac{1}{1-s^{k-1}} \sum_{i=0}^{k-1} P_{ki}(0)c_i, \ k \geq 2.$$
Thus by Theorem 4.6, (4.40) has a solution ϕ which is analytic on a neighborhood of the origin.

We remark that the analytic solution ϕ just found satisfies the interesting property
$$\phi(z) = \eta \lim_{n \to \infty} \frac{f^{[n]}(z)}{s^n}, \tag{4.41}$$
a proof of this fact can be found in Kuczma [104].

We remark further that the condition $f(0) = 0$ asserts that 0 is a fixed point of f. When $\phi'(0) = \eta \neq 0$, by the Analytic Inverse Function Theorem 4.2, we know that the inverse function $\psi(z) = \phi^{-1}(z)$ exists and is analytic on a neighborhood of the origin. By substituting $\phi(z) = w$ and $z = \psi(w)$ into the Schröder equation, we see that
$$\phi(f(\psi(w)))) = sw,$$

or
$$f(\psi(w)) = \psi(sw),$$
which is called the Poincaré equation. For obvious reasons, we will write this equation as
$$\psi(sz) = f(\psi(z)).$$

Next we consider the case where $|s| = 1$. This case is more complicated. Under the assumption on f in the previous Theorem, we may write
$$f(z) = sz + \sum_{n=2}^{\infty} a_n z^n.$$

If $|s| = 1$ but s is not a root of unity, then (4.22) has a solution $\{c_k\}_{k \in \mathbf{N}}$. Note that c_1 is arbitrary as before, in other words, (4.40) has a one parameter family of formal solutions of the form
$$\phi(z) = \sum_{n=1}^{\infty} \frac{c_n}{n!} z^n. \qquad (4.42)$$

However, it is possible that for each c_1, $\phi(z)$ is divergent for any $z \neq 0$. Thus it is of great interest to find a set of points in the unit circle such that the corresponding Schröder equation has analytic solutions.

Theorem 4.10. *Suppose s is a Siegel number. Suppose further that*
$$f(z) = sz + \sum_{n=2}^{\infty} a_n z^n$$
is analytic at 0. Then (4.40) has an analytic solution defined over a neighborhood of 0.

Proof. In view of the above remark, it suffices to derive an analytic solution of the Poincaré equation
$$\psi(sz) = f(\psi(z)) \qquad (4.43)$$
in a neighborhood of the origin. To accomplish this, note that by applying Cauchy's Estimation (Theorem 3.26) to the power series function
$$\sum_{n=0}^{\infty} a_{n+2} z^n,$$
we see that there is $\rho > 0$ such that
$$|a_{n+2}| \leq \rho^{n+1}, \; n \in \mathbf{N}.$$
Introducing new functions $\Psi(z) = \rho \psi(\rho^{-1} z)$ and $F(z) = \rho f(\rho^{-1} z)$, we obtain from (4.43) that
$$\Psi(sz) = F(\Psi(z)),$$

which is again an equation of the form (4.43). Here F is of the form

$$F(z) = sz + \sum_{n=2}^{\infty} A_n z^n,$$

but $|A_n| = |a_n \rho^{1-n}| \leq 1$ for $n \geq 2$. Consequently, we may assume that

$$|a_n| \leq 1, \ n \geq 2.$$

Let us assume an analytic solution of (4.43) in the form

$$\psi(z) = \sum_{n=0}^{\infty} b_n z^n,$$

where $b_0 = 0$ and b_1 is an arbitrary nonzero number. By substituting ψ into (4.43), we obtain

$$\psi(sz) - s\psi(z) = f(\psi(z)) - s\psi(z),$$

and

$$\sum_{n=2}^{\infty} (s^n - s) b_n z^n = \sum_{n=2}^{\infty} a_n (\psi(z))^n = \sum_{n=2}^{\infty} \sum_{i=2}^{n} a_i b_n^{\langle i \rangle} z^n.$$

By the Unique Representation Theorem 3.8, we see that

$$b_n = (s^{n-1} - 1)^{-1} \sum_{i=2}^{n} a_i b_n^{\langle i \rangle}.$$

It now suffices to show that $\psi(z)$ is analytic over a neighborhood of the origin. To see this, consider

$$Q(z) = \sum_{n=0}^{\infty} u_n z^n,$$

where $u_0 = 0$, $u_1 = |b_1|$ and

$$u_n = \frac{1}{|s^{n-1} - 1|} \sum_{i=2}^{n} u_n^{\langle i \rangle}, \ n \geq 2.$$

We assert that

$$|b_n| \leq u_n, \ n \in \mathbf{Z}^+.$$

Indeed, $|b_1| \leq u_1$. Assume by induction that $|b_k| \leq u_k$ for $k = 2, 3, ..., n-1$, then

$$|b_n| \leq \frac{1}{|s^{n-1} - 1|} \sum_{i=2}^{n} u_n^{\langle i \rangle} = u_n$$

as asserted. In other words, $Q(z)$ is a majorant series of $\psi(z)$.

Next, we show that Q has a positive radius of convergence. To see this, note that by Example 4.8, the solution

$$\Phi(z) = \sum_{n=0}^{\infty} v_n z^n$$

of the implicit relation

$$\Phi - |b_1| z - \frac{\Phi^2}{1 - \Phi} = 0$$

is analytic on a neighborhood of the origin and $v = \{v_n\}_{n \in \mathbf{N}}$ is given by $v_0 = 0$, $v_1 = |b_1|$ and

$$v_n = \sum_{k=2}^{n} v_n^{(k)}, \ n \geq 2.$$

By Cauchy's Estimation (Theorem 3.26), there is a positive number A such that

$$|v_n| \leq A^{n+1}, \ n \in \mathbf{Z}^+.$$

Hence by Theorem 3.32, there is a positive number δ such that

$$u_n \leq A^{n+1} \left(2^{5\delta+1}\right)^{n-1} n^{-2\delta}, \ n \geq 2,$$

which shows that $Q(z)$ has a positive radius of convergence. The proof is complete.

We remark that since the coefficient b_1 of $\psi(z)$ in the above proof can be taken as any nonzero number, we have actually shown that (4.40) has a one-parameter family of nontrivial analytic solutions defined over a neighborhood of 0.

4.5 Nonlinear Equations

We have considered some functional equations of the form

$$F(z, \phi(z), \phi(f_1(z)), ..., \phi(f_m(z))) = 0,$$

where F is linear. We now consider some equations where F is nonlinear. Such equations are difficult to handle. We will only give two examples.

Recall the Poincaré equation derived in the previous section. Although it is obtained from the linear Schröder equation, it can be regarded as nonlinear. Here let us illustrate the technique of finding analytic solutions further by considering the nonhomogeneous Poincaré equation

$$f(\psi(z)) = \psi(\alpha z) + F(z), \qquad (4.44)$$

where f and F are given functions, and α is a given number.

Theorem 4.11. *Suppose*

$$f(z) = \alpha z + \sum_{n=2}^{\infty} f_n z^n,$$

where $\alpha \geq (1 + \sqrt{5})/2$, is analytic on $B(0; \delta)$. Suppose further that the power series function

$$F(z) = \sum_{n=2}^{\infty} F_n z^n \qquad (4.45)$$

is analytic on $B(0; \tau)$. Then for any $\eta \neq 0$, (4.44) has a solution $\psi(z)$ which is analytic on a neighborhood of the origin and satisfies $\psi(0) = 0$ and $\psi'(0) = \eta$.

Proof. In view of Cauchy's Estimation (Theorem 3.26), there is $p > 0$ such that $|f_n| \leq p^{n-1}$ for $n \geq 2$. Introducing new functions ϕ and g defined by

$$\phi(z) = p\psi\left(\frac{z}{p}\right), \quad g(z) = pf\left(\frac{z}{p}\right),$$

it is easily seen from (4.44) that

$$g(\phi(z)) = \phi(\alpha z) + pF\left(\frac{z}{p}\right)$$

which is again an equation of the form (4.44). Here $g(z) = pf(z/p)$ is of the form

$$g(z) = \alpha z + \sum_{n=2}^{\infty} \frac{f_n}{p^{n-1}} z^n,$$

but

$$|g_n| \leq \frac{|f_n|}{p^{n-1}} \leq 1, \ n = 2, 3, \ldots.$$

Consequently, we may assume that

$$|f_n| \leq 1, \ n = 2, 3, \ldots.$$

Next, by Cauchy's Estimation (Theorem 3.26), there is $M > 0$ such that

$$|F_n| \leq \frac{M}{r^n}, \ 0 < r < \tau, \ n = 2, 3, \ldots.$$

Suppose (4.44) has an analytic solution

$$\psi(z) = \sum_{n=0}^{\infty} b_n z^n \tag{4.46}$$

where $b_0 = 0$. By substituting ψ into (4.44), we obtain

$$\alpha b_1 = b_1 \alpha, \tag{4.47}$$

and

$$(\alpha^n - \alpha) b_n = \sum_{i=2}^{n} f_i b_n^{\langle i \rangle} - F_n, \ n \geq 2. \tag{4.48}$$

Since (4.47) is satisfied by taking b_1 in an arbitrary manner, we will take $b_1 = \eta$, and then $b_2, b_3, \ldots$ can then be determined by (4.48) in a unique manner. It suffices now to show that the subsequent series (4.46) converges on a neighborhood of the origin. To see this, note that $\alpha \geq (1 + \sqrt{5})/2$ implies $\alpha^2 - \alpha - 1 \geq 0$ for $n \geq 2$ and hence

$$|b_n| \leq |\alpha^n - \alpha| |b_n| \leq \sum_{i=2}^{n} |b|_n^{\langle i \rangle} + |F_n| \leq \sum_{i=2}^{n} |b|_n^{\langle i \rangle} + \frac{M}{r^n}$$

for $n \geq 2$. If we now define a sequence $q = \{q_n\}_{n \in \mathbf{N}}$ by $q_0 = 0$, $q_1 = |\eta|$ and

$$q_n = \sum_{i=2}^{n} q_n^{\langle i \rangle} + \frac{M}{r^n}, \ n \geq 2,$$

then it is easily seen that
$$|b_n| \le q_n, \ n \ge 1.$$
In other words, the series $Q(z) = \hat{q}(z)$ is a majorant series of ψ. But by Example 4.13, $Q(z)$ is a solution of the implicit relation
$$F(z, Q) \equiv Q - \eta z - \frac{Q^2}{1-Q} - \frac{Mz^2}{r^2 - rz} = 0$$
which is analytic on a neighborhood of the origin, thus $\psi(z)$ is also analytic there. The proof is complete.

Next, we consider the more difficult nonlinear equation
$$\phi(z) = h(z, \phi(f(z))), \ z \in \mathbf{C}, \tag{4.49}$$
where (i) $f(z)$ is analytic on $B(\xi; \sigma_0)$, $f(\xi) = \xi$ and $0 < |f'(\xi)| < 1$, and (ii) $h(z, w)$ is analytic over the dicylinder
$$\Omega = \{(z, w) \in \mathbf{C}^2 | \ |z - \xi| < \sigma_0, |w - \eta| < \tau\} \tag{4.50}$$
with dicenter (ξ, η) and $h(\xi, \eta) = \eta$.

In view of the above assumptions, we may write
$$f(z) = \xi + \sum_{n=1}^{\infty} b_n (z - \xi)^n, \ |z - \xi| < \sigma_0, 0 < |b_1| < 1. \tag{4.51}$$
and
$$h(z, w) = \sum_{n,m=0}^{\infty} a_{nm}(z - \xi)^n (w - \eta)^m, \ a_{00} = \eta, (z, w) \in \Omega. \tag{4.52}$$
We will seek an analytic solution of (4.49) of the form
$$\phi(z) = \eta + \sum_{n=1}^{\infty} c_n (z - \xi)^n. \tag{4.53}$$
After substituting (4.51), (4.52) and (4.53) into (4.49) and comparing coefficients, we see that
$$(1 - a_{01} b_1) c_1 = a_{10},$$
and
$$(1 - b_1^n a_{01}) c_n = F_n(c_1, ..., c_{n-1}), \ n = 2, 3, ..., \tag{4.54}$$
where F_n is a $(n-1)$-variate polynomial, with coefficients depending on a_{ij} and b_k. If
$$b_1^n a_{01} \ne 1, \ n \in \mathbf{Z}^+, \tag{4.55}$$
then $c_1, c_2, ...$ can be uniquely determined and hence a formal solution of (4.49) is found. If
$$b_1 a_{01} = 1 \text{ and } a_{10} = 0,$$

then we may let c_1 to be an arbitrary number, and then $c_2, c_3, ...$, can be determined. Or if for some $l \geq 2$,

$$b_1^l a_{01} = 1 \text{ and } F_l(c_1, ..., c_{l-1}) = 0, \tag{4.56}$$

then c_l can be chosen in an arbitrary manner and the subsequent $c_{l+1}, c_{l+2}, ...$ can be determined.

Next, observe that since $\phi(z)$ is an analytic solution of (4.49) satisfying $\phi(\xi) = \eta$, in view of (4.49), we see that

$$\phi'(z) = h'_z(z, w) + h'_w(z, w)\phi'(f(z))f'(z)$$

where $w = \phi(f(z))$. Let us write $w_1 = \phi'(f(z))$ and

$$H_1(z, w, w_1) = h'_z(z, w) + h'_w(z, w)f'(z)w_1,$$

the applying another differentiation, we see that

$$\phi''(z) = \frac{\partial H_1(z, w, w_1)}{\partial z} + \frac{\partial H_1(z, w, w_1)}{\partial w}\phi'(f(z))f'(z) + \frac{\partial H_1(z, w, w_1)}{\partial w_1}\phi''(f(z))f'(z)$$

$$= \frac{\partial H_1}{\partial z} + f'(z)\left(\frac{\partial H_1}{\partial w}w_1 + \frac{\partial H_1}{\partial w_1}w_2\right)$$

where we have set $w_1 = \phi''(f(z))$. In general, let $f = f(z)$ and $h = h(z, w)$ be C^r functions and let

$$H_1(z, w, w_1) = h'_z(z, w) + h'_w(z, w)f'(z)w_1,$$

and

$$H_{k+1}(z, w, w_1, ..., w_{k+1}) = \frac{\partial H_k}{\partial z} + f'(z)\left(\frac{\partial H_k}{\partial w}w_1 + \cdots + \frac{\partial H_k}{\partial w_k}w_{k+1}\right)$$

for $k \geq 2$. Then $H_k(z, w, w_1, ..., w_k)$ is a C^{r-k} function and

$$H_k(z, w, w_1, ..., w_k) = P_k(z, w, w_1) + Q_k(z, w, w_k) + R_k(z, w, w_1, ..., w_{k-1}), \tag{4.57}$$

where

$$P_k(z, w, w_1) = \sum_{i=0}^{k} C_i^{(k)} \frac{\partial^k h(z, w)}{\partial z^{k-i} \partial w^i} [f'(z)]^i w_1^i, \tag{4.58}$$

$$Q_k(z, w, w_k) = \frac{\partial h(z, w)}{\partial w} [f'(z)]^k w_k, \tag{4.59}$$

and $R_k(z, w, w_1, ..., w_{k-1})$ is a $(k+1)$-variate polynomial with coefficients which are C^{r-k} functions in z and w. Furthermore, if $\phi(z)$ is a C^r solution of (4.49), then

$$\phi^{(k)}(z) = H_k(z, \phi(f(z)), \phi'(f(z)), ..., \phi^{(k)}(f(z))), \ k = 1, 2, ..., r. \tag{4.60}$$

Under the additional analyticity assumptions on h, f, it is easy to see that for every positive integer k, $H_k(z, w, w_1, ..., w_k)$ and $R_k(z, w, w_1, ..., w_{k-1})$ are analytic over $\{(z, w, w_1, ..., w_k)|\ (z, w) \in \Omega\}$.

Now that $\phi(z)$ in (4.53) is a formal solution of (4.49), we infer from (4.60) that

$$\eta_1 = c_1, \quad \eta_2 = 2!c_2, \quad ..., \quad \eta_k = k!c_k, \quad ... \tag{4.61}$$

and

$$\eta_k = H_k(\xi, \eta, \eta_1, ..., \eta_k).$$

Theorem 4.12. *Suppose $f(z)$ defined by (4.51) is analytic on $B(\xi; \sigma_0)$, and $h(z, w)$ defined by (4.52) is analytic on the dicylinder Ω defined by (4.50). Suppose further that (4.55) or (4.56) is satisfied. Then the formal solution $\phi(z)$ given by (4.53) is analytic on a disk $B(\xi; \delta)$, and $\phi(z)$ is unique in case (4.55) while $\phi(z)$ depends on a parameter in case (4.56).*

Proof. Since $0 < |b_1| < 1$, there is a positive integer $r \geq 1$ such that

$$\left| h'_w(\xi, \eta) \left[f'(\xi) \right]^r \right| < 1. \tag{4.62}$$

We can also find numbers $\nu \in (0,1)$, $\sigma_1 > 0$ and $d \in (0, \tau)$ such that for $|z - \xi| < \sigma_1$ and $|w - \eta| \leq d$,

$$\left| h'_w(z, w) \left[f'(z) \right]^r \right| < \nu. \tag{4.63}$$

Furthermore, for any positive numbers $M_1, ..., M_r$, we can find $L_0, L_1, ..., L_{r-1}$ and $L_r = \nu$ such that for any $(z, w', w'_1, ..., w'_r)$, $(z, w'', w''_1, ..., w''_r)$ in $\overline{B}(\xi; \sigma_1) \times \overline{B}(\eta; d) \times \overline{B}(\eta_1; M_1) \times \cdots \times \overline{B}(\eta_r; M_r)$ (where $\eta_1, ..., \eta_k$ are given in (4.61)),

$$|H_r(z, w', w'_1, ..., w'_r) - H_r(z, w'', w''_1, ..., w''_r)| \leq L_0 |w' - w''| + \sum_{k=1}^{r} L_k |w'_k - w''_k|. \tag{4.64}$$

Let

$$P(z) = \eta + \sum_{i=1}^{r} c_i(z - \xi)^i$$

and pick a positive number K such that for $z \in \overline{B}(\xi; \sigma_1)$,

$$|H_r(z, \eta, \eta_1, ..., \eta_r) - H_r(\xi, \eta, \eta_1, ..., \eta_r)| \leq \frac{1-\nu}{2} K.$$

Also pick $\sigma_2 > 0$ such that when $|z - \xi| \leq \sigma \leq \sigma_2$,

$$|P(z) - \eta| < d - \frac{\sigma^r K}{r!}. \tag{4.65}$$

In view of $0 < |b_1| < 1$, there is some positive number σ_3 such that for $|z| \leq \sigma_3$,

$$|f(z) - \xi| \leq |z - \xi|.$$

Finally, pick $\lambda \in (0,1)$ and σ which satisfies

$$0 < \sigma \leq \min\{1, \sigma_0, \sigma_1, \sigma_2, \sigma_3\},$$

$$\sum_{k=1}^{r} |\eta_k| \frac{\sigma^k}{k!} + K \frac{\sigma^k}{r!} < d,$$

$$\sum_{k=0}^{r-1}\sum_{i=1}^{r-k}L_k\,|\eta_{k+i}|\,\frac{\sigma^i}{i!}+K\sum_{k=1}^{r}L_{r-k}\frac{\sigma^k}{k!}<\frac{1-\nu}{2}K$$

and

$$\sum_{k=0}^{r-1}L_k\frac{\sigma^{r-k}}{(r-k)!}+\nu<\lambda<1.$$

Let S be the set of functions of the form

$$\phi(z)=P(z)+\sum_{n=r+1}^{\infty}u_n(z-\xi)^n \qquad (4.66)$$

which is analytic in $B(\xi;\sigma)$, with $\phi^{(r)}(z)$ continuous on $\overline{B}(\xi;\sigma)$, and

$$\left|\phi^{(r)}(z)-\eta_r\right|\le K,\ z\in\overline{B}(\xi;\sigma).$$

The set S is nonempty since P belongs to it, and is a complete metric space when equipped with the usual linear structure and metric

$$\rho(\phi_1,\phi_2)=\sup_{z\in\overline{B}(\xi;\sigma)}\left|\phi_1^{(r)}(z)-\phi_2^{(r)}(z)\right|.$$

Note that in view of Example 3.9,

$$\sup_{z\in\overline{B}(\xi;\sigma)}\left|\phi_1^{(k)}(z)-\phi_2^{(k)}(z)\right|\le\frac{\sigma^{r-k}}{(r-k)!}\rho(\phi_1,\phi_2). \qquad (4.67)$$

Define a mapping T on S by

$$(T\phi)(z)=h(z,\phi(f(z))).$$

Since for $z\in\overline{B}(\xi;\sigma)$ and $\phi\in S$,

$$|\phi(z)-\eta|=|\phi(z)-P(z)|+|P(z)-\eta|$$
$$\le\frac{\sigma^r}{r!}\sup_{z\in\overline{B}(\xi;\sigma)}\left|\phi^{(r)}(z)-\eta_r\right|+\sup_{z\in\overline{B}(\xi;\sigma)}|P(z)-\eta|$$
$$<d,$$

$T\phi$ is analytic in $B(\xi;\sigma)$. In view of the properties of $H_r, \phi(f(z)), ..., \phi^{(r)}(f(z))$, the function

$$(T\phi)^{(r)}(z)=H_r(z,\phi(f(z)),...,\phi^{(r)}(f(z)))$$

is continuous on $\overline{B}(\xi;\sigma)$. Also, $T\phi$ can be expressed in the form (4.66). Finally, we will show that for $z\in\overline{B}(\xi;\sigma)$,

$$\left|(T\phi)^{(r)}(z)-\eta_r\right|\le K.$$

To see this, note that in view of Example 3.9, the function

$$\left(\phi(z)-\eta-\sum_{j=1}^{r-1}c_j(z-\xi)^j\right)^{(k)}=\phi^{(k)}(z)-\sum_{j=k}^{r-1}\frac{\eta_j}{(j-k)!}(z-\xi)^{j-k}$$

satisfies

$$\phi^{(k)}(z) - \eta_k = \sum_{i=1}^{r-k-1} \frac{\eta_{i+k}}{i!}(z-\xi)^i + \frac{(z-\xi)^{r-k}}{(r-k-1)!}\int_0^1 (1-t)^{r-k-1}\phi^{(r)}(\xi+t(z-\xi))\,dt,$$

so that

$$\left|\phi^{(k)}(z) - \eta_k\right| \le \sum_{i=1}^{r-k}|\eta_{i+k}|\frac{\sigma^i}{i!} + K\frac{\sigma^{r-k}}{(r-k)!}, \quad k=0,1,...,r-1. \qquad (4.68)$$

Hence,

$$\begin{aligned}
\left|(T\phi)^{(r)}(z) - \eta_r\right| &= \left|H_r(z,\phi(f(z)),\phi'(f(z)),...,\phi^{(r)}(f(z))) - H_r(\xi,\eta,\eta_1,...,\eta_r)\right| \\
&\le \left|H_r(z,\phi(f(z)),\phi'(f(z)),...,\phi^{(r)}(f(z))) - H_r(z,\eta,\eta_1,...,\eta_r)\right| \\
&\quad + |H_r(z,\eta,\eta_1,...,\eta_r) - H_r(\xi,\eta,\eta_1,...,\eta_r)| \\
&\le L_0|\phi(f(z))-\eta| + \sum_{k=1}^{r-1} L_k\left|\phi^{(k)}(f(z)) - \eta_k\right| \\
&\quad + \nu\left|\phi^{(r)}(f(z)) - \eta_r\right| + \frac{1-\nu}{2}K \\
&\le \sum_{k=0}^{r-1}\sum_{i=1}^{r-k} L_k|\eta_{k+i}|\frac{\sigma^i}{i!} + K\sum_{k=1}^r L_{r-k}\frac{\sigma^k}{k!} + \nu K + \frac{1-\nu}{2}K \\
&< \frac{1-\nu}{2}K + \frac{1-\nu}{2}K + \nu K \\
&= K.
\end{aligned}$$

These show that $T\phi \in S$.

Now by means of (4.64), (4.67) and

$$\sum_{k=0}^{r-1} L_k \frac{\sigma^{r-k}}{(r-k)!} + \nu < \lambda < 1,$$

we see that

$$\begin{aligned}
\rho(T\phi_1,T\phi_2) &= \sup_{z\in B(\xi,\sigma)}\left|(T\phi_1)^{(r)}(z) - (T\phi_2)^{(r)}(z)\right| \\
&\le \sup_{z\in B(\xi,\sigma)}\left|L_0|\phi_1(f(z)) - \phi_2(f(z))| + \sum_{k=1}^n L_k\left|\phi_1^{(k)}(f(z)) - \phi_2^{(k)}(f(z))\right|\right| \\
&\le L_0\left\{\sup_{z\in B(\xi,\sigma)}|\phi_1(z)-\phi_2(z)| + \sum_{k=1}^r L_k \sup_{z\in B(\xi,\sigma)}\left|\phi_1^{(k)}(z) - \phi_2^{(k)}(z)\right|\right\} \\
&\le \sum_{k=0}^{r-1} L_k \frac{\sigma^{r-k}}{(r-k)!}\rho(\phi_1,\phi_2) + L_r\rho(\phi_1,\phi_2) \\
&\le \lambda\rho(\phi_1,\phi_2).
\end{aligned}$$

By means of the Banach contraction theorem, (4.49) has a solution in S. The proof is complete.

Example 4.18. Consider the equation

$$\psi(f(z)) = (\psi(z))^p, \qquad (4.69)$$

where p is a positive integer greater than or equal to 2. Suppose $f(z) = z^p F(z)$ where F is analytic in $B(0; \delta)$ and $F(0) \neq 0$. Then (4.69) has a solution $\psi(z)$ which is analytic on a neighborhood of the origin and satisfies $\psi(0) = 0$ and $\psi'(0) \neq 0$. Indeed, pick c such that $c^{p-1} = F(0)$ and consider

$$\phi(z) = (F(z)\phi(f(z)))^{1/p} \qquad (4.70)$$

where $u^{1/p}$ denotes the branch of the root function in a neighborhood of $u = c^p$ for which $(c^p)^{1/p} = c$. By Theorem 4.12, equation (4.70) has an analytic solution $\phi(z)$ defined over a neighborhood of 0 and $\phi(0) = c$.

4.6 Notes

The Analytic Implicit Function Theorem 4.1 is well known and is believed to be due to Cauchy. The presentation here is based on Fichtenholz [63]. More information can be found in Krantz and Parks [101]. It is interesting to note that the basic idea of the proof is to make use of the Newton binomial series as a majorant!

The results in the section on polynomial and rational functional equations are obtained in the processes of deriving analytic solutions of other functional equations (see later discussions).

The linear functional equation (4.14) is studied by Li [112], in which Theorem 4.3 is obtained.

The linear functional equation (4.15) is also studied by Li [112], in which Theorem 4.4 is also obtained.

Equation (4.16) and the corresponding Theorem 4.5 are in [145].

Equation (4.26) has been studied by Smajdor and Smajdor [210], Myrberg [145], Li and Si [125] and Kuczma [105, 106]. In Kuczma [105], Theorem 4.7 is given.

A Chapter on Schröder equation (see Schröder [173]) can be found in Kuczma [104], in which more references can be found. A new reference is Smajdor [209]. Theorem 4.10 is due to Siegel [203].

The same method of proof of Theorem 4.11 found in [125] can be used to deal with a more general equation of the form

$$f(\psi(z)) = \sum_{i=1}^{m} \psi(\alpha_i z) + F(z).$$

Equation (4.69) is the Böttcher equation which is a special conjugacy equation. Suppose f is analytic on the unit disk D, maps D into itself, and can be expanded as $f(z) = a_k z^k + a_{k+1} z^{k+1} + \cdots$, where $a_k \neq 0$ and $k \geq 2$. Cowen [43] gives a necessary and sufficient condition for the existence of single-valued analytic solutions defined on all of D to the Böttcher's equation. It is seen in [43] that the only non-zero solutions occur when $k = m$. There is always a solution of the equation $\psi(f(z)) = (\psi(z))^k$ that is analytic and univalent in a neighborhood of the origin (see Valiron [216]).

Implicit function theorems for equations (4.20), (4.49) and more general equations such as

$$\Phi(z) = H\left(z, \Phi(f_1(z)), ..., \Phi(f_m(z))\right), \ z \in C,$$

where Φ is the unknown function, and $f_1, ..., f_m$ as well as H are given complex valued functions, can be handled by the Banach contraction theorem as can be seen in Smajdor [205–208], Matkowski [135], Baron et al. [15].

Chapter 5

Functional Equations with Differentiation

5.1 Introduction

A differential equation is an equation that involves an unknown function and its derivatives. Differential equations play an extremely important and useful role in applied mathematics since they are used to model natural evoluntionary processes in which the unknown functions and their rate of changes are involved.

It may occur in some natural processes that the unknown function is involved in an indirect manner. For instance, the following relation

$$f'(t) = f(t-1)$$

states that the derived function of the unknown function f is equal to the function f translated one unit to the right. Such an equation or similar ones arise when the rate of changes at time t of the unknown processes are influenced by the past histories of the processes. We will grossly call such equations functional differential equations. Nowadays, functional equations with differentiation can be much more complicated than the functional differential equations and may involve operations of iteration, composition, integration, etc., besides the usual algebraic operations.

Recall our very first example in this book which is concerned with finding a solution $y = y(t)$ to the equation

$$\frac{dy}{dt} = ry,$$

and the condition $y(0) = 1$. If we assume that y is an analytic function on a neighborhood of 0, then $y(t) = \widehat{a}(t)$ for t in the open interval $B(0; \rho(a))$ for some sequence $a \in l^{\mathbf{N}}$. By Theorem 3.7, $y'(t) = \widehat{a}'(t) = \widehat{Da}(t)$ for $t \in B(0; \rho(a))$. Thus, from $\widehat{Da}(t) = ry(t) = \widehat{ra}(t)$ and Theorem 3.18, we have $Da = ra$, that is,

$$(k+1)a_{k+1} = ra_k, \ k \in \mathbf{N}.$$

Since $y(0) = a_0 = 1$, we see that

$$a = \left\{\frac{r^k}{k!}\right\}_{k \in \mathbf{N}}.$$

Finally, by the ratio test, we see that
$$\lim_{k\to\infty}\left|\frac{a_{k+1}}{a_k}\right| = \lim_{k\to\infty}\left|\frac{r}{k+1}\right| = 0.$$
Thus $\rho(a) = \infty$. This means $y = \widehat{a}(x)$ is an analytic function on $\mathbf{R}$, and since $Da = ra$ implies $y'(t) = ry(t)$ for $t \in \mathbf{R}$, it is the unique analytic solution of our equation satisfying $y(0) = 1$.

Based on similar ideas, we may obtain many existence theorems for different types of ordinary and partial differential equations. Some of these equations have been discussed quite extensively and systematically in standard texts (see e.g. Hille [78], Balser [13]), while others including the functional differential equations are only reported in different research papers. We will present a variety of existence results in this Chapter, but more attention will be paid to the less known and more recent results in the literature for obvious reasons.

5.2 Linear Systems

Recall that a matrix function $B(t)$ of one variable is analytic at $t = t_0$ if all its component functions are analytic at t_0. The initial value problem
$$x'(t) = A(t)x(t), \quad x(t_0) = x_0 \tag{5.1}$$
is now studied under the condition that the m by m matrix function $A(t)$ is analytic at $t = t_0$. Such a point is also called an ordinary point of the differential system $x' = A(t)x$.

Assume that $x(t)$ and $A(t)$ are respectively analytic functions of the form
$$x(t) = \sum_{k=0}^{\infty} x_k(t - t_0)^k, \tag{5.2}$$
and
$$A(t) = \sum_{k=0}^{\infty} A_k(t - t_0)^k \tag{5.3}$$
in a neighborhood of the point t_0, say $B(t_0; \rho)$, where we interpret the above notations as an abbreviation for simultaneously writing component functions. Then inserting these expressions and employing the Unique Representation Theorem 3.8, we see that
$$(k+1)x_{k+1} = \sum_{m=0}^{k} A_{k-m}x_m, \; k \in \mathbf{N}. \tag{5.4}$$

Hence, given x_0, we can recursively compute $x_1, x_2, ...$ in a unique manner which proves the uniqueness of the analytic solution $x(t)$. To see that (5.2) is indeed an analytic solution, we may assume without loss of generality that x_0 is not 0 for otherwise $x_n = 0$ for all n. We only need to show that it converges in a neighborhood of t_0. This is accomplished by finding majorant functions for the components of $x(t)$. To this end, first observe that the convergence of the series in (5.3) implies, in view of Cauchy's Estimation (Theorme 3.26), that for $0 < r < \rho$, there is some $M_r > 0$ such that

$$\|A_k\| \leq \frac{M_r}{r^k}, \ k \in \mathbf{N},$$

where $\|\cdot\|$ stands for a natural norm for matrices. Let the sequence $c = \{c_k\}_{k \in \mathbf{N}}$ be defined by $c_0 = \|x_0\|$ and

$$(k+1)c_{k+1} = M_r \sum_{m=0}^{k} \frac{1}{r^{k-m}} c_m, \ k \in \mathbf{N}.$$

Then we may conclude by induction that $c_k \neq 0$ and $\|x_k\| \leq c_k$ for $k \in \mathbf{N}$, and

$$(k+1)c_{k+1} = \left(M_r + \frac{k}{r}\right) c_k, \ k \in \mathbf{N}.$$

By means of the ratio test, we see that $\{r' \cdot c\}$ is absolutely summable for any $r' \in (0, r)$. Since r is an arbitrary number satisfying $0 < r < \rho$, we see further that $\rho(c) = \rho$. Finally, since each component function of $x(t)$ is majorized by $\widehat{c}(t)$, we see that $x(t)$ defined by (5.2) is an analytic solution on $B(t_0; \rho)$.

Theorem 5.1. *Suppose the m by m matrix function $A(t)$ is analytic at each point t in an open set S of $\mathbf{R}$. Then for every $t_0 \in S$, there exists a unique vector function $x(t)$ analytic on the largest open ball $B(t_0; \rho)$ with center t_0 contained in S such that $x'(t) = A(t)x(t)$ for $t \in B(t_0; \rho)$ and $x(t_0) = x_0$.*

We remark that with the help of a monodromy theorem (see e.g. page 225 in Balser [13]), we can extend the above local existence theorem into a global existence theorem.

We remark further that the above theorem holds if the functions involved are defined on subsets of the complex plane. The proof is almost the same with minor modifications.

Example 5.1. Consider the differential system

$$\begin{pmatrix} y'(t) \\ z'(t) \end{pmatrix} = \begin{pmatrix} 0 & 1 \\ 1 & 1 \end{pmatrix} \begin{pmatrix} y(t) \\ z(t) \end{pmatrix}.$$

Since the coefficient matrix is analytic on $\mathbf{R}$, by Theorem 5.1, it has an unique solution $(y(t), z(t))^\dagger$ which is analytic on $\mathbf{R}$ and satisfies $y(0) = a_0$ and $z(0) = b_0$. Furthermore, if we let $y(t) = \widehat{a}(t)$ and $z(t) = \widehat{b}(t)$, then in view of (5.4), we see that

$$(k+1) \begin{pmatrix} a_{k+1} \\ b_{k+1} \end{pmatrix} = \begin{pmatrix} 0 & 1 \\ 1 & 1 \end{pmatrix} \begin{pmatrix} a_k \\ b_k \end{pmatrix}, \ k \in \mathbf{N}.$$

Since $y(0) = a_0$ and $z(0) = b_0$, we see that

$$\begin{pmatrix} a_k \\ b_k \end{pmatrix} = \left\{ \frac{1}{k!} \begin{pmatrix} 0 & 1 \\ 1 & 1 \end{pmatrix}^k \begin{pmatrix} a_0 \\ b_0 \end{pmatrix} \right\}_{k \in \mathbf{N}}.$$

Hence,
$$\begin{pmatrix} y(t) \\ z(t) \end{pmatrix} = \sum_{k=0}^{\infty} \frac{t^k}{k!} \begin{pmatrix} 0 & 1 \\ 1 & 1 \end{pmatrix}^k \begin{pmatrix} a_0 \\ b_0 \end{pmatrix}, \quad t \in \mathbf{R}.$$

We remark that when a scalar linear differential equation can be expressed as a linear differential system, then the above theorem can be applied. For instance, the following equation
$$y''(t) - y(t) = 0, \tag{5.5}$$
can be written as
$$\begin{pmatrix} y'(t) \\ z'(t) \end{pmatrix} = \begin{pmatrix} 0 & 1 \\ 1 & 0 \end{pmatrix} \begin{pmatrix} y(t) \\ z(t) \end{pmatrix}.$$

Thus an analytic solution of (5.5) exists. However, to find the explicit solution, it is sometimes easier to proceed in a direct manner illustrated as follows.

We first assume that $y(t) = \widehat{a}(t)$ is an analytic solution of (5.5) on $B(0; \rho(a))$ for some $a \in l^{\mathbf{N}}$ and $y(0) = a_0 = \alpha$ and $y'(0) = a_1 = 0$. By Theorem 3.7, $y''(t) = \widehat{a}''(t) = \widehat{D^2 a}(t)$ for $t \in B(0; \rho(a))$. Thus, from $y''(t) = y(t)$ we have $D^2 a = a$, that is,
$$(k+1)(k+2)a_{k+2} = a_k, \quad k \in \mathbf{N}.$$

Since $a_0 = \alpha$ and $a_1 = 0$, we may calculate
$$a_{2k} = \frac{\alpha}{(2k)!}, \quad a_{2k+1} = 0, \quad k \in \mathbf{Z}^+.$$

This shows the uniqueness of the analytic solution
$$y(t) = \sum_{k=0}^{\infty} \frac{\alpha}{(2k)!} t^{2k}.$$

If $\alpha = 0$, then $y(x) = 0$ is analytic on $\mathbf{R}$. If $\alpha \neq 0$, note that for $t \neq 0$,
$$\lim_{k \to \infty} \left| \frac{\alpha}{(2(k+1))!} t^{2(k+1)} \right| \left| \frac{(2k)!}{\alpha} \frac{1}{t^{2k}} \right| = \lim_{k \to \infty} \frac{|t^2|}{(2k+2)(2k+1)} = 0.$$

Thus $y(x)$ is analytic on $\mathbf{R}$.

Similarly, if we assume that $z(t) = \widehat{b}(t)$ is an analytic solution in $B(0; \rho(b))$ for some $a \in l^{\mathbf{N}}$ and $z(0) = b_0 = 0$ and $z'(0) = b_1 = \beta$, then $D^2 b = b$ is valid and
$$b_{2k} = 0, \quad b_{2k+1} = \frac{\beta}{(2k+1)!}, \quad k \in \mathbf{Z}^+.$$

As before, we may show that
$$z(t) = \sum_{k=0}^{\infty} \frac{\beta}{(2k+1)!} t^{2k+1}$$
is analytic on $\mathbf{R}$.

Therefore the radius of convergence of $a + b$ is ∞, which follows from the fact that the sum
$$\sum_{k=0}^{\infty} a_{2k} t^{2k} + \sum_{k=0}^{\infty} b_{2k+1} t^{2k+1}$$
is convergent. Finally, the analytic function $w(t) = \widehat{a+b}(t)$ over $\mathbf{R}$ is the unique analytic solution of (5.5) satisfying $w(0) = \alpha$ and $w'(0) = \beta$.

The same method can be used to find analytic solutions for linear ordinary differential equations with nonconstant analytic coefficient functions. For instance, consider the second-order equation
$$y'' + p(x) y' + q(x) y = 0. \qquad (5.6)$$
An important fact about (5.6) is that the behavior of its solutions near a point x_0 is determined by the behavior of $p(x)$ and $q(x)$ near this point. If $p(x)$ and $q(x)$ are analytic at x_0, then $x = x_0$ is called an ordinary point of the equation. A solution is sought of the form
$$y(x) = \widehat{a}(x - x_0), \qquad (5.7)$$
where $a \in l^{\mathbf{N}}$. For instance, if we consider equation (5.6) with $p(x) = 3x/(x^2 + 4)$ and $q(x) = 1/(x^2 + 4)$, or in working form
$$(x^2 + 4) y'' + 3xy' + y = 0, \qquad (5.8)$$
then 0 is an ordinary point and the substitution $y(x) = \widehat{a}(x)$ leads to
$$a_{n+2} = -\frac{(n+1)}{4(n+2)} a_n, \quad n \in \mathbf{N}.$$
Given a_0 and a_1, we may then find the terms in the series. To treat convergence, we first find the radius of convergence of the sequence $\rho(a)$ from the recurrence formula involving two terms such as
$$a_{n+p} = f(n) a_n, \quad n \in \mathbf{N}. \qquad (5.9)$$

Example 5.2. Suppose the sequence $a = \{a_k\}_{k \in \mathbf{N}}$ is generated by the recurrence formula (5.9) where $p \in \mathbf{Z}^+$. Then
$$\frac{1}{\rho(a)} = \left[\lim_{k \to \infty} |f(pk)| \right]^{1/p} \qquad (5.10)$$
(with the convention that $\frac{1}{+\infty} = 0$). The idea of proof has been explained. More specifically, note that
$$\lim_{k \to \infty} \left[\left| \frac{a_{pk+p+s}}{a_{pk+s}} \right| |x|^p \right] = \lim_{k \to \infty} \left[|f(pk+s)| |x|^p \right] = |x|^p \lim_{k \to \infty} |f(pk)|.$$
If $\lim_{k \to \infty} |f(pk)| = f_p < \infty$, then when $|x| < 1/f_p^{1/p}$, we have
$$\sum_{k=0}^{\infty} a_{pk+s} x^{pk+s} < \infty, \ s = 0, 1, 2, ..., p-1.$$
Thus $\underline{x} \cdot a$ is summable, which shows that $\rho(a) \geq f_p^{-1/p}$. If $|x| > 1/f_p^{1/p}$, then $\underline{x} \cdot a$ is not summable, thus $\rho(a) \leq f_p^{-1/p}$. The case where $\lim_{k \to \infty} |f(pk)|$ is either 0 or ∞ is similarly proved.

Example 5.3. Consider equation (5.8). Application of (5.10) now gives

$$\rho(a) = \lim_{k\to\infty} \left| \frac{4(2k+2)}{2k+1} \right|^{1/2} = 2.$$

Note that the radius of convergence of the series solutions are equal to the distance of the ordinary point ($x = 0$) to the nearest singularity $\pm 2i$ of the rational function $1/(z^2 + 4)$.

Example 5.4. Consider the Airy's equation $y'' - xy = 0$. Substitution of solution y to Airy's equation about the ordinary point $x_0 = 0$ in the form of (5.7) gives

$$a_{n+3} = \frac{a_n}{(n+2)(n+3)}, \quad n \in \mathbf{N},$$

with $a_2 = 0$, which is of the form (5.9). Application of relation (5.10) gives

$$\rho(a) = \lim_{k\to\infty} |(3k+2)(3k+3)|^{1/3} = \infty.$$

Example 5.5. Consider the Chebyshev's equation

$$\left(1 - x^2\right) y'' - xy' + \frac{1}{4}y = 0.$$

The point $x = 0$ is also an ordinary point for Chebyshev's equation so that the substitution (5.7) with $x_0 = 0$ leads to

$$a_{n+2} = \frac{n^2 - 1/4}{(n+2)(n+1)} a_n, \quad n \in \mathbf{N}. \tag{5.11}$$

The application of (5.10) gives

$$\rho(a) = \lim_{k\to\infty} \left| \frac{(2k+2)(2k+1)}{(2k)^2 - 1/4} \right|^{1/2} = 1.$$

5.3 Neutral Systems

Let J be an interval in $[0, \infty)$. Consider the system [35]

$$tx'(t) + cx'(t/\alpha) = Ax(t) + F(t), \quad t \in J \subseteq [0, \infty), \tag{5.12}$$

under the condition

$$x(0) = x_0 \in \mathbf{R}^k \tag{5.13}$$

where $F: J \to \mathbf{R}^k$ is analytic and is of the form

$$F(t) = \sum_{n=0}^{\infty} F_n t^n, \quad F_n \in \mathbf{R}^k, \ t \in J, \tag{5.14}$$

A is a real k by k nonsingular matrix, $c, \alpha \in \mathbf{R}$ and $\alpha > 1$, $c \neq 0$.

Let
$$x(t) = \sum_{n=0}^{\infty} x_n t^n, \quad x_n \in \mathbf{R}^k \tag{5.15}$$
be a formal power series solution of (5.12) and (5.13). Then
$$x'(t) = \sum_{n=0}^{\infty} n x_n t^{n-1}, \quad x'(t/\alpha) = \sum_{n=0}^{\infty} n x_n \alpha^{1-n} t^{n-1}. \tag{5.16}$$
Substituting $x(t)$ into (5.12), we obtain
$$(n-1) x_{n-1} + cn\alpha^{1-n} x_n = A x_{n-1} + F_{n-1}, \quad n \in \mathbf{Z}^+,$$
so that
$$x_n = \frac{1}{nc} \alpha^{n-1} \{A - (n-1)E x_{n-1} + F_{n-1}\}, \quad n \in \mathbf{Z}^+, \tag{5.17}$$
where E is the identity matrix. Thus
$$x_1 = \frac{1}{c}(A x_0 + F_0),$$
$$x_2 = \frac{\alpha}{2c}[(A - E)x_1 + F_1] = \frac{\alpha}{2c^2}(A - E)A x_0 + \frac{\alpha}{2c^2}(A - E)F_0 + \frac{\alpha}{2c}F_1,$$
$$x_3 = \frac{\alpha^2}{3c}[(A - 2E)x_2 + F_2] = \frac{\alpha^3}{3!c^2}(A - 2E)F_1 + \frac{\alpha^2}{3c}F_2$$
$$= \frac{\alpha^3}{3!c^3}\left(\prod_{j=1}^{3}[A - (3-j)E]\right)x_0 + \sum_{l=0}^{2} \frac{l!\alpha^{3-l(l-1)/2}}{3!c^{3-l}}\left(\prod_{j=1}^{2-l}[A - (3-j)E]\right)F_l,$$
and by induction,
$$x_n = \frac{\alpha^{n(n-1)/2}}{n!c^n}\left(\prod_{j=1}^{n}[A - (n-j)E]\right)x_0$$
$$+ \sum_{l=0}^{n-1} \frac{l!\alpha^{(n-l)(n+l-1)/2}}{n!c^{n-1}}\left(\prod_{j=1}^{n-1-l}[A - (n-j)E]\right)F_l. \tag{5.18}$$

Theorem 5.2. *Suppose* $\det(A - nE) \neq 0$ *for* $n \in \mathbf{Z}^+$. *Then for any nonzero* x_0 *in* $\mathbf{R}^k$, *the homogeneous system*
$$tx'(t) + cx'(t/\alpha) = Ax(t), \quad t \in J \subseteq [0, \infty), \tag{5.19}$$
does not have any nontrivial analytic solution that satisfies $x(0) = x_0$.

Proof. Any formal solution $x(z)$ of (5.19) that satisfies $x(0) = x_0$ is of the form
$$x_n = \frac{\alpha^{n(n-1)/2}}{n!c^n}\left(\prod_{j=1}^{n}[A - (n-j)E]\right)x_0. \tag{5.20}$$

In view of (5.17), we see that
$$x_n = -\left(\alpha^{n-1}/c\right)\left(E - (A+E)/n\right)x_{n-1}.$$
Since $\alpha > 1$ and $\|A+E\|$ is finite, there exists a $m \in \mathbf{Z}^+$ such that
$$r := \alpha\left(1 - \frac{1}{m}\|A+E\|\right) > 1.$$
Set $t = c\alpha^{-m}$. Then
$$\alpha\left(1 - \frac{1}{m+k}\|A+E\|\right) \geq r,\ k \in \mathbf{N},$$
and
$$\bar{x}_n = -\alpha^{n-m-1}\left(E - (A+E)/n\right)\bar{x}_{n-1}, \tag{5.21}$$
where $\bar{x}_n = x_n t^n$. We have
$$\|\bar{x}_{m+2}\| = |\alpha|\left\|\bar{x}_{m+1} - \frac{1}{m+2}(A+E)\bar{x}_{m+1}\right\|$$
$$\geq |\alpha|\left(\|\bar{x}_{m+1}\| - \frac{1}{m+2}\|A+E\|\,\|\bar{x}_{m+1}\|\right)$$
$$\geq |\alpha|\left(1 - \frac{1}{m+2}\|A+E\|\right)\|\bar{x}_{m+1}\|$$
$$\geq r\|\bar{x}_{m+1}\|.$$
By induction, we see that
$$\|\bar{x}_{m+1+k}\| \geq r^k \|\bar{x}_{m+1}\|,\ k \in \mathbf{Z}^+$$
and hence $\lim_{n\to\infty}\|\bar{x}_n\| = +\infty$. This shows that $x(t) = \sum_{n=0}^{\infty} x_n t^n = \sum_{n=0}^{\infty} \bar{x}_n$ diverges for any t different from 0. The proof is complete.

Next, let A has real and simple eigenvalues $\lambda_1, \lambda_2, ..., \lambda_k$ only. Let $\bar{A}$ be the Jordan canonical form of the matrix A so that
$$\bar{A} = TAT^{-1} \tag{5.22}$$
for some nonsingular $T \in \mathbf{R}^{k\times k}$. Let
$$z(t) = Tx(t) \tag{5.23}$$
and
$$z(0) = Tx(0) = Tx_0 = z_0.$$
Then (5.19) can be written as
$$tT^{-1}z'(t) + cT^{-1}z'(t/\alpha) = AT^{-1}z(t),\ t \in J,$$
so that
$$tz'(t) + cz'(t/\alpha) = \bar{A}z(t),\ t \in J. \tag{5.24}$$

Let
$$z(t) = (z_1(t), z_2(t), ..., z_k(t))^\dagger = \sum_{m=0}^{\infty} z_m t^m,$$

where $z_m = (z_{m1}, z_{m2}, ..., z_{mk})^\dagger$, be a formal power series solution of (5.24). Then

$$z_m = T x_m, \; m \in \mathbf{Z}^+, \qquad (5.25)$$

and

$$z_{mi} = \frac{\alpha^{m-1}}{mc} [\lambda_i - (m-1)] z_{m-1,i} = \frac{\alpha^{m(m-1)/2}}{m! c^m} \prod_{j=1}^{m} [\lambda_i - (m-j)] z_{0i} \qquad (5.26)$$

for $i = 1, 2, ..., k$ and $m \in \mathbf{Z}^+$.

Note that when one of the eigenvalues $\lambda_1, ..., \lambda_k$, say, λ_ξ, is a positive integer, then

$$z_{m\xi} = \frac{\alpha^{m(m-1)/2}}{m! c^m} \prod_{j=1}^{m} [\lambda_\xi - (m-j)] z_{0\xi} = 0$$

for $m = \lambda_\xi + 1, \lambda_\xi + 2, ...$. This shows that

$$z_\xi(t) = \sum_{m=0}^{\lambda_\xi} z_{m\xi} t^m, \; t \in J,$$

which is a polynomial in t.

As a consequence, when the eigenvalues of the matrix A are simple positive integers, then in view of (5.23), we see that the homogeneous system (5.23) has a unique analytic solution that satisfies $x(0) = x_0$. Each component of this solution is a polynomial.

Next, we consider the nonhomogeneous equation (5.12) when $F(t)$ is a polynomial.

Theorem 5.3. *Suppose*

$$F(t) = \sum_{n=0}^{M} F_n t^n, \; t \in J.$$

Suppose further that $\det(A - nE) \neq 0$ *for* $n = M+2, M+3, ...$, *and that for any* $x_0 \in R^k$,

$$F_M \neq (ME - A) \left[\frac{\alpha^{M(M-1)/2}}{M! c^M} \left(\prod_{j=1}^{M} [A - (M-j) E] \right) x_0 + \sum_{l=0}^{M-1} \frac{l! \alpha^{(M-l)(M+l-1)/2}}{M! c^{M-l}} \left(\prod_{j=1}^{M-1-l} [A - (M-j) E] \right) F_l \right].$$

Then there does exist any analytic solution $x(z)$ *of (5.12) on* J *that also satisfies* $x(0) = x_0$.

Proof. In view of (5.17), we see that a formal power series solution $x(t) = \sum_{n=0}^{\infty} x_n t^n$ of (5.12) satisfies
$$x_n = \frac{1}{nc}\alpha^{n-1}\left\{[A - (n-1)E]x_{n-1} + F_{n-1}\right\}, \ n = 1, ..., M+1,$$
and
$$x_n = \frac{1}{nc}\left\{[A - (n-1)E]x_{n-1}\right\}, \ n = M+2, ...$$
Thus, for any $x_0 \in \mathbf{R}^k$, x_n is uniquely determined up to $n = M+1$, and
$$x_{M+1} = \frac{1}{Mc}\alpha^M\left[(A - ME)x_M + F_M\right].$$
Furthermore, in view of (5.18) and the condition on F_M, we see further that $F_M \neq (ME - A)x_M$ and $x_{M+1} \neq 0$. For $n \geq M+2$, we can determine x_n from
$$x_n = -\frac{\alpha^{n-1}}{c}\left(E - \frac{A+E}{n}\right)x_{n-1}.$$
We may now proceed as in the proof of Theorem 5.2 to show that the corresponding formal power series solution diverges for all nonzero t. The proof is complete.

Let us investigate the existence of analytic solutions for the nonhomogeneous problem (5.12), based on the analysis of structures of the spectrum of the matrix A. A substitution
$$z(t) = Tx(t), \tag{5.27}$$
where $T \in \mathbf{R}^{k \times k}$ is a real nonsingular matrix, allows us to write
$$z(0) = Tx(0) = Tx_0 = z_0$$
and (5.12) in the form
$$tT^{-1}z'(t) + cT^{-1}z'(t/\alpha) = AT^{-1}z(t) + F(t), \ t \in J,$$
or, equivalently,
$$tz'(t) + cz'(t/\alpha) = \bar{A}z(t) + B(t), \ t \in J, \ z(0) = z_0, \tag{5.28}$$
where $\bar{A} = TAT^{-1}$ and $B(t) = TF(t)$. Let
$$z(t) = (z_1(t), z_2(t), ..., z_k(t))^\dagger = \sum_{n=0}^{\infty} z_n t^n, \ z_n \in \mathbf{R}^k, \ t \in J, \tag{5.29}$$
be a formal solution of (5.28), and let
$$B(t) = (B_1(t), ..., B_k(t))^\dagger = \sum_{n=0}^{\infty} B_n t^n, B_n = (B_{n1}, ...B_{nk})^\dagger,$$
then
$$z_i(t) = \sum_{n=0}^{\infty} z_{ni} t^n, \ z_i \in \mathbf{R}, \ i = 1, 2, ..., k \tag{5.30}$$
and
$$z_{ni} = \frac{\alpha^{n-1}}{nc}\left\{[\lambda_i - (n-1)]z_{n-1,i} + B_{n-1,i}\right\}, n \in \mathbf{Z}^+; \ i = 1, 2, ..., k. \tag{5.31}$$

Theorem 5.4. *Suppose the eigenvalues of the matrix A are simple positive integers $\lambda_1, ..., \lambda_k$. If each component $B_i(t)$ of $B(t)$ is a polynomial of degree less than or equal to $\lambda_i - 1$, then for each $x_0 \in \mathbf{R}^k$, equation (5.28) has a unique solution $x(t)$ that satisfies $x(0) = x_0$ and is a polynomial of degree no greater than $\max\{\lambda_1, ..., \lambda_k\}$.*

Proof. It suffices to look at $i = 1$. Since
$$z_{n1} = \frac{\alpha^{n-1}}{nc}\left\{[\lambda_1 - (n-1)]z_1^{n-1} + B_i^{n-1}\right\}, \ 1 \leq n \leq \lambda_1,$$
and
$$z_{n1} = \frac{\alpha^{n-1}}{nc}[\lambda_1 - (n-1)]z_1^{n-1}, \ n = \lambda_1 + 1, ...,$$
thus,
$$z_{n1} = 0, \ n \geq \lambda_1 + 1.$$
It follows that
$$z_1(t) = \sum_{n=0}^{\lambda_1} z_{n1} t^n, \ t \in J.$$
The proof is complete.

As a direct consequence, if the eigenvalues of A are simple positive integers, and if each component $F_i(t)$ of $F(t)$ is a polynomial of degree not exceeding $\min\{\lambda_1, ..., \lambda_k\} - 1$, then for each $x_0 \in \mathbf{R}^k$, equation (5.28) has a unique solution $x(t)$ that satisfies $x(0) = x_0$ and is a polynomial of degree no greater than $\max\{\lambda_1, ..., \lambda_k\}$.

By (5.15) and (5.18), the solution of (5.12) can be written as
$$x(t) = X(t)x_0 + Y(t),$$
where the 'fundamental matrix' $X(t)$ and the vector $Y(t)$ are determined as
$$X(t) = \sum_{n=0}^{\max_{1 \leq i \leq k} \lambda_i} \frac{\alpha^{n(n-1)/2}}{n!c^n}\left(\prod_{j=1}^{n}[A - (n-j)E]\right)t^n,$$
and
$$Y(t) = \sum_{n=0}^{\max_{1 \leq i \leq k} \lambda_i} \sum_{l=0}^{n-1} \frac{l!\alpha^{(n-l)(n+l-1)/2}}{n!c^{n-l}}\left(\prod_{j=1}^{n-1-l}[A - (n-j)E]\right) F_l t^n.$$

5.4 Nonlinear Equations

We consider the simple nonlinear differential equation
$$\frac{dy}{dx} = f(x, y). \tag{5.32}$$
which includes the following nonhomogeneous equation
$$F'(t) = G(F(t)) + H(t), \tag{5.33}$$
where G and H are known functions.

Theorem 5.5. *Suppose $G(x) = \widehat{g}(x)$ and $H(t) = \widehat{h}(t)$ are analytic over some $B(0;\delta)$. Then equation (5.33) under the condition $F(0) = 0$ has an analytic solution of the form $F(t) = \widehat{b}(t)$ over some $B(0;\varepsilon)$, where $b_0 = 0$ and $b_1 = g_0 + h_0$.*

Proof. Let $F(t) = \widehat{b}(t)$, where $b_0 = 0$, be a formal solution of (5.33). After substituting it into (5.33), we see that
$$Db = g \circ b + h.$$
Hence, $b_1 = g_0 b_0^{\langle 0 \rangle} + h_0 = g_0 + h_0$ and
$$(n+1)b_{n+1} = (g \circ b)_n + h_n = \sum_{k=0}^{n} g_k b_n^{\langle k \rangle} + h_n \qquad (5.34)$$
for $n \geq 2$. The sequence $\{b_n\}_{n=2}^{\infty}$ can be uniquely determined by (5.34), and thus the formal solution $F(t)$ is found. Next, we will show that $F(t)$ is analytic on some $B(0; \varepsilon)$. To this end, note that by Cauchy's Estimation (Theorem 3.26), there is some $p > 0$ such that
$$|g_n| \leq p^n, \ n \in \mathbf{Z}^+. \qquad (5.35)$$
Since (5.33) is invariant under the transformations $\bar{F}(t) = pF\left(p^{-1}t\right)$ and $\bar{G}(x) = pG\left(p^{-1}x\right)$:
$$\bar{F}'(x) = \bar{G}\left(\bar{F}(qt)\right) + H\left(p^{-1}t\right),$$
and
$$\bar{G}(x) = G\left(p^{-1}x\right) = \sum_{n=0}^{\infty} \bar{g}_n x^n = \sum_{n=0}^{\infty} \frac{g_n}{p^n} x^n,$$
where $|\bar{g}_n| = |g_n/p^n| \leq 1$ for $n \in \mathbf{Z}^+$, we may thus assume without loss of generality that $|g_n| \leq 1$ for $n \geq 1$. Next, we consider the power series function $\gamma(t) = \widehat{u}(t)$, where $u_0 = 0$,
$$u_1 = |g_0| + |h_0|,$$
and
$$u_{n+1} = \sum_{k=0}^{n} u_n^{\langle k \rangle} + |h_n|, \ n \in \mathbf{Z}^+.$$
By induction, we may easily show that
$$|b_n| \leq u_n, \ n \in \mathbf{Z}^+,$$
thus $F(t)$ is majorized by $\gamma(t)$. Note that, by Example 4.11, the power series function $\gamma(t)$ is an analytic solution of the implicit relation
$$P(t, \gamma) \equiv \gamma - (|g_0| + |h_0|) t - t \left(\frac{\gamma}{1-\gamma} + \bar{H}(t) \right) = 0$$
near the origin. Hence $F(t)$ is also analytic in some $B(0; \varepsilon)$. The proof is complete.

As a corollary, if $G(x)$ is analytic on some disk $B(0; \delta)$, then the equation
$$F'(t) = G(F(t)) \qquad (5.36)$$

under the condition $F(0) = 0$ has an analytic solution of the form

$$F(t) = g_0 t + \sum_{n=2}^{\infty} g_n t^n$$

on some $B(0; \varepsilon)$.

Theorem 5.6. *Suppose $f = f(x, y)$ is analytic in a neighborhood of the point (x_0, y_0). Then there is a unique analytic solution $y = g(x)$ of (5.32) in a neighborhood of x_0 which satisfies $y_0 = g(x_0)$.*

Proof. Assume without loss of generality that $x_0 = y_0 = 0$ and

$$f(x, y) = \widehat{a}(x, y)$$

for $|x| < \alpha$, $|y| < \beta$. Assume further that $g(x) = \widehat{b}(x)$ for $|x| < \mu \le \alpha$. Then $b_0 = 0$. Furthermore, in view of the Substitution Theorems 3.11 and 3.25, the composite function $\widehat{a}(x, \widehat{b}(x))$ is analytic on a neighborhood of 0, so that

$$\widehat{a}(x, \widehat{b}(x)) = \sum_{m=0}^{\infty} \sum_{n=0}^{\infty} a_{mn} x^m \left(\widehat{b}(x)\right)^n = \sum_{m=0}^{\infty} \sum_{n=0}^{\infty} \sum_{k=0}^{\infty} a_{mn} b_k^{\langle n \rangle} x^{m+k} = \sum_{i=0}^{\infty} c_i x^i$$

for x in a neighborhood of 0, where $\{c_i\} \in l^{\mathbf{N}}$ with $\rho(c) > 0$. The first few terms of the sequence c can be easily determined. To this end, let us write

$$a^{(m)} = \{a_{mn}\}_{n \in \mathbf{N}}, \ n \in \mathbf{N}.$$

Then

$$c_0 = \sum_{n=0}^{\infty} a_{0n} b_0^{\langle n \rangle} = (a^{(0)} \circ b)_0 = a_{00} + a_{01} b_0 + a_{02} b_0^2 + \cdots = a_{00},$$

and

$$c_1 = \sum_{n=0}^{\infty} a_{0n} b_1^{\langle n \rangle} + \sum_{n=0}^{\infty} a_{1n} b_0^{\langle n \rangle}$$
$$= (a^{(0)} \circ b)_1 + (a^{(1)} \circ b)_0$$
$$= \left\{a_{00} b_1^{\langle 0 \rangle} + a_{01} b_1^{\langle 1 \rangle} + a_{02} b_2^{\langle 2 \rangle} + \cdots\right\} + \left\{a_{10} + a_{11} b_0 + a_{12} b_0^2 + \cdots\right\}$$
$$= a_{01} b_1 + a_{10},$$

and in general,

$$c_t = (a^{(0)} \circ b)_t + (a^{(1)} \circ b)_{t-1} + \cdots + (a^{(t)} \circ b)_0 = \sum_{p=0}^{t} (a^{(p)} \circ b)_{t-p}.$$

In view of (5.32) and the Unique Representation Theorem 3.8, we see from

$$\widehat{Db}(x) = \widehat{a}(x, \widehat{b}(x))$$

that
$$b_1 = a_{00},$$
$$b_2 = \frac{1}{2}(a_{10} + a_{01}a_{00}),$$

and by induction that
$$b_t = \sum_{p=0}^{t}(a^{(p)} \circ b)_{t-p} = P_t(a_{00}, a_{10}, a_{01}, ..., a_{0,t-1}), \ t \geq 3,$$

where each P_n is a nontrivial polynomial with nonnegative coefficients. This proves the uniqueness of the analytic function $g(x)$. To see that $g(x)$ is indeed an analytic solution, we only need to show that it converges absolutely in a neighborhood of 0. This is accomplished by finding a majorant function for $g(x)$. For this purpose, let $\tau \in (0, \alpha)$ and $\rho \in (0, \beta)$. Then by Example 3.15, there is some positive constant M such that
$$f(x,y) \ll \frac{M}{(1-x/\tau)(1-y/\rho)} = \sum_{m=0}^{\infty}\sum_{n=0}^{\infty} A_{mn} x^m y^n,$$
where
$$A_{mn} = \frac{M}{\tau^m \rho^n}, \ m, n \in \mathbf{N}.$$

Note that any analytic solution $h(x) = \widehat{d}(x)$ of the differential equation
$$\frac{dy}{dx} = \frac{M}{(1-x/\tau)(1-y/\rho)} \tag{5.37}$$
satisfying the initial condition $h(0) = 0$ is a majorizing function of $g(x)$:
$$d_1 = A_{00} \geq |a_{00}| = b_1,$$
$$d_2 = \frac{1}{2}(A_{10} + A_{01}A_{00}) \geq \frac{1}{2}(|a_{10}| + |a_{01}||a_{00}|) \geq b_2,$$
and by induction,
$$d_n = P(A_{00}, A_{10}, A_{01}, ..., A_{0,n-1}) \geq b_n, \ n \geq 3.$$

Furthermore, this solution $h(x)$ can be found by rewriting (5.37) as
$$\left(1 - \frac{y}{\rho}\right)\frac{dy}{dx} = \frac{M}{(1-\frac{x}{\tau})},$$
then integrating and substituting $(x,y) = (0,0)$, we obtain
$$y = \rho\left(1 - \sqrt{1 + \frac{2M\tau}{\rho}\ln\left(1 - \frac{x}{\tau}\right)}\right).$$

Note that the function
$$\sqrt{1 + \frac{2M\tau}{\rho}\ln\left(1 - \frac{x}{\tau}\right)}$$

is a composite function of the form $F(G(x))$ where $G(x) = \ln(1 - x/\tau)$ and $F(u) = (1 + 2M\tau u/\rho)^{1/2}$. Since G is analytic for $|x| < \tau$ and F is analytic for $(2M\tau/\rho)|u| < 1$, we see that $F(G(x))$, and hence, $g(x)$ are analytic on a neighborhood of 0 by Theorem 3.16. The proof is complete.

We remark that since it can easily be checked that $F(G(x))$ is analytic for $|x| < \tau \left(1 - e^{-\rho/(2M\tau)}\right)$, the above theorem can be stated in a more precise manner, namely, if $f(x, y)$ is analytic for $|x| < \alpha$ and $|y| < \beta$, then the equation $y' = f(x, y)$ has a solution $y = g(x)$ which satisfies $g(0) = 0$ and is analytic for $|x| < \tau \left(1 - e^{-\rho/(2M\tau)}\right)$, where $\tau \in (0, \alpha)$, $\rho \in (0, \beta)$ and $|f(x,y)| \le M$ for $|x| < \tau$ and $|y| < \rho$.

We remark further that if $f(x, y) = \widehat{a}(x, y)$ where a is a nonnegative double sequence, then the analytic solution found above is of the form $\widehat{b}(x)$ and b is a nonnegative sequence.

Example 5.6. Consider the equation

$$\frac{1}{\alpha}\frac{du}{dt} = \frac{M}{\left(1 - \frac{u+t}{r}\right)\left(1 - \frac{1}{\rho}\frac{du}{dt}\right)} - M \tag{5.38}$$

in the unknown function $u = u(t)$, where $\alpha \in (0, 1)$ and M, r, ρ are fixed positive numbers. If we rewrite (5.38) as

$$\frac{1}{\alpha\rho}\left(\frac{du}{dt}\right)^2 - \left(\frac{1}{\alpha} - \frac{M}{\rho}\right)\frac{du}{dt} + \frac{M}{1 - (u+t)/r} - M = 0,$$

then

$$\frac{du}{dt} = \frac{\alpha\rho}{2}\left\{\left(\frac{1}{\alpha} - \frac{M}{\rho}\right) - \sqrt{\left(\frac{1}{\alpha} - \frac{M}{\rho}\right)^2 - \frac{4M}{\alpha\rho}\left(\frac{t+u}{r}\right)\left(1 - \frac{t+u}{r}\right)^{-1}}\right\}$$

$$= \frac{\alpha\rho}{2}\left(\frac{1}{\alpha} - \frac{M}{\rho}\right)\left(1 - \sqrt{1 - W}\right), \tag{5.39}$$

where

$$W = \frac{4M\alpha\rho}{(\rho - \alpha M)^2}\left(\frac{t+u}{r}\right)\left(1 - \frac{t+u}{r}\right)^{-1}.$$

Since

$$1 - \sqrt{1 - W} = \frac{1}{2}W + \frac{1}{2 \times 4}W^2 + \frac{1 \times 3}{2 \times 4 \times 6}W^3 + \cdots$$

$$= \sum_{k=1}^{\infty} \frac{1 \times 3 \times \cdots \times (2k-1)}{2 \times 4 \times \cdots \times (2k)} W^k$$

for $|W| < 1$, we see that the function $1 - \sqrt{1 - W}$ is analytic at 0 and generated by a nonnegative sequence. Thus, for α sufficiently small, the composite function on the right hand side of (5.39) is also analytic at $(0, 0)$ and is generated by a nonnegative double sequence. In view of the preceding remark, we see that (5.39), and hence (5.38), have a solution

$$u(t) = \sum_{k=1}^{\infty} b_k t^k$$

which satisfies the additional condition $u(0) = 0$ and $b_k \geq 0$ for $k \in Z^+$ and is analytic at 0.

There is a natural extension of Theorem 5.6: Suppose $f_1(t, x_1, ..., x_n), ..., f_n(t, x_1, ..., x_n)$ are analytic in a neighborhood of the point $(t_0, u_0, ..., u_n)$. Then there is a unique analytic solution $(x_1, ..., x_n) = (x_1(t), ..., x_n(t))$ of

$$x_1'(t) = f_1(t, x_1(t), ..., x_n(t)),$$
$$x_2'(t) = f_2(t, x_1(t), ..., x_n(t)),$$
$$... = ...$$
$$x_n'(t) = f_n(t, x_1(t), ..., x_n(t)),$$

in a neighborhood of t_0 satisfying $(x_1(t_0), ..., x_n(t_0)) = (u_0, ..., u_n)$. The proof, in spite of the more complicated technical detail, is similar to that of Theorem 5.6 and hence is omitted. Instead, we consider a more specific example as follows.

Example 5.7. Consider the nonlinear differential equation

$$x'''(t) = e^{-t} x^2(t), \qquad (5.40)$$

subject to the boundary conditions

$$x(0) = 1, \ x'(0) = 1, \ x(1) = e.$$

Our problem is to try to find an 'approximate' solution. Since (5.40) is equivalent to a system of differential equations just described, we see that a unique analytic solution $x(t) = \hat{a}(t)$ of (5.40) exists which satisfies $x(0) = a_0 = 1$, $x'(0) = a_1 = 1$ and $x''(0) = a_2$. Substituting this solution into (5.40), we see that

$$D^3 a = (\underline{-1} \cdot \varpi) * a^{(2)}.$$

Hence

$$a_{k+3} = \frac{1}{(k+3)(k+2)(k+1)} \sum_{j=0}^{k} \sum_{i=0}^{k-j} \frac{(-1)^j}{j!} a_i a_{k-j-i}, \ k \in \mathbf{N},$$

which yields

$$a_3 = \frac{1}{60} a_2 - \frac{1}{80},$$
$$a_4 = \frac{1}{6300} a_2 - \frac{23}{25200},$$
$$a_5 = a_2^2 - \frac{3149}{1058400} a_2 + \frac{1301}{2116800},$$

etc. These lead us to

$$x(t) = 1 + t + a_2 t^2 + \left(\frac{1}{60}a_2 - \frac{1}{80}\right)t^3 + \left(\frac{1}{6300}a_2 - \frac{23}{25200}\right)t^4 + \cdots.$$

If we now solve

$$e = x(1) \approx 1 + 1 + a_2 1^2 + \left(\frac{1}{60}a_2 - \frac{1}{80}\right)1^3 + \left(\frac{1}{6300}a_2 - \frac{23}{25200}\right)1^4,$$

we obtain

$$a_2 \approx \frac{3150}{3203}e - \frac{25031}{12812} \approx 0.71959.$$

Then

$$\tilde{x}(t) = 1 + t + a_2 t^2 + \left(\frac{1}{60}a_2 - \frac{1}{80}\right)t^3 + \left(\frac{1}{6300}a_2 - \frac{23}{25200}\right)t^4$$

is a candidate for an approximate solution of our problem. To check our assertion, note that $x(t) = e^t$ is the unique analytic solution of our problem (as can be checked by direct verification). Now we may obtain the following supporting data:

$$\tilde{x}(0.5) = 1.6798 \cdots \approx 1.6487 \cdots = e^{0.5},$$

$$\tilde{x}(0.1) = 1.09020509 \cdots \approx 1.1052 \cdots = e^{0.1},$$

etc.

5.5 Cauchy-Kowalewski Existence Theorem

As an application of Theorem 5.6, let us consider the following partial differential equation in the unknown function $u = u(x, y)$,

$$\frac{\partial u}{\partial x} = f\left(x, y, u, \frac{\partial u}{\partial y}\right), \qquad (5.41)$$

subject to the initial data

$$u(0, y) = 0. \qquad (5.42)$$

We will assume that f is an analytic function so that $f(0,0,0,0) = 0$ and

$$f(x, y, u, v) = \sum_{(i,j,k,l) \in \mathbf{N}^4} a_{ijkl} x^i y^j u^k v^l, \quad |x|, |y|, |u| \leq r; |v| \leq \rho.$$

The Cauchy-Kowalewski Theorem asserts the existence of a solution $u = u(x, y)$ which is analytic at $(0, 0)$. To this end, we first compute the partial derivatives of $u(x, y)$. Note that $u(0, 0) = 0$ and

$$\frac{\partial u}{\partial x}(0,0) = f\left(0, 0, u(0,0), \frac{\partial u}{\partial y}(0,0)\right) = f(0,0,0,0) = 0.$$

Furthermore, since

$$\frac{\partial^2 u}{\partial y \partial x} = f_2(x, y, u, u_y) + f_3(x, y, u, u_y)z_y + f_4(x, y, u, u_y)u_{yy}$$
$$= f_2(x, y, u, u_y),$$

we see that
$$u_{xy}(0,0) = f_2(0,0,0,0) = a_{0100}.$$

Similarly,
$$\frac{\partial^2 u}{\partial x^2} = f_1(x, y, u, u_y) + f_3(x, y, u, u_y)u_x + f_4(x, y, u, u_y)u_{yx},$$

so that
$$\frac{\partial^2 u}{\partial x^2}(0,0) = a_{1000} + a_{0001}a_{0100}.$$

It should be clear by now that we can calculate all the partial derivatives of $u(x, y)$ at $(0,0)$, and

$$\frac{\partial^{\alpha+\beta} u}{\partial x^\alpha \partial y^\beta}(0,0) = P_{\alpha+\beta}\left(a_{0000}, a_{1000}, ..., a_{000(\alpha+\beta-1)}\right).$$

where P_n is a nontrivial polynomial with a finite number of independent variables and nonnegative coefficients. We have thus found a unique formal power series solution $u(x, y)$.

To see that the power series function

$$u(x, y) = \sum_{(\alpha,\beta) \in \mathbf{N}^2} \frac{1}{\alpha! \beta!} \frac{\partial^{\alpha+\beta} u}{\partial x^\alpha \partial y^\beta}(0,0) x^\alpha y^\beta$$

is analytic at $(0,0)$, we first infer from the Cauchy's Estimation (Theorem 3.26) that there is some positive constant M such that

$$|a_{ijkl}| \leq \frac{M}{r^{i+j+k}\rho^l}, \quad (i, j, k, l) \in \mathbf{N}^4.$$

Thus

$$|a_{ijkl}| \leq \frac{(i+j+k)!M}{i!j!k!\delta^i r^{i+j+k}\rho^l} := c_{ijkl}$$

for some $\delta \in (0,1)$. As a consequence,

$$f(x, y, u, v) \ll \sum_{(i,j,k,l) \in \mathbf{N}^4} \frac{(i+j+k)!M}{i!j!k!\delta^i r^{i+j+k}\rho^l} x^i y^j u^k v^l - M$$

$$= M \left\{ \sum_{(i,j,k) \in \mathbf{N}^3} \frac{(i+j+k)!}{i!j!k!} \left(\frac{x}{\delta r}\right)^i \left(\frac{y}{r}\right)^j \left(\frac{u}{r}\right)^k \sum_{l \in \mathbf{N}} \left(\frac{v}{\rho}\right)^l \right\} - M$$

$$= M \left\{ \frac{1}{\left(1 - \frac{x/\delta + y + u}{r}\right)\left(1 - \frac{v}{\rho}\right)} - 1 \right\}.$$

We now consider the problem

$$\frac{\partial U}{\partial x} = M \left\{ \frac{1}{\left(1 - \frac{x/\delta + y + U}{r}\right)\left(1 - \frac{1}{\rho}\frac{\partial U}{\partial y}\right)} - 1 \right\}. \tag{5.43}$$

In view of Example 5.6, for sufficiently small $\delta > 0$, there is a nontrivial analytic solution

$$W(t) = \sum_{k=1}^{\infty} b_k t^k$$

of (5.38):

$$\frac{1}{\alpha}\frac{dW}{dt} = \frac{M}{\left(1 - \frac{W+t}{r}\right)\left(1 - \frac{1}{\rho}\frac{dW}{dt}\right)} - M$$

which is analytic at 0 and $b_k \geq 0$ for $k \in \mathbf{Z}^+$. If we let

$$U(x,y) = W\left(\frac{x}{\delta} + y\right),$$

then $U(x,y)$ is a solution of (5.43) which is analytic at $(0,0)$ and satisfies

$$U(0,y) = W(y) = \sum_{k=1}^{\infty} b_k y^k \geq 0$$

for y in a neighborhood of 0. Furthermore, since

$$\frac{\partial^{\alpha+\beta} U}{\partial x^\alpha \partial y^\beta}(0,0) = P_{\alpha+\beta}\left(c_{0000}, c_{1000}, ..., c_{000(\alpha+\beta-1)}\right)$$

$$\geq P_{\alpha+\beta}\left(|a_{0000}|, |a_{1000}|, ..., |a_{000(\alpha+\beta-1)}|\right)$$

$$\geq \frac{\partial^{\alpha+\beta} u}{\partial x^\alpha \partial y^\beta}(0,0),$$

we see that $U(x,y)$ majorizes $u(x,y)$ in a neighborhood of $(0,0)$. This shows that $u(x,y)$ is analytic near $(0,0)$. The proof is complete.

5.6 Functional Equations with First Order Derivatives

In this section, we consider functional differential equations involving first order derivatives of the unknown function. Sometimes it is relatively easy to find analytic solutions. For example, we first consider a simple equation

$$x'(z) = x(az),$$

where a is a fixed complex number different from 0. We may easily show that it has a solution of the form

$$x(z) = \sum_{n=0}^{\infty} \frac{a^{(n(n-1)/2)}}{n!} \eta z^n$$

which is analytic at each $z \in \mathbf{C}$. Indeed, if we seek a power series solution of the form

$$x(z) = \sum_{n=0}^{\infty} b_n z^n,$$

then substituting it into the above equation and comparing coefficients, we see that

$$Db = \underline{a} \cdot b,$$

or

$$(n+1)b_{n+1} = a^n b_n, \ n \in \mathbf{N}.$$

Let b_0 be an arbitrary number η, then the sequence $\{b_n\}$ is uniquely determined by

$$b_n = \frac{a^{(n(n-1)/2)}}{n!} \eta, \ n \in \mathbf{N}.$$

If $\eta = 0$, our original assertion clearly holds. If $\eta \neq 0$, then

$$\lim_{n \to \infty} \frac{b_{n+1}}{b_n} = \lim_{n \to \infty} \frac{a^n}{n+1} = 0,$$

so that $x(z)$ is analytic at each $z \in \mathbf{C}$.

5.6.1 Equation I

We consider the equation [118]

$$x'(z) = G(x(qz)) + H(z), \ q \in \mathbf{C}. \tag{5.44}$$

Theorem 5.7. *Suppose $G(z) = \widehat{g}(z)$, where $g_0 = 0$, and $H(z) = \widehat{h}(z)$ are analytic on a neighborhood of the origin. If $|q| < 1$, then the equation (5.44) has a solution $x(z)$ which is analytic on a neighborhood of the origin.*

Proof. By Cauchy's Estimation (Theorem 3.26), there is $p > 0$ such that

$$|g_m| \leq p^m, \ m \geq 1. \tag{5.45}$$

Let $x(z) = \widehat{b}(z)$, where $b_0 = 0$, be a formal power series solution of (5.44). Substituting it into (5.44), we see that

$$Db = \underline{q} \cdot (g \circ b) + h.$$

Hence

$$b_1 = h_0$$

and

$$(n+1)b_{n+1} = q^n \sum_{t=1}^{n} g_t b_n^{\langle t \rangle} + h_n, \ n \in \mathbf{N}. \tag{5.46}$$

Note that Example 4.11 asserts that the implicit relation

$$F(z,w) \equiv w - z\left(\frac{p\,|q|\,w}{1-p\,|q|\,w} + \bar{h}(z)\right) = 0 \qquad (5.47)$$

where $h(z) = \sum_{n=0}^{\infty} |h_n| z^n$, has a solution $w(z)$ which is analytic on a neighborhood of the origin and

$$w(z) = \sum_{n=0}^{\infty} u_n z^n,$$

where the sequence $u = \{u_n\}_{n \in \mathbf{N}}$ is defined by $u_0 = 0$, $u_1 = |h_0|$ and

$$u_{n+1} = \sum_{t=1}^{n} p^t |q|^t u_n^{\langle t \rangle} + |h_n|, \ n \in \mathbf{N}. \qquad (5.48)$$

We assert that b is majorized by u. Indeed, $b_0 = u_0 = 0$, $|b_1| = |h_0| = u_1$. Assume by induction that $|b_i| \leq u_i$ for $i = 0, 1, ..., n$ where $n \geq 1$. Then

$$|b_{n+1}| = \left|\frac{q^n}{n+1}\sum_{t=1}^{n} g^t b_n^{\langle t \rangle} + h_n\right| \leq \sum_{t=1}^{n} |q|^t p^t u_n^{\langle t \rangle} + |h_n| = u_{n+1}$$

as required. Now that $b \ll u$, thus $x(z)$ has a positive radius of convergence. The proof is complete.

5.6.2 Equation II

We consider the equation

$$G(x) F'(x) = G(F(x)), \qquad (5.49)$$

where G and H are known functions.

Theorem 5.8. *Suppose $G(x) = \widehat{a}(x)$, where $a_0 \neq 0$, is analytic on some $B(0; \delta)$. Then the equation (5.49) under the condition $F(0) = 0$ has the unique analytic solution $F(x) = x$ on $B(0; \delta)$.*

Proof. Let $F(x) = \widehat{c}(x)$, where $c_0 = 0$, be a formal power series solution of (5.49). Substituting it into (5.49), we see that

$$a * Dc = a \circ c$$

or,

$$\sum_{k=0}^{n} (k+1)c_{k+1} a_{n-k} = \sum_{k=0}^{n} a_k c_n^{\langle k \rangle}, \ n \in \mathbf{N}. \qquad (5.50)$$

Since $a_0 c_1 = a_0$, $2a_0 c_2 + c_1 a_1 = a_1 c_1$ and since $a_0 \neq 0$, we see that $c_1 = 1$ and $c_2 = 0$. Assume by induction that $c_k = 0$ for $k = 2, ..., n$, then in view of (5.50),

$$(n+1) a_0 c_{n+1} = a_n c_1^n - a_n c_1 = 0,$$

so that $c_{n+1} = 0$. In other words, $F(x) = x$, which is analytic for $|x| < \delta$. The proof is complete.

Theorem 5.9. *Suppose $G(x) = \widehat{a}(x)$, where $a_0 = 0$ and $a_1 \neq 0$, is analytic on some $B(0;\delta)$. Then the equation (5.49) under the condition $F(0) = 0$ has an analytic solution of the form*

$$F(x) = \eta x + \sum_{n=2}^{\infty} c_n x^n \qquad (5.51)$$

on some $B(0;\varepsilon)$, where η is arbitrary.

Proof. Let $F(x) = \widehat{c}(x)$, where $c_0 = 0$, be a formal power series solution of (5.49). As in the proof of Theorem 5.8, we see that (5.50) holds. Since $a_0 c_1 = a_0$, and $a_0 = 0$, we may choose c_1 to be any number η. Then $c_2, c_3, \ldots$ can be determined from (5.50):

$$n a_1 c_n = \sum_{k=0}^{n} a_k c_n^{\langle k \rangle} - \sum_{k=1}^{n-1} k c_k a_{n+1-k} \qquad (5.52)$$

for $n \geq 2$. Since $a_1 \neq 0$, c_1 can be chosen in an arbitrary manner. Let $c_1 = \eta$. Then the sequence $\{c_n\}_{n=1}^{\infty}$ can be uniquely determined by (5.52). In other words, we have determined the formal power series solution $F(x)$. We will show that $F(x)$ is analytic in some $B(0;\varepsilon)$. To this end, note that by Cauchy's Estimation (Theorem 3.26), there is $p > 0$ such that $|a_n| \leq p^{n-1}$ for $n \geq 2$. Since the equation (5.49) is invariant under the transformations $F(x) = f(px)/p$ and $G(x) = g(px)/p$:

$$g(x) f'(x) = g(f(x)),$$

and

$$g(x) = \sum_{n=1}^{\infty} g_n x^n = pG\left(\frac{x}{p}\right) = \sum_{n=1}^{\infty} \frac{a_n}{p^{n-1}} x^n,$$

where $|a_n/p^{n-1}| \leq 1$ for $n \geq 1$, we may assume without loss of generality that

$$|a_n| \leq 1, \; n \in \mathbf{Z}^+. \qquad (5.53)$$

Next, let the power series function $h(x) = \sum_{n=0}^{\infty} h_n x^n$ be defined by $h_0 = 0$, $h_1 = |\eta|$ and

$$h_n = \frac{1}{|a_1|} \left(\sum_{k=0}^{n} h_n^{\langle k \rangle} + \sum_{k=1}^{n-1} h_k \right), \; n \geq 1. \qquad (5.54)$$

Then $|c_1| = |\eta| \leq h_1$. Assume by induction that $|c_k| \leq h_k$ for $k = 2, 3, \ldots, n-1$. Then in view of (5.52),

$$|c_n| \leq \frac{1}{|a_1| n} \left(\sum_{k=0}^{n} |a_k| |c|_n^{\langle k \rangle} + \sum_{k=1}^{n-1} k |c_k| |a_{n+1-k}| \right)$$

$$\leq \frac{1}{|a_1|} \left(\sum_{k=0}^{n} \frac{1}{n} |c|_n^{\langle k \rangle} + \sum_{k=1}^{n-1} \frac{k}{n} |c_k| \right)$$

$$\leq \frac{1}{|a_1|} \left(\sum_{k=0}^{n} h_n^{\langle k \rangle} + \sum_{k=1}^{n-1} h_k \right)$$

$$\leq h_n.$$

In other words, $F(x)$ is majorized by $h(x)$. Next, by Example 4.9, $h = h(x)$ is an analytic solution of the implicit relation

$$h - |\eta|x - \frac{1}{|a_1|}\left(\frac{h^2}{1-h} + \frac{x}{1-x}h\right) = 0$$

near the origin. Hence $F(x)$ is also analytic near the origin. The proof is complete.

5.6.3 Equation III

We consider a simple equation

$$f(2z) = 2f'(z)f(z). \qquad (5.55)$$

Assume $f(z) = \hat{a}(z)$ is a solution of (5.55) which is analytic on a neighborhood of 0 and generated by the sequence $a = \{a_k\}_{k \in \mathbf{N}}$. Substituting it into (5.55) and comparing coefficients, we obtain

$$a \cdot \underline{2} = 2(Da) * a,$$

or

$$a_0 = 2a_1 a_0,$$
$$2a_1 = 4a_2 a_0 + 2a_1^2,$$

and

$$2^n a_n = 2(n+1)a_0 a_{n+1} + \sum_{k=1}^{n} 2k a_k a_{n+1-k}, \ n \geq 2. \qquad (5.56)$$

In view of the first two equations, there are three cases: (i) $a_0 = 0$, $a_1 = 0$; (ii) $a_0 = 0$, $a_1 = 1$; and (iii) $a_0 \neq 0$. In the first case, we may show by induction that

$$a_k = 0, \ k \in \mathbf{N}.$$

Thus the trivial analytic function is an analytic solution of (5.55). In the second case, we substitute $n = 2$ and $n = 3$ into (5.56) to obtain

$$4a_2 = 2a_1 a_2 + 4a_2 a_1 = 6a_2,$$

and

$$8a_3 = 2a_1 a_3 + 4a_2^2 + 6a_3 a_1 = 8a_3 + 4a_2^2$$

respectively. Thus $a_2 = 0$ and a_3 is arbitrary, say, $a_3 = \alpha/3!$. Then by induction, we may show that

$$a_{2n} = 0, \ n \in \mathbf{N},$$

and

$$a_{2n+1} = \frac{\alpha^n}{(2n+1)!}, \ n \in \mathbf{N}.$$

Indeed, the fact that $a_{2n} = 0$ for $n \in \mathbf{N}$ is easily seen by induction. Next, assume by induction that $a_{2k+1} = \alpha^k/(2k+1)!$ for $k = 1, 2, ..., n$, then in view of (5.56),

$$\{2^{2n+2} - (2n+3) - 1\} a_{2n+3}$$
$$= \frac{3\alpha^{n+1}}{3!(2n+1)!} + \frac{5\alpha^{n+1}}{5!(2n-1)!} + \cdots + \frac{(2n+1)\alpha^{n+1}}{(2n+1)!3!}$$
$$= \left\{ \frac{(2n+3)!}{2!(2n+1)!} + \frac{(2n+3)!}{4!(2n-1)!} + \cdots + \frac{(2n+3)!}{(2n)!3!} \right\} \frac{\alpha^{n+1}}{(2n+3)!},$$

that is,

$$\left\{ 2^{2n+2} - C_{2n+2}^{(2n+3)} - C_0^{(2n+3)} \right\} a_{2n+3} = \left\{ C_2^{(2n+3)} + C_4^{(2n+3)} + \cdots + C_{2n}^{(2n+3)} \right\} \frac{\alpha^{n+1}}{(2n+3)!}.$$

Since

$$2^{2n+2} = C_0^{(2n+2)} + C_2^{(2n+2)} + \cdots + C_{2n}^{(2n+2)} + C_{2n+2}^{(2n+2)},$$

we see further that

$$a_{2n+3} = \frac{\alpha^{n+1}}{(2n+3)!}$$

as desired.

It is easily checked that the radius of convergence of the sequence a is $\rho(a) = \infty$. Thus the power series function

$$\widehat{a}(z) = z + \sum_{k=1}^{\infty} \frac{\alpha^k}{(2k+1)!} z^{2k+1}, \quad z \in \mathbf{C},$$

is an analytic solution of (5.55). Note that when $\alpha = 0$, $\widehat{a}(z) = 0$ for $z \in \mathbf{C}$; when $\alpha > 0$,

$$\widehat{a}(z) = \frac{1}{\alpha^{1/2}} \sum_{k=0}^{\infty} \frac{\left(\alpha^{1/2} z\right)^{2k+1}}{(2k+1)!} = \frac{1}{\alpha^{1/2}} \sinh \alpha^{1/2} z, \quad z \in \mathbf{C},$$

and when $\alpha < 0$,

$$\widehat{a}(z) = \frac{1}{(-\alpha)^{1/2}} \sum_{k=0}^{\infty} (-1)^k \frac{\left((-\alpha)^{1/2} z\right)^{2k+1}}{(2k+1)!} = \frac{1}{(-\alpha)^{1/2}} \sin(-\alpha)^{1/2} z, \quad z \in \mathbf{C}.$$

In the third case, let $a_0 = \beta \neq 0$, then $a_1 = 1/2$. We may show that

$$a_n = \frac{1}{n! 2^n \beta^{n-1}}, \quad n \in \mathbf{Z}^+.$$

Indeed, assume by induction that $a_k = 1/(k!2^k \beta^{k-1})$ for $k = 2, 3, ..., n$. Then in view of (5.56),

$$a_{n+1} = \frac{1}{2(n+1)\beta} \left\{ 2^n a_n - \sum_{k=1}^{n} 2k a_k a_{n+1-k} \right\}$$

$$= \frac{1}{2(n+1)\beta} \left\{ (2^n - 1 - n) a_n - \sum_{k=2}^{n-1} 2k a_k a_{n+1-k} \right\}$$

$$= \left\{ 2^n - 1 - n - \sum_{k=2}^{n-1} C_{k-1}^{(n)} \right\} \frac{1}{(n+1)!2^{n+1}\beta^n}$$

$$= \left\{ 2^n - \sum_{k=0}^{n} C_k^{(n)} + 1 \right\} \frac{1}{(n+1)!2^{n+1}\beta^n}$$

$$= \frac{1}{(n+1)!2^{n+1}\beta^n}.$$

It is again easily checked that the radius of convergence of the sequence $a = \{a_k\}_{k \in \mathbf{N}}$ is $\rho(a) = \infty$. Thus the power series function

$$\hat{a}(z) = \beta \sum_{k=0}^{\infty} \frac{1}{k!} \left(\frac{z}{2\beta}\right)^k = \beta \exp\left\{\frac{z}{2\beta}\right\}, \quad z \in \mathbf{C},$$

is an analytic solution of (5.55).

Theorem 5.10. *Analytic solutions of (5.55) on $\mathbf{C}$ exist and are uniquely determined by the values of their zeroth, first, second and/or third derivatives at the origin. More specifically, (1) if $f(0) = 0$ and $f'(0) = 0$, then the trivial function $f(x) = 0$ is the only analytic solution of (5.55), (2) if $f(0) = 0$, $f'(0) = 1$ and $f'''(0) = 0$, then the identity function $f(z) = z$ is the only analytic solution of (5.55), (3) if $f(0) = 0$, $f'(0) = 1$ and $f'''(0) = \alpha > 0$, then $f(z) = \alpha^{-1/2} \sinh \alpha^{1/2} z$ is the only analytic solution of (5.55), (4) if $f(0) = 0$, $f'(0) = 1$ and $f'''(0) = \gamma < 0$, then the function $f(x) = (-\gamma)^{-1/2} \sin(-\gamma)^{1/2} z$ is the only analytic solution of (5.55), and (5) if the additional condition $f(0) = \beta \neq 0$ is imposed, then $f(z) = \beta \exp\{z/(2\beta)\}$ is the only analytic solution of (5.55).*

5.6.4 Equation IV

Let us now consider functional differential equations of the form

$$y'(z) = \sum_{i=1}^{p} a_i y'(\lambda_i z) + \sum_{j=1}^{q} b_j y(\mu_j z) + c y(z). \tag{5.57}$$

Theorem 5.11. *Suppose $0 < \lambda_1, ..., \lambda_p, \mu_1, ..., \mu_q < 1$. Suppose $a_1, ..., a_p, b_1, ..., b_q, c$ are nonnegative numbers and $a_1 + \cdots + a_p < 1$. Then (5.57) has a solution $y = y(z)$ which is analytic on $\mathbf{C}$ and satisfies $y(0) = 1$.*

Proof. We first assume that
$$y(z) = \sum_{n=0}^{\infty} d_n z^n, \tag{5.58}$$
where $d_0 = 1$, is a formal power series solution of (5.57). Substituting it into (5.57), we obtain
$$\left(1 - \sum_{i=1}^{p} a_i\right) d_1 = \sum_{j=1}^{q} b_j + c,$$
and
$$(n+1) d_{n+1} = (n+1) \left(\sum_{i=1}^{p} a_i \lambda_i^n\right) d_{n+1} + \left(\sum_{j=1}^{q} b_j u_j^n\right) d_n + c d_n$$
for $n \geq 1$. Thus
$$d_1 = \frac{\sum_{j=1}^{q} b_j + c}{1 - \sum_{i=1}^{p} a_i}, \tag{5.59}$$
and
$$d_{n+1} = \frac{\sum_{j=1}^{q} b_j u_j^n + c}{(n+1)\left(1 - \sum_{i=1}^{p} a_i \lambda_i^n\right)} d_n \tag{5.60}$$
for $n \geq 1$. Since $a_1 + \cdots + a_p < 1$, we see that
$$\sum_{i=1}^{p} a_i \lambda_i^n < 1, \ n \in \mathbf{Z}^+.$$
Thus $\{d_n\}_{n=1}^{\infty}$ can be uniquely determined by (5.59) and (5.60), and $d_n \geq 0$ for $n \geq 0$. Furthermore, from our assumptions on λ_i and μ_j,
$$\lim_{n \to \infty} \frac{d_{n+1}}{d_n} = \lim_{n \to \infty} \frac{\sum_{j=1}^{q} b_j u_j^n + c}{(n+1)\left(1 - \sum_{i=1}^{p} a_i \lambda_i^n\right)} = 0.$$
This shows that the series (5.58) converges for all $z \in C$. The proof is complete.

5.6.5 Equation V

Consider the equation [184]
$$x'(z) = \sum_{i=1}^{l} A_i(z) x(F_i(z)) + H(z), \tag{5.61}$$
a special case of which is
$$x'(z) = A(z) x(F(z)) + H(z). \tag{5.62}$$

Theorem 5.12. *Suppose $A(z) = \widehat{a}(z)$, $F(z) = \widehat{f}(z)$ and $H(z) = \widehat{h}(z)$ are analytic over a neighborhood of the origin. Suppose further that $F(0) = f_0 = 0$ and $|F'(0)| = |s| \in (0,1)$. Then (5.62) has a solution of the form $x(z) = \widehat{b}(z)$, where $b_0 = 0$ and $b_1 = h_0$, which is analytic over a neighborhood of the origin.*

Proof. Let $x(z) = \widehat{b}(z)$ be a formal power series solution of (5.62). Then substituting it into (5.62), we see that

$$Db = a * (b \circ f) + h. \tag{5.63}$$

Hence

$$(n+1)b_{n+1} = \sum_{k=0}^{n} a_{n-k} (b \circ f)_k + h_k = \sum_{k=0}^{n} a_{n-k} \left(\sum_{j=0}^{k} b_j f_k^{\langle j \rangle} \right) + h_k, \ k \in \mathbf{N}.$$

If we let $b_0 = 0$, then $b_1 = h_0$, and $\{b_n\}_{n=2}^{\infty}$ can be uniquely determined. To show that the formal solution is also analytic, we consider

$$y(z) = z\overline{A}(z)y\left(\overline{F}(z)\right) + z\overline{H}(z), \tag{5.64}$$

where

$$\overline{A}(z) = \widehat{|a|}(z), \ \overline{F}(z) = \widehat{|f|}(z), \ \overline{H}(z) = \widehat{|h|}(z).$$

In view of Theorem 4.6, (5.64) has a solution $y(z) = \widehat{u}(z)$, where $u_0 = 0$, which is analytic over a neighborhood of the origin. Furthermore, by substituting $y(z) = \widehat{u}(z)$ into (5.64), we see that

$$u = \hbar * |a| * (u \circ |f|) + \hbar * |h|.$$

Now it suffices to show that b is majorized by y. Indeed, $b_0 = 0 \le u_0$. Furthermore, from (5.63),

$$\hbar * Db = \{0, b_1, 2b_2, 3b_3, ...\} = \hbar * a * (b \circ f) + \hbar * h.$$

Hence,

$$b_1 = (\hbar * a * (b \circ f) + \hbar * h)_1 \le (\hbar * |a| * (u \circ |f|) + \hbar * |h|)_1 = u_1,$$

and by induction, $b_n \le u_n$ for all $n \ge 2$. The proof is complete.

By similar methods, we may also deduce analytic solutions for the more general equation (5.61).

Theorem 5.13. *Suppose $A_1(z), ..., A_l(z), f_1(z), ..., f_l(z)$ and $h(z)$ are analytic for $|x - \xi| < R$, $f_1(\xi) = f_2(\xi) = \cdots = f_l(\xi) = \xi$ and $f_i'(\xi) = s_i$ for $i = 1, ..., l$, where $0 < |s_i| < 1$. Then (5.61) has a solution of the form*

$$x(z) = \eta(z - \xi) + \sum_{n=2}^{\infty} b_n (z - \xi)^n, \ \eta = h(\xi), \tag{5.65}$$

which is analytic on a neighborhood of the point ξ.

5.6.6 Equation VI

Consider the equation [178]

$$G(z)F'(z) = \sum_{j=1}^{m} p_j G(F(q_j z)), \qquad (5.66)$$

where $p_1, ..., p_m, q_1, ..., q_m$ are complex numbers. Note that (5.66) is an extension of (5.49).

Theorem 5.14. *Suppose $\sum_{j=1}^{m} |p_j| \leq 1$ and $|q_1|, ..., |q_m| \leq 1$. Suppose further that $G(z) = \hat{a}(z)$, where $a_0 = 0$ and $a_1 \neq 0$, is analytic on a neighborhood of the origin. Then for any complex number η, equation (5.66) has a solution $F(z)$ which is analytic on a neighborhood of the origin and satisfies $F(0) = 0$ and $F'(0) = \eta + \sum_{j=1}^{m} p_j$.*

Proof. In view of the Cauchy Estimation (Theorem 3.26), there exists a positive β such that

$$|a_n| \leq \beta^{n-1}, \; n \geq 2. \qquad (5.67)$$

Introducing new functions $f(z) = \beta F(\beta^{-1} z)$ and $g(z) = \beta G(\beta^{-1} z)$, we may transform (5.66) into an equation of the form

$$g(z) f'(z) = \sum_{j=1}^{m} p_j g(f(q_j z))$$

where $g(z)$ is of the form

$$g(z) = \sum_{n=1}^{\infty} g_n z^n$$

with

$$g_n = \frac{a_n}{\beta^{n-1}}, \; n \geq 2.$$

Since $|g_n| \leq 1$ for $n \geq 2$, we may assume without loss of generality that the original sequence $\{a_n\}_{n \in \mathbb{N}}$ satisfies $|a_n| \leq 1$ for $n \geq 2$.

Let

$$F(z) = \sum_{n=0}^{\infty} b_n z^n, \qquad (5.68)$$

where $b_0 = 0$, be a solution of equation (5.66) which is analytic at 0. Inserting $G(z) = \hat{a}(z)$ and (5.68) into (5.66) and comparing coefficients we obtain

$$a_0 \left(b_1 - \sum_{j=1}^{m} p_j \right) = 0,$$

and
$$\left(n - \sum_{j=1}^{m} p_j q_j^n\right) a_1 b_n = \sum_{j=1}^{m} p_j q_j^n \sum_{t=2}^{n} a_t b_n^{\langle t \rangle} - \sum_{k=0}^{n-2} (k+1) a_{n-k} b_{k+1}, \qquad (5.69)$$
for $n \geq 2$. Since $a_0 = G(0) = 0$, we may choose $b_1 - \sum_{j=1}^{m} p_j = \eta$. Since $a_1 = G'(0) \neq 0$ and
$$\left|\sum_{j=1}^{m} p_j q_j^n\right| \leq \sum_{j=1}^{m} |p_j| |q_j|^n \leq \sum_{j=1}^{m} |p_j| \leq 1,$$
we see that the sequence $\{b_n\}_{n=2}^{\infty}$ is successively determined by the relation (5.69) in a unique manner. To complete our proof, it suffices now to show that the sequence $\{b_n\}_{n \in N}$ has a positive radius of convergence. To this end, first note that
$$\left|\frac{k+1}{n - \sum_{j=1}^{m} p_j q_j^n}\right| \leq 1 \qquad (5.70)$$
for $0 \leq k \leq n-2$ and $n \geq 2$. Furthermore, since
$$\lim_{n \to \infty} \frac{\sum_{j=1}^{m} p_j q_j^n}{n - \sum_{j=1}^{m} p_j q_j^n} = 0,$$
there exists a positive number M, such that
$$\left|\frac{\sum_{j=1}^{m} p_j q_j^n}{n - \sum_{j=1}^{m} p_j q_j^n}\right| \leq M, \; n \geq 2. \qquad (5.71)$$
Next note that Example 4.9 asserts that the equation
$$W(z) - \left(|\eta| + \sum_{j=1}^{m} |p_j|\right) z - \frac{M}{|a_1|} \frac{[W(z)]^2}{1 - W(z)} - \frac{1}{|a_1|} \frac{z}{1-z} W(z) = 0$$
has a solution
$$W(z) = \sum_{n=0}^{\infty} B_n z^n$$
which is analytic on a neighborhood of the origin and satisfies $B_0 = 0$, $B_1 = |\eta| + \sum_{j=1}^{m} |p_j|$ and
$$B_n = \frac{M}{|a_1|} \sum_{t=1}^{n} B_n^{\langle t \rangle} + \frac{1}{|a_1|} \sum_{k=0}^{n-2} B_{k+1}$$
for $n \geq 2$. In view of (5.69) and the inequalities (5.70) and (5.71), it is clear that
$$|b_n| \leq B_n, \; n \in \mathbf{Z}^+. \qquad (5.72)$$
This implies that $\{b_n\}_{n \in N}$ has a positive radius of convergence. The proof is complete.

Theorem 5.15. Suppose $\sum_{j=1}^{m} |p_j| \leq 1$ and $|q_1|, ..., |q_m| \leq 1$. Suppose further that $G(z) = \widehat{a}(z)$ is analytic on a neighborhood of zero and $G(0) = a_0 \neq 0$, Then equation (5.66) has a solution $F(z)$ which is analytic on a neighborhood of the origin and satisfies $F(0) = 0$ and $F'(0) = \sum_{j=1}^{m} p_j$.

Proof. As in the previous proof, we seek a power series solution of the form (5.68). Since $a_0 = G(0) \neq 0$, by defining $b_1 = \sum_{j=1}^{m} p_j$ and then substituting $G(z)$ and (5.68) into (5.66), we see that the sequence $\{b_n\}_{n=2}^{\infty}$ is successively determined by the condition

$$(n+1)a_0 b_{n+1} + \left(n - \sum_{j=1}^{m} p_j q_j^n\right) a_1 b_n = \sum_{j=1}^{m} p_j q_j^n \sum_{t=2}^{n} a_t b_n^{\langle t \rangle} - \sum_{k=0}^{n-2}(k+1)a_{n-k}b_{k+1},$$
(5.73)

for $n \geq 2$, in a unique manner. Furthermore, it is easy to see from (5.70) and (5.71) that

$$|b_{n+1}| \leq \left|\frac{n - \sum_{j=1}^{m} p_j q_j^n}{n+1}\right| \left|\frac{a_1}{a_0}\right| |b_n|$$

$$+ \frac{\left|\sum_{j=1}^{m} p_j q_j^n\right|}{(n+1)|a_0|} \sum_{t=2}^{n} |a_t| |b|_n^{\langle t \rangle} + \frac{1}{|a_0|} \sum_{k=0}^{n-2} \frac{k+1}{n+1} |a_{n-k}| |b_{k+1}|$$

$$\leq \left|\frac{a_1}{a_0}\right| |b_n| + \frac{1}{|a_0|} \sum_{t=2}^{n} |b|_n^{\langle t \rangle} + \frac{1}{|a_0|} \sum_{k=0}^{n-2} |b_{k+1}|. \quad (5.74)$$

Note that Example 4.10 asserts that the equation

$$\Phi(z) = \left(\sum_{j=1}^{m} |p_j|\right) z + \mu z^2 + \left|\frac{a_1}{a_0}\right| z \left(\Phi(z) - \left(\sum_{j=1}^{m} |p_j|\right) z\right)$$

$$+ \left|\frac{1}{a_0}\right| \frac{z[\Phi(z)]^2}{1 - \Phi(z)} + \left|\frac{1}{a_0}\right| \frac{z^2}{1-z} \Phi(z).$$

has a solution

$$\Phi(z) = \sum_{n=0}^{\infty} B_n z^n$$

which is analytic on a neighborhood of the origin and satisfies $B_0 = 0$, $B_1 = \sum_{j=1}^{m}|p_j|$, $B_2 = \sum_{j=1}^{m}|p_j||a_1/a_0| + \mu$ and

$$B_{n+1} = \left|\frac{a_1}{a_0}\right| B_n + \frac{1}{|a_0|}\sum_{t=2}^{n} B_n^{\langle t \rangle} + \frac{1}{|a_0|}\sum_{k=0}^{n-2} B_{k+1} \quad (5.75)$$

for $n \geq 2$. By choosing μ so that $|b_2| \leq B_2$, it is then easily seen from (5.74) and (5.75) that $|b_n| \leq B_n$ for $n \geq 1$. Thus the sequence $\{b_n\}_{n \in \mathbb{N}}$ has a positive radius of convergence. The proof is complete.

5.7 Functional Equations with Higher Order Derivatives

In this section, we consider several functional differential equations involving higher order derivatives of the unknown functions.

5.7.1 Equation I

Consider first the equation

$$x''(z) = G(x(qz)) + H(z), \quad q \in \mathbf{C}. \tag{5.76}$$

Theorem 5.16. *Suppose $G(z) = \widehat{g}(z)$ and $H(z) = \widehat{h}(z)$ are analytic on a neighborhood of the origin. If $|q| < 1$, then for any complex number η, equation (5.76) has a solution $x(z)$ which is analytic on a neighborhood of the origin and satisfies $x(0) = 0$ and $x'(0) = \eta$.*

Proof. By Cauchy's Estimation (Theorem 3.26), there is $p > 0$ such that

$$|g_n| \le p^n, \quad n \ge 1.$$

Let

$$x(z) = \sum_{n=0}^{\infty} b_n z^n \tag{5.77}$$

be a formal power series solution of (5.76) generated by the sequence $b = \{b_n\}_{n \in \mathbf{N}}$ that satisfies $b_0 = 0$. Then substituting it into (5.76), we obtain

$$D^2 b = g \circ (b \cdot q) + h,$$

so that

$$2b_2 = g_0 + h_0,$$

and

$$(n+1)(n+2) b_{n+2} = q^n \sum_{m=1}^{n} g_m b_n^{\langle m \rangle} + h_n, \quad n \in \mathbf{Z}^+. \tag{5.78}$$

By imposing the condition $b_1 = \eta$, where η is an arbitrary number, we see that $\{b_n\}_{n=1}^{\infty}$ is then uniquely determined. Note that Example 4.12 asserts that the equation

$$u - |\eta| z - \frac{1}{2}(|g_0| + |h_0|) z^2 - z^2 \left[\frac{p|q|u}{1 - p|q|u} + \tilde{H}(z) \right] = 0,$$

where $\tilde{H}(z) = \sum_{n=0}^{\infty} |h_n| z^n$, has a solution $u(z)$ which is analytic on a neighborhood of the origin and

$$u(z) = \sum_{n=0}^{\infty} u_n z^n$$

where the sequence $u = \{u_n\}_{n \in \mathbf{N}}$ is defined by $u_0 = 0$, $u_1 = |\eta|$, $u_2 = (|g_0| + |h_0|)/2$ and

$$u_{n+2} = \sum_{m=1}^{n} p^m |q|^m u_n^{\langle m \rangle} + |h_n|, \quad n \in \mathbf{N}.$$

We assert that $|b_n| \leq u_n$ for $n \in \mathbf{N}$. Indeed, $|b_0| = 0 = u_0$, $|b_1| = |\eta| = u_1$, $|b_2| = |g_0 + h_0|/2 \leq (|g_0| + |h_0|)/2 = u_2$. Assume by induction that $|b_i| \leq u_i$ for $i = 0, 1, ..., n+1$ where $n \geq 1$. Then

$$|b_{n+2}| \leq \left|\frac{q^n}{(n+1)(n+2)}\right| \sum_{m=1}^{n} |g_m| |b|_n^{\langle m \rangle} + |h_n|$$

$$\leq \sum_{m=1}^{n} p^m |q|^m u_n^{\langle m \rangle} + |h_n|$$

$$= u_{n+2}$$

as required. Now that we have proved that b is majorized by u, we see that

$$x(z) = \eta z + \sum_{n=2}^{\infty} b_n z^n$$

is convergent for each z near 0. The proof is complete.

5.7.2 Equation II

Consider the equation [184]

$$z^n x^{(n)}(z) = \sum_{i=1}^{l} \beta_i x(q_i z) + G(x(qz)) + H(z), \quad q, q_1, ..., q_l, \beta_1, ..., \beta_l \in \mathbf{C}. \quad (5.79)$$

Theorem 5.17. *Suppose (i) $G(z) = \hat{g}(z)$, where $g_0 = 0$ and $g_1 = \alpha$, is analytic on a neighborhood of the origin, (ii) $H(z) = \hat{h}(z)$, where $h_0 = h_1 = 0$, is analytic for $|z| < R$, and (iii) $|q| \leq 1$, $\left|\sum_{i=1}^{l} \beta_i q_i^m + \alpha q^m\right| \geq 1$ for $m = 1, ..., n-1$, as well as*

$$\left|\frac{m!}{(m-n)!} - \sum_{i=1}^{l} \beta_i q_i^m - \alpha q^m\right| \geq 1$$

for $m \geq n$. Then the equation (5.79) has an analytic solution of the form

$$x(z) = \sum_{m=1}^{\infty} b_m z^m, \quad b_i = \eta_i, \quad i = 1, ..., n-1, \quad (5.80)$$

in a neighborhood of the origin, where $\eta_1, ..., \eta_{n-1}$ satisfy

$$\sum_{i=1}^{l} \beta_i q_i^m \eta_m + q^m \sum_{t=1}^{m} \sum_{l_1 + \cdots + l_t = m; l_1, l_2, ..., l_t \in \mathbf{Z}^+} g_t \eta_{l_1} \cdots \eta_{l_t} + h_m = 0 \quad (5.81)$$

for $m = 1, 2, ..., n-1$.

Proof. In view of the condition on $G(z)$, by Cauchy's Estimation (Theorem 3.26), there is $p > 0$ such that

$$|g_m| \leq p^{m-1}, \quad m = 2, 3, ... \,. \quad (5.82)$$

It is easily checked that (5.79) is invariant with respect to the transformations
$$y(z) = px(p^{-1}z), \ \psi(z) = pG(p^{-1}z),$$
and
$$\psi(z) = \alpha z + \sum_{m=2}^{\infty} \frac{g_m}{p^{m-1}} z^m$$
with
$$\left|\frac{g_m}{p^{m-1}}\right| \leq 1, \ m \geq 2.$$
Consequently we may assume that
$$|g_m| \leq 1, \ m \geq 2. \tag{5.83}$$
Next, in view of (ii) and Cauchy's Estimation (Theorem 3.26), for any $r \in (0, R)$, there is $M > 0$ such that,
$$|h_m| \leq \frac{M}{r^m}. \tag{5.84}$$
Assume that $x(z) = \widehat{b}(z)$, where $b_0 = 0$, is a formal power series solution of (5.79), then substituting it into (5.79), we see that
$$\hbar * D^n b = \sum_{i=1}^{l} \beta_i b \cdot q_i + g \circ (b \cdot q) + h,$$
so that
$$\left(\sum_{i=1}^{l} \beta_i q_i^m - \alpha q^m\right) b_m + q^m \sum_{t=1}^{m} g_t b_m^{\langle t \rangle} - h_m, \ 1 \leq m \leq n-1, \tag{5.85}$$
and
$$\left(\frac{m!}{(m-n)!} - \sum_{i=1}^{l} \beta_i q_i^m - \alpha q^m\right) b_m = q^m \sum_{t=1}^{m} g_t b_m^{\langle t \rangle} + h_m, \ m \geq n. \tag{5.86}$$

Set $b_i = \eta_i$ for $i = 1, ..., n-1$, where $\eta_1, ..., \eta_{n-1}$ satisfy (5.81). Then (5.85) is satisfied. Furthermore, in view of (5.86), $\{b_m\}_{m=0}^{\infty}$ can then be uniquely determined.

Now we need to show that $x(z)$ is analytic in a neighborhood of the origin. To this end, consider the implicit relation
$$y = |\eta_1| z + \frac{y^2}{1-y} + \frac{Mz^2}{r^2 - rz}, \tag{5.87}$$
which, in view of Example 4.13 has the analytic solution $y = y(z)$ near 0. If we write $y(z) = \sum_{m=1}^{\infty} u_m z^m$, then substituting it into (5.87), we see that $u_1 = |\eta_1|$ and
$$u_m = \sum_{t=1}^{m} u_m^{\langle t \rangle} + \frac{M}{r^m} \tag{5.88}$$

for $m \geq 1$. In view of (iii), (5.83), (5.85) and (5.86),

$$|b_m| \leq \sum_{t=1}^{m} |b|_m^{\langle t \rangle} + |h_m| \qquad (5.89)$$

for $m \geq 1$. We assert that

$$|b_m| \leq u_m, \; m \geq 1. \qquad (5.90)$$

Indeed, $|b_1| \leq u_1$. Assume by induction that $|b_k| \leq u_k$ for $k = 1, ..., m-1$, then in view of (5.89) and (5.84),

$$|b_m| \leq \sum_{t=1}^{m} |b|_m^{\langle t \rangle} + |h_m| \leq \sum_{t=1}^{m} u_m^{\langle t \rangle} + \frac{M}{r^m} = u_m$$

as desired.

Finally, since $x(z)$ is majorized by $y(z)$, $x(z)$ is also analytic for $|z| < \delta$. The proof is complete.

5.7.3 Equation III

Consider the equation [174]

$$f^{(k)}(z) + \sum_{i=1}^{k} \varphi_i(z) f^{(k-i)}(z) + \sum_{j=1}^{n} \alpha_j(z) f'(p_j z) + \sum_{l=1}^{s} \beta_l(z) f(q_l z) = g(z) \quad (5.91)$$

where $k \geq 1$, $|p_j| \leq 1$ and $|q_l| \leq 1$ for $j = 1, 2, ..., n$ and $l = 1, 2, ..., s$.

In case $k = 1$, we have the following existence result.

Theorem 5.18. *Suppose* $k = 1$. *Suppose further that (i)* $\varphi_1(z)$, $\alpha_1(z), ..., \alpha_n(z)$, $\beta_1(z), ..., \beta_s(z)$ *are analytic on* $B(0; \rho)$, *and (ii)* $|p_1|, ..., |p_n| < 1$ *and* $|q_1|, ..., |q_s| \leq 1$. *Then (5.91) has a solution* $y = y(z)$ *which is analytic on* $B(0; \rho)$.

Proof. We may assume that

$$\varphi_1(z) = \sum_{m=0}^{\infty} \varphi_{1m} z^m,$$

$$\alpha_j(z) = \sum_{m=0}^{\infty} \alpha_{jm} z^m, \; j = 1, 2, ..., n, \qquad (5.92)$$

$$\beta_l(z) = \sum_{m=0}^{\infty} \beta_{lm} z^m, \; l = 1, 2, ..., s, \qquad (5.93)$$

and

$$g(z) = \sum_{m=0}^{\infty} g_m z^m. \qquad (5.94)$$

By Cauchy's Estimation (Theorem 3.26), there is some $M > 0$ such that for any $r \in (0, \rho)$,

$$|\varphi_{1m}| \leq \frac{M}{r^m}, \quad |\alpha_{jm}| \leq \frac{M}{r^m}, \quad |\beta_{lm}| \leq \frac{M}{r^m}, \quad |g_m| \leq \frac{M}{r^m}, \quad (5.95)$$

where $m \geq 0$, $1 \leq j \leq n$ and $1 \leq l \leq s$. Let

$$f(z) = \sum_{m=0}^{\infty} f_m z^m \quad (5.96)$$

be a formal power series solution of (5.91), then

$$g_m = \left(1 + \sum_{j=1}^{n} \alpha_{j0} p_j^m\right)(m+1) f_{1+m} + \sum_{u=0}^{m} \varphi_{1u} f_{m-u}$$
$$+ \sum_{j=1}^{n} \sum_{u=0}^{m} \alpha_{ju}(m+1-u) f_{m+1-u} p_j^{m-u} + \sum_{l=1}^{s} \sum_{u=0}^{m} \beta_{lu} q_l^{m-u} f_{m-u} \quad (5.97)$$

for $m \geq 0$. Since $|p_j| < 1$, there exists T such that for $m > T$, we have

$$\sum_{j=1}^{n} |\alpha_{j0}| |p_j|^{m-1} < 1.$$

Let $\{A_m\}$ be defined by $A_m = B_m$ for $0 \leq m \leq T$ and

$$A_m = \frac{1}{m\left(1 - \sum_{j=1}^{n} |\alpha_{j0}| |p_j|^{m-1}\right)} \times \left(\sum_{u=0}^{m-1} \frac{M}{r^u} A_{m-1-u}\right.$$
$$+ \sum_{j=1}^{n} \sum_{u=0}^{m-1} \frac{M}{r^u}(m-u) A_{m-u} |p_j|^{m-1-u}$$
$$\left. + \sum_{l=1}^{s} \sum_{u=0}^{m-1} \frac{M}{r^u} |q_l|^{m-1-u} A_{m-1-u} + \frac{M}{r^{m-1}}\right) \quad (5.98)$$

for $m > T$, where $B_m > 0$ satisfies $|f_m| \leq B_m$ for $0 \leq m \leq T$. In view of (5.98), $A_m > 0$ for $m \geq 0$. Furthermore, in view of (5.95), (5.97) and (5.98), we may show by induction that

$$|f_m| \leq A_m. \quad (5.99)$$

Thus, when $m > T$,

$$(m+1)\left(1 - \sum_{j=1}^{n} |\alpha_{j0}||p_j|^m\right) A_{m+1}$$

$$= \sum_{u=0}^{m} \frac{M}{r^u} A_{m-u} + \sum_{j=1}^{n}\sum_{u=1}^{m} \frac{M}{r^u} (m+1-u) A_{m+1-u} |p_j|^{m-u}$$

$$+ \sum_{l=1}^{s}\sum_{u=0}^{m} \frac{M}{r^u} |q_l|^{m-u} A_{m-u} + \frac{M}{r^m}$$

$$= MA_m + \frac{M}{r} m A_m \sum_{j=1}^{n} |p_j|^{m-1} + MA_m \sum_{l=1}^{s} |q_l|^m$$

$$+ \frac{1}{r}\left(\sum_{u=1}^{m-1} \frac{M}{r^u} A_{m-1-u} + \sum_{j=1}^{n}\sum_{u=1}^{m-1} \frac{M}{r^u} (m-u) A_{m-u} |p_j|^{m-1-u}\right.$$

$$\left. + \sum_{l=1}^{s}\sum_{u=0}^{m-1} \frac{M}{r^u} |q_l|^{m-1-u} A_{m-1-u} + \frac{M}{r^{m-1}}\right)$$

$$= MA_m + \frac{M}{r} m A_m \sum_{j=1}^{n} |p_j|^{m-1} + MA_m \sum_{l=1}^{s} |q_l|^m$$

$$+ \frac{1}{r} m \left(1 - \sum_{j=1}^{n} |\alpha_{j0}||p_j|^{m-1}\right) A_m,$$

so that

$$\frac{A_{m+1}}{A_m} = \frac{M + \frac{M}{r} m \sum_{j=1}^{n} |p_j|^{m-1} + M \sum_{l=1}^{s} |q_l|^m + \frac{m}{r}\left(1 - \sum_{j=1}^{n} |\alpha_{j0}||p_j|^{m-1}\right)}{(m+1)\left(1 - \sum_{j=1}^{n} |\alpha_{j0}||p_j|^m\right)}.$$

By taking limits on both sides, we may easily see that

$$\lim_{m \to \infty} \frac{A_{m+1}}{A_m} = \frac{1}{r}.$$

Since the radius of convergence of $\sum_{m=0}^{\infty} A_m z^m$ is r and since r is an arbitrary number in $(0, \rho)$, we see that $\sum_{m=0}^{\infty} A_m z^m$ is analytic on $B(0; \rho)$. Finally, since $|f_m| \le A_m$ for $m \ge 0$, we see that $\sum_{m=0}^{\infty} f_m z^m$ is analytic on $B(0; \rho)$. The proof is complete.

In case $k \ge 2$, we have the following result.

Theorem 5.19. *Suppose $k \ge 2$. Suppose further that (i) $\varphi_1(z), ..., \varphi_k(z)$, $\alpha_1(z), ..., \alpha_n(z)$ and $\beta_1(z), ..., \beta_s(z)$ are analytic on $B(0; \rho)$, and (ii) $|p_1|, ..., |p_n|$, $|q_1|, ..., |q_s| \le 1$. Then (5.91) has a solution which is analytic on $B(0; \rho)$.*

Proof. We may assume (5.92), (5.93) and (5.94) as well as

$$\varphi_i(z) = \sum_{m=0}^{\infty} \varphi_{im} z^m, \quad i = 1, ..., k.$$

By Cauchy's Estimation (Theorem 3.26), there is some $M > 0$ such that for any $r \in (0, \rho)$,

$$|\varphi_{im}| \leq \frac{M}{r^m}, \quad |\alpha_{jm}| \leq \frac{M}{r^m}, \quad |\beta_{lm}| \leq \frac{M}{r^m}, \quad |g_m| \leq \frac{M}{r^m}, \quad (5.100)$$

where $m \geq 0, 1 \leq i \leq k, 1 \leq j \leq n$ and $1 \leq l \leq s$. Let

$$f(z) = \sum_{m=0}^{\infty} f_m z^m \quad (5.101)$$

be a formal power series solution of (5.91), then

$$f_k \cdot k! + \sum_{i=1}^{k} \varphi_{i0} f_{k-i}(k-i)! + \sum_{j=1}^{n} \alpha_{j0} f_1 + \sum_{l=1}^{s} \beta_{l0} f_0 = g_0, \quad (5.102)$$

and

$$g_m = f_{k+m} \frac{(k+m)!}{m!} + \sum_{i=1}^{k} \sum_{u=0}^{m} \varphi_{iu} f_{k-i+m-u} \frac{(k-i+m-u)!}{(m-u)!}$$

$$+ \sum_{i=1}^{k} \sum_{u=0}^{m} \alpha_{ju}(m+1-u) f_{m+1-u} p_j^{m-u+1} + \sum_{l=1}^{s} \sum_{u=0}^{m} \beta_{lu} q_l^{m-u} f \quad (5.103)$$

for $m \geq 1$. Let $\{A_m\}$ be defined by $A_m = B_m$ for $0 \leq m \leq k - 1$ and

$$A_m = \frac{(m-k)!}{m!} \times \left(\sum_{i=1}^{k} \sum_{u=0}^{m-k} \frac{M}{r^u} A_{m-i-u} \frac{(m-i-u)!}{(m-k-u)!} \right.$$

$$+ \sum_{j=1}^{n} \sum_{u=0}^{m-k} \frac{M}{r^u}(m-k+1-u) A_{m-k+1-u} |p_j|^{m-k-u+1}$$

$$\left. + \sum_{l=1}^{s} \sum_{u=0}^{m-k} \frac{M}{r^u} |q_l|^{m-k-u} A_{m-k-u} + \frac{M}{r^{m-k}} \right) \quad (5.104)$$

for $m \geq k$, where $B_m > 0$ satisfies $|f_m| \leq B_m$ for $0 \leq m \leq k-1$. In view of (5.104), we see that $A_m > 0$ for $m \geq 0$. Furthermore, in view of (5.100), (5.102), (5.103) and (5.104), we see that $|f_k| \leq A_k$. Then by induction, we may show that

$$|f_m| \leq A_m, \quad m \in \mathbf{N}.$$

Thus, for $m \geq k$, we have

$$A_{m+1} = \frac{(m+1-k)!}{(m+1)!} \Biggl(\sum_{i=1}^{k} \sum_{u=0}^{m+1-k} \frac{M}{r^u} A_{m+1-i-u} \frac{(m+1-i-u)!}{(m+1-k-u)!}$$

$$+ \sum_{j=1}^{n} \sum_{u=0}^{m+1-k} \frac{M}{r^u} (m+2-k-u) A_{m+2-k-u} |p_j|^{m+1-k-u}$$

$$+ \sum_{l=1}^{s} \sum_{u=0}^{m+1-k} \frac{M}{r^u} |q_l|^{m+1-k-u} A_{m+1-k-u} + \frac{M}{r^{m+1-k}} \Biggr)$$

$$= \frac{(m+1-k)!}{(m+1)!} \Biggl(\sum_{i=1}^{k} M A_{m+1-i} + \frac{1}{r} \sum_{i=1}^{k} \sum_{u=0}^{m-k} \frac{M}{r^u} A_{m-i-u} \frac{(m-i-u)!}{(m-k-u)!}$$

$$+ \sum_{j=1}^{n} M A_{m+2-k} (m+2-k) |p_j|^{m+1-k}$$

$$+ \frac{1}{r} \sum_{j=1}^{n} \sum_{u=0}^{m-k} \frac{M}{r^u} (m-k+1-u) A_{m-k+1-u} |p_j|^{m-k-u+1}$$

$$+ \sum_{l=1}^{s} M A_{m+1-k} |q_l|^{m+1-k} + \frac{1}{r} \sum_{l=1}^{s} \sum_{u=0}^{m-k} \frac{M}{r^u} |q_l|^{m-k-u} A_{m-k-u} + \frac{M}{r^{m+1-k}} \Biggr)$$

$$= \frac{(m+1-k)!}{(m+1)!} \Biggl(\sum_{i=1}^{k} M A_{m+1-i} + \sum_{j=1}^{n} M A_{m+2-k} (m+2-k) |p_j|^{m+2-k}$$

$$+ \sum_{l=1}^{s} M A_{m+1-k} |q_l|^{m+1-k} + \frac{1}{r} \frac{m!}{(m-k)!} A_m \Biggr),$$

so that

$$\frac{A_{m+1}}{A_m} = \frac{M}{(m+1)_{\langle k \rangle}} + \frac{M}{(m+1)_{\langle k \rangle}} \sum_{i=2}^{k} \frac{A_{m+1-i}}{A_m} + \frac{M \sum_{j=1}^{n} |p_j|^{m+2-k}}{(m+1)_{\langle k-1 \rangle}} \cdot \frac{A_{m+2-k}}{A_m}$$

$$+ \frac{M \sum_{l=1}^{s} |q_l|^{m+1-k}}{(m+1)_{\langle k \rangle}} \cdot \frac{A_{m+1-k}}{A_m} + \frac{m+1-k}{r(m+1)}. \qquad (5.105)$$

Let $C_m = \frac{A_{m+1}}{A_m}$. Then (5.105) becomes

$$C_m = \frac{M}{(m+1)_{\langle k \rangle}} + \frac{M}{(m+1)_{\langle k \rangle}} \sum_{i=2}^{k} \frac{1}{C_{m-1} C_{m-2} \cdots C_{m+2-i}}$$

$$+ \frac{M \sum_{j=1}^{n} |p_j|^{m+2-k}}{(m+1)_{\langle k-1 \rangle}} \cdot \frac{1}{C_{m-1} C_{m-2} \cdots C_{m+2-k}}$$

$$+ \frac{M \sum_{l=1}^{s} |q_l|^{m+1-k}}{(m+1)_{\langle k \rangle}} \cdot \frac{1}{C_{m-1} C_{m-2} \cdots C_{m+1-k}} + \frac{m+1-k}{r(m+1)}. \qquad (5.106)$$

In view of (5.106), $C_m \geq \frac{m+1-k}{r(m+1)}$, thus $\lim_{m \to \infty} C_m \geq \frac{1}{r} > 0$. As a consequence,

$$\limsup_{m \to \infty} \frac{1}{C_{m-1}C_{m-2}\cdots C_{m+1-i}} = \left\{ \liminf_{m \to \infty} C_{m-1}C_{m-2}\cdots C_{m+1-i} \right\}^{-1}$$
$$\leq \left\{ \liminf_{m \to \infty} C_{m-1} \right\}^{-1} \cdots \left\{ \liminf_{m \to \infty} C_{m+1-i} \right\}^{-1}$$
$$< +\infty$$

for $i = 1, 2, ..., k$. For similar reasons,

$$\limsup_{m \to \infty} \frac{1}{C_{m-1}C_{m-2}\cdots C_{m+2-i}} < +\infty.$$

Thus from (5.106),

$$\limsup_{m \to \infty} C_m$$
$$\leq \limsup_{m \to \infty} \frac{M}{(m+1)_{\langle k \rangle}} + \limsup_{m \to \infty} \frac{M}{(m+1)_{\langle k \rangle}} \sum_{i=2}^{k} \limsup_{m \to \infty} \frac{1}{C_{m-1}C_{m-2}\cdots C_{m+1-i}}$$
$$+ \limsup_{m \to \infty} \frac{M \sum_{j=1}^{n} |p_j|^{m+2-k}}{(m+1)_{\langle k-1 \rangle}} \cdot \limsup_{m \to \infty} \frac{1}{C_{m-1}C_{m-2}\cdots C_{m+2-k}}$$
$$+ \limsup_{m \to \infty} \frac{M \sum_{l=1}^{s} |q_l|^{m+1-k}}{(m+1)_{\langle k \rangle}} \cdot \limsup_{m \to \infty} \frac{1}{C_{m-1}C_{m-2}\cdots C_{m+1-k}}$$
$$+ \limsup_{m \to \infty} \frac{m+1-k}{r(m+1)}$$
$$= \frac{1}{r}.$$

This implies

$$\limsup_{m \to \infty} \frac{A_{m+1}}{A_m} \leq \frac{1}{r},$$

and hence

$$\limsup_{m \to \infty} \sqrt[m]{A_m} \leq \limsup_{m \to \infty} \frac{A_{m+1}}{A_m} \leq \frac{1}{r},$$

or

$$\frac{1}{\limsup_{m \to \infty} \sqrt[m]{A_m}} \geq \frac{1}{r}.$$

Since r is an arbitrary number in $(0, \rho)$, we now see that $\sum_{m=0}^{\infty} A_m z^m$ converges for $|z| < \rho$. Since $|f_m| \leq A_m$ for $m \geq 0$, we may now conclude that $f(z)$ is analytic on $B(0; \rho)$. The proof is complete.

Theorem 5.20. *Suppose $k \geq 1$ and $g(z) \equiv 0$. Suppose further that (i) $|p_1|$, ..., $|p_n|$, $|q_1|$, ..., $|q_s| \leq 1$, (ii) $z = 0$ is a 'regular singular point' of (5.91), so that*

$z^i \varphi_i(z)$, $z^{k-1}\alpha_j(z)$, $z^k \beta_l(z)$, where $1 \leq i \leq k, 1 \leq j \leq n$ and $1 \leq l \leq s$, are analytic on $B(0; \rho)$, and (iii) d is a root of the 'indicial' equation

$$\omega(z) \equiv z_{\langle k \rangle} + \sum_{i=1}^{k} \varphi_{i0} z_{\langle k-i \rangle} + \sum_{j=i}^{n} \alpha_{j0} z p_j^{z-1} + \sum_{l=1}^{s} \beta_{l0} q_l^z = 0$$

and $\omega(m+d) \neq 0$ for $m \in \mathbf{N}$. Then (5.91) has a solution of the form

$$f(z) = z^d \sum_{m=0}^{\infty} f_m z^m, \quad f_0 \neq 0, \tag{5.107}$$

which is analytic on $B(0; \rho)$.

Proof. In view of the conditions imposed in (i), we may assume that

$$z^i \varphi_i(z) = \sum_{m=0}^{\infty} \varphi_{im} z^m, \quad i = 1, ..., k,$$

$$z^{k-1} \alpha_j(z) = \sum_{m=0}^{\infty} \alpha_{jm} z^m, \quad j = 1, ..., n,$$

and

$$z^k \beta_l(z) = \sum_{m=0}^{\infty} \beta_{lm} z^m, \quad l = 1, ..., s,$$

are analytic near 0. By Cauchy's Estimation (Theorem 3.26), there is some $M > 0$ such that for any $r \in (0, \rho)$,

$$|\varphi_{im}| \leq \frac{M}{r^m}, \quad |\alpha_{jm}| \leq \frac{M}{r^m}, \quad |\beta_{lm}| \leq \frac{M}{r^m}, \tag{5.108}$$

where $m \geq 0$, $1 \leq i \leq k$, $1 \leq j \leq n$ and $1 \leq l \leq s$. Substituting the formal solution (5.107) into (5.91), we obtain

$$\omega(d) f_0 = 0 \tag{5.109}$$

and

$$w(m+d) f_m = -\sum_{i=1}^{k} \sum_{u=1}^{m} \varphi_{iu}(m-u+d)_{\langle k-i+2 \rangle} f_{m-u}$$

$$-\sum_{j=1}^{k} \sum_{u=1}^{m} \alpha_{iu}(m-u+d) p_j^{m-u+d-1} f_{m-u} - \sum_{l=1}^{s} \sum_{u=1}^{m} \beta_{lu} q_l^{m-u+d} f_{m-u},$$

for $m \geq 1$. Furthermore, in view of (iii), we may also see that $\{f_m\}_{m=0}^{\infty}$ can be uniquely determined by the above recurrence relations. Let the sequence $\{A_m\}_{m=0}^{\infty}$ be determined by $A_0 = |f_0|$ and A_m is equal to

$$\frac{1}{|w(m+d)|} \Bigg(\sum_{i=1}^{k} \sum_{u=1}^{m} \frac{M}{r^u} (m-u+d)_{\langle k-i+2 \rangle} A_{m-u}$$

$$+ \sum_{j=1}^{k} \sum_{u=1}^{m} \frac{M}{r^u} (m-u+d) |p_j|^{m-u+d-1} + \sum_{l=1}^{s} \sum_{u=1}^{m} \frac{M}{r^u} |q_l|^{m-u+d} A_{m-u} \Bigg) \tag{5.110}$$

for $m \geq 1$. Then

$$|w(m+1+d)|A_{m+1}$$
$$= \sum_{i=1}^{k}\sum_{u=1}^{m} \frac{M}{r^u}(m+1-u+d)_{\langle k-i+2\rangle} A_{m+1-u}$$
$$+ \sum_{j=1}^{k}\sum_{u=1}^{m} \frac{M}{r^u}(m-u+d)|p_j|^{m-u+d} A_{m+1-u} + \sum_{l=1}^{s}\sum_{u=1}^{m} \frac{M}{r^u}|q_l|^{m+1-u+d} A_{m+1-u}$$
$$= \frac{M}{r}\sum_{i=1}^{k}(m+d)_{\langle k-i\rangle} A_m + \frac{M}{r}\sum_{j=1}^{k}(m+d)|p_j|^{m+d-1} A_m$$
$$+ \frac{M}{r}\sum_{l=1}^{s}|q_l|^{m+d} A_m + \frac{1}{r}|w(m+d)|A_m.$$

Thus
$$\frac{A_{m+1}}{A_m} = \frac{1}{r|w(m+1+d)|}\Bigg(M\sum_{i=1}^{k}(m+d)_{\langle k-i\rangle}$$
$$+ M\sum_{j=1}^{k}(m+d)|p_j|^{m+d-1} + M\sum_{l=1}^{s}|q_l|^{m+d} + |w(m+d)|\Bigg),$$
so that
$$\lim_{m\to\infty} \frac{A_{m+1}}{A_m} = \frac{1}{r}.$$

In other words, the series $\sum_{m=0}^{\infty} A_m z^m$ converges for $|z| < r$. Since r is an arbitrary number in $(0,\rho)$, we see that $\sum_{m=0}^{\infty} A_m z^m$ converges for $|z| < \rho$. On the other hand, from (5.108)-(5.110), we may prove by induction that $|f_m| \leq A_m$ for $m \geq 0$. As a consequence, $\sum_{m=0}^{\infty} f_m z^m$ converges for $|z| < \rho$. This shows that the formal solution (5.107) is analytic on $B(0;\rho)$. The proof is complete.

We remark that the same techniques can be used to handle the following equation

$$f^{(k)}(z) + \sum_{i=1}^{k}\varphi_i f^{(k-i)}(z) + \sum_{n=1}^{M}\sum_{j=0}^{k}\psi_{nj}(z) f^{(k-j)}(p_{nj}z) = g(z) \tag{5.111}$$

under the initial condition

$$f^{(t)}(0) = \xi_t, \quad t = 0, 1, 2, .., k-1. \tag{5.112}$$

Theorem 5.21. *Suppose* $|p_{nj}| \leq u < 1$ *for* $j = 0, ..., K$ *and* $n = 1, 2, ..., T$ *and suppose* $\phi_1, ..., \phi_k, \psi_{10}, \psi_{11}, ..., \psi_{TK}$, *as well as* g *are analytic functions of the form*

$$\phi_i(z) = \sum_{m=0}^{\infty}\phi_{im}z^m, \quad i = 1, 2, ..., K,$$

$$\psi_{nj}(z) = \sum_{m=0}^{\infty}\psi_{njm}z^m, \quad j = 0, ..., K; n = 1, 2, ..., T,$$

$$g(z) = \sum_{m=0}^{\infty} g_m z^m$$

on $B(0;\gamma)$, and
$$1 + \sum_{n=1}^{T} \psi_{n0}(0) p_{n0}^m \neq 0, \ m \in \mathbf{N}.$$

Then the initial value problem (5.111)-(5.112) has an unique analytic solution on $B(0;\gamma)$.

We now consider several examples.

Example 5.8. Consider
$$f'(z) = af(\lambda z) + bf(z) + cf'(\beta z) + g(z) \tag{5.113}$$
where $g(z)$ is analytic for $|z| < R$, $|\beta| < 1$ and $|\lambda| \leq 1$. First in view of Theorem 5.18, (5.113) has an analytic solution for $|z| < R$. Let
$$g(z) = \sum_{m=0}^{\infty} g_m z^m$$
and
$$f(z) = \sum_{m=0}^{\infty} d_m z^m.$$

Substituting these into (5.113), we see that
$$(m+1)(1 - c\beta^m) d_{m+1} = (a\lambda^m + b) d_m + g_m, \ m = 0, 1, 2, \dots. \tag{5.114}$$

If $1 - c\beta^m \neq 0$ for $m \geq 0$, then
$$d_{m+1} = \frac{\prod_{i=0}^{m}(a\lambda^i + b)}{\prod_{i=0}^{m}(i+1)(1 - c\beta^i)} \eta + \sum_{k=0}^{m} \frac{\prod_{i=0}^{m}(a\lambda^i + b)}{\prod_{i=0}^{m}(i+1)(1 - c\beta^i)} g_k,$$

and
$$f(z) = \eta + \sum_{k=0}^{m} \left[\frac{\prod_{i=0}^{m}(a\lambda^i + b)}{\prod_{i=0}^{m}(i+1)(1 - c\beta^i)} \eta + \sum_{k=0}^{m} \frac{\prod_{i=0}^{m}(a\lambda^i + b)}{\prod_{i=0}^{m}(i+1)(1 - c\beta^i)} g_k \right] z^{m+1},$$

where $f(0) = \eta$. If $1 - c\beta^m = 0$ for some $m \in \mathbf{N}$, since $|\beta| < 1$, there exists some N such that $|c\beta^m| < 1$ for $m > N$. Thus there are $m_1, m_2, \dots, m_r \leq N$ such that $1 - c\beta^{m_j} = 0$ for $j = 1, \dots, r$. It is then not difficult to see that
$$f(z) = \sum_{j=1}^{r} c_j z^{m_j+1} + \sum_{j=1}^{r} \tilde{c}_j z^{m_j} + \sum_{m=0; m \neq m_1, \dots, m_r}^{N} \eta_m z^m$$
$$+ \sum_{m=N}^{\infty} \left[\frac{\prod_{i=0}^{m}(a\lambda^i + b)}{\prod_{i=0}^{m}(i+1)(1 - c\beta^i)} \eta + \sum_{k=0}^{m} \frac{\prod_{i=0}^{m}(a\lambda^i + b)}{\prod_{i=0}^{m}(i+1)(1 - c\beta^i)} g_k \right] z^{m+1},$$

where $c_1, ..., c_r$ are arbitrary and $\tilde{c}_1, ..., \tilde{c}_r$ satisfy

$$(a\lambda^{m_i} + b)\tilde{c}_i + g_{m_i} = 0, \quad i = 1, ..., r,$$

and η_m is determined by

$$(m+1)(1 - c\beta^m)\eta_{m+1} = (a\lambda^m + b)d_m + g_m.$$

Example 5.9. Consider the Bessel equation

$$z^2 y''(z) + zy'(vz) + \left(z^2 - v^{-\frac{1}{2}}\right) y(vz) = 0 \qquad (5.115)$$

where $v \in (0, 1]$. We may rewrite (5.115) in the form

$$y''(z) + \frac{1}{z} y'(vz) + \frac{z^2 - v^{-\frac{1}{2}}}{z^2} y(vz) = 0,$$

where $\alpha(z) = \frac{1}{z}$ and $\beta(z) = \frac{z^2 - v^{-\frac{1}{2}}}{z^2}$ have regular singular points at $z = 0$. Since $z = 1$ is a root of the indicial equation

$$z(z-1) + zv^{z-1} - v^{z-1} = 0,$$

in view of Theorem 5.20, (5.115) has an analytic solution of the form

$$y(z) = z \sum_{m=0}^{\infty} a_m z^m, \quad a_0 \neq 0. \qquad (5.116)$$

After substituting it into (5.115), we see that

$$a_m = \frac{v^{m-1}}{m(m+1+v^m)} a_{m-2}, \quad m \in \mathbf{Z}^+.$$

Thus $a_{2k+1} = 0$ for $k \geq 1$ and

$$a_{2k} = \frac{(-1)^k v^{2k-1} a_0}{\prod_{i=1}^{k} [2i(2i+1+v^{2i})]}, \quad k \in \mathbf{Z}^+.$$

This shows that

$$y(z) = a_0 + \sum_{k=1}^{\infty} \frac{(-1)^k v^{2k-1} a_0}{\prod_{i=1}^{k} [2i(2i+1+v^{2i})]} z^{2k},$$

where a_0 is an arbitrary number different from 0.

Example 5.10. Consider the equation

$$f^{(k)}(z) = af^{(k-1)}(\lambda z) + bf^{(k-1)}(z) + cf^{(k)}(\beta z) + g(z), \qquad (5.117)$$

under the condition

$$f^{(m)}(0) = \xi_t, \quad m = 0, 1, .., k-1, \qquad (5.118)$$

where $|\beta| < 1, |\lambda| \leq 1$ and $1 - c\beta^m \neq 0$ for $m \geq 0$. We will assume that
$$g(z) = \sum_{m=0}^{\infty} g_m z^m$$
is analytic for $|z| < \gamma$. By Theorem 5.21, our problem has a unique solution of the form
$$f(z) = \sum_{m=0}^{\infty} f_m z^m$$
which is analytic for $|z| < \gamma$. To find this solution, we substitute $f(z)$ into (5.117) and find the difference equation
$$\frac{(m+1)!}{m!}(1 - c\beta^m) f_{m+k} = (a\lambda^m + b)\frac{(m+k-1)!}{m!} f_{m+k-1} + g_m, \ m \in \mathbf{N}.$$
Since this equation is of the form
$$h_{m+1} = \rho_m h_m + \chi_m, \ m \in \mathbf{N},$$
we easily see that
$$f_{m+K} = \frac{\prod\limits_{i=0}^{m}(a\lambda^i + b)}{\prod\limits_{i=0}^{m}(i+k)(1 - c\beta^i)} \frac{\xi_{k-1}}{(k-1)!} + \sum_{n=0}^{m} \frac{n!}{(m+k)!} \frac{\prod\limits_{i=n+1}^{m}(a\lambda^i + b)}{\prod\limits_{i=K}^{m}(1 - c\beta^i)} g_n$$
for $m \in \mathbf{N}$.

5.7.4 Equation IV

Consider the equation [174]
$$f^{(k)}(z) + \sum_{i=1}^{k} \varphi_i(z) f^{(k-i)}(z) + \sum_{j=1}^{n} \alpha_j(z) f'(z - \tau_j) + \sum_{l=1}^{s} \beta_s(z) f(z - w_l) = g(z)$$
(5.119)

where $k \geq 1$ and $|p_1|, \ldots |p_n|, |q_1|, \ldots, |q_s| \leq 1$.

Theorem 5.22. *Suppose there is a positive number D such that for all negative b that satisfies $|b| > D$, the functions $\varphi_1(z), \ldots, \varphi_k(z), \alpha_1(z), \ldots, \alpha_n(z), \beta_1(z), \ldots, \beta_s(z)$ are analytic for $|z - b| < |b| - D$:*

$$\varphi_i(z) = \sum_{m=0}^{\infty} \varphi_{im}(z-b)^m, \ i = 1, 2, \ldots, k, \tag{5.120}$$

$$\alpha_j(z) = \sum_{m=0}^{\infty} \alpha_{jm}(z-b)^m, \ j = 0, \ldots, n, \tag{5.121}$$

$$\beta_l = \sum_{m=0}^{\infty} \beta_{lm}(z-b)^m, \ l = 0, \ldots, s, \tag{5.122}$$

and
$$g(z) = \sum_{m=0}^{\infty} g_m (z-b)^m. \quad (5.123)$$

Then the following equation
$$\varphi^{(k)}(z) = \sum_{i=1}^{k} \tilde{\varphi}_i \varphi^{(k-i)}(z) + \left(\sum_{j=1}^{n} \tilde{\alpha}_j(z)\right) \varphi'(z) + \left(\sum_{l=1}^{s} \tilde{\beta}_l(z)\right) \varphi(z) + \tilde{g}(z) \quad (5.124)$$

where
$$\tilde{\varphi}_i(z) = \sum_{m=0}^{\infty} |\varphi_{im}| (z-b)^m, \quad i = 1, 2, ..., k, \quad (5.125)$$

$$\tilde{\alpha}_j(z) = \sum_{m=0}^{\infty} |\alpha_{jm}| (z-b)^m, \quad j = 0, ..., n, \quad (5.126)$$

$$\tilde{\beta}_l = \sum_{m=0}^{\infty} |\beta_{lm}| (z-b)^m, \quad l = 0, ..., s, \quad (5.127)$$

and
$$\tilde{g}(z) = \sum_{m=0}^{\infty} |g_m| (z-b)^m, \quad (5.128)$$

has a solution $\varphi(z)$ which is analytic for $|z-b| < |b| - D$ and satisfies
$$0 \le \varphi^{(i)}(b) < +\infty, \quad i = 0, 1, ..., k-1. \quad (5.129)$$

Proof. Let
$$\varphi(z) = \sum_{m=0}^{\infty} b_m (z-b)^m \quad (5.130)$$

be a formal power series solution of (5.124). After substituting it into (5.124), we obtain
$$k! b_k = \sum_{i=1}^{k} |\varphi_{i0}| b_{k-i} (k-i)! + \sum_{j=1}^{n} |\alpha_{j0}| b_1 + \sum_{l=1}^{s} |\beta_{l0}| b_0 + |g_0|,$$

and
$$b_{k+m} \frac{(k+m)!}{m!} = \sum_{i=1}^{k} \sum_{u=0}^{m} |\varphi_{iu}| b_{k-i+m-u} \frac{(k-i+m-u)!}{(m-u)!}$$
$$+ \sum_{j=1}^{n} \sum_{u=0}^{m} |\alpha_{ju}| (m+1-u) b_{m+1-u}$$
$$+ \sum_{l=1}^{s} \sum_{u=0}^{m} |\beta_{lu}| b_{m-u} + |g_m|, \quad (5.131)$$

for $m \geq 1$. By Cauchy's Estimation (Theorem 3.26), there is some $M > 0$ such that for any $r \in (0, |b| - D)$,

$$|\varphi_{im}| \leq \frac{M}{r^m}, \quad |\alpha_{jm}| \leq \frac{M}{r^m}, \quad |\beta_{lm}| \leq \frac{M}{r^m}, \quad |g_m| \leq \frac{M}{r^m}, \tag{5.132}$$

where $m \geq 0, 1 \leq i \leq k, 1 \leq j \leq n, 1 \leq l \leq s$. Let $\{A_m\}_{m=0}^{\infty}$ be defined by $A_m = B_m$ for $0 \leq m \leq k-1$ and

$$A_m = \frac{(m-k)!}{m!} \times \left(\sum_{i=1}^{k} \sum_{u=0}^{m-k} \frac{M}{r^u} A_{m-i-u} \frac{(m-i-u)!}{(m-k-u)!} \right.$$

$$+ \sum_{j=1}^{n} \sum_{u=0}^{m-k} \frac{M}{r^u} (m+1-k-u) A_{m+1-k-u}$$

$$\left. + \sum_{l=1}^{s} \sum_{u=0}^{m-k} \frac{M}{r^u} A_{m-k-u} + \frac{M}{r^{m-k}} \right), \tag{5.133}$$

for $m \geq k$, where $B_m \geq 0$ satisfies $b_m \leq B_m$ for $0 \leq m \leq k-1$ (since $0 \leq \varphi^{(i)}(b) < +\infty$ for $i = 0, 1, ..., k-1$). In view of (5.131)-(5.133), we may show by induction that

$$0 \leq b_m \leq A_m, \ m \geq 0. \tag{5.134}$$

As in the proof of Theorem 5.21, we may show that $\sum_{m=0}^{\infty} A_m (z-b)^m$ converges for $|z - b| < |b| - D$. This then implies $\varphi(z)$ converges for $|z - b| < |b| - D$. The proof is complete.

We now turn to equation (5.119). In order to find an analytic solution, we need to consider "approximating equations". Let

$$b < 0, \ p_j = 1 + z_j b^{-1}, \ q_l = 1 + w_l b^{-1}, j = 1, ..., n; \ l = 1, ..., s.$$

Construct equation

$$f^{(k)}(z) + \sum_{i=1}^{k} \varphi_i(z) f^{(k-i)}(z) + \sum_{j=1}^{n} \alpha_j(z) f'(p_j z - \tau_j) + \sum_{l=1}^{s} \beta_s(z) f(q_l z - w_l) = g(z).$$

$$\tag{5.135}$$

When $b \to -\infty$, we have $p_j \to 1$ and $q_l \to 1$ for $j = 1, ..., n$ and $l = 1, ..., s$. Thus (5.135) 'tends' to (5.119) in a formal manner.

Theorem 5.23. *Suppose $k \geq 2$. Suppose further that there is a positive number D such that for all negative b that satisfies $|b| > D$, the functions $\varphi_1(z), ..., \varphi_k(z), \alpha_1(z), ..., \alpha_n(z), \beta_1(z), ..., \beta_s(z)$ are analytic for $|z - b| < |b| - D$. If the analytic solution found in Theorem 5.21 is bounded, i.e., for $b < 0$ with sufficiently large $|b|$, there is $K > 0$ such that*

$$\left| \sum_{m=0}^{\infty} b_m (z-b)^m \right| \leq K, \ |z - b| < |b| - D \tag{5.136}$$

then there is $Q < 0$ such that $|Q| > D$ and (5.119) has a solution $f(z)$ which is analytic for $\Re(z) < Q$ and

$$f(z) = \lim_{b \to -\infty} f(b, z), \tag{5.137}$$

where $f(b, z)$ is an analytic solution of (5.135), and the convergence is interpreted as uniform convergence on compact subsets of the region defined by $\Re(z) < Q$.

Proof. By Theorem 5.22, there is a solution of the form (5.130) for equation (5.124) which is analytic for $|z - b| < |b| - D$. We assert that (5.135) has an analytic solution $f(z)$ which satisfies

$$\left| f^{(i)}(b) \right| \leq \varphi^{(i)}(b), \quad i = 0, 1, ..., k - 1. \tag{5.138}$$

To this end, let

$$f(b, z) = \sum_{m=0}^{\infty} f_m (z - b)^m \tag{5.139}$$

be a formal power series solution of (5.135). Substituting it into (5.135), we obtain

$$g_0 = k! f_k = \sum_{i=1}^{k} \varphi_{i0} f_{k-i} (k - i)! + \sum_{j=1}^{n} \alpha_{j0} f_1 p_j + \sum_{l=1}^{s} \beta_{l0} f_0,$$

$$g_m = f_{k+m} \frac{(k+m)!}{m!} + \sum_{i=1}^{k} \sum_{u=0}^{m} \varphi_{iu} f_{k-i+m-u} \frac{(k-i+m-u)!}{(m-u)!}$$
$$+ \sum_{j=1}^{n} \sum_{u=0}^{m} \alpha_{ju} (m + 1 - u) f_{m+1-u} p_j^{m-u+1} + \sum_{l=1}^{s} \sum_{u=0}^{m} \beta_{lu} f_{m-u},$$

for $m \in \mathbf{Z}^+$. In view of (5.131) and (5.138), we may show by induction that $|f_m| \leq b_m$ for $m \geq 0$. Thus, (5.135) has a solution of the form (5.139) which is analytic for $|z - b| < |b| - D$ and satisfies (5.138). Furthermore, in view of our assumptions,

$$\left| \sum_{m=0}^{\infty} f_m (z-b)^m \right| \leq \left| \sum_{m=0}^{\infty} b_m (z-b)^m \right| \leq K,$$

that is, $f(b, z)$ is bounded for $|z - b| < |b| - D$. Let B be a bounded region whose closure is contained in $\Re(z) < Q$. Then there exists $b_0 < 0$ such that for $b < b_0$, we have

$$B \subset \{z : |z - b| < |b| - |Q|\}.$$

Let $\{b_k\}_{k=0}^{\infty}$ be a sequence which is decreasing and tends to $-\infty$. The family $\{f(b_j, z)\}_{j=0}^{\infty}$ of solutions is a sequence of bounded functions on B. Thus there is a subsequence of $\{b_j\}$ which we may, without loss of generality, denote by $\{b_j\}$ such that the limiting function $\lim_{j \to \infty} f(b_j, z)$ is analytic on B, and it is the uniform limit of $\{f(b_j, z)\}_{j=0}^{\infty}$ on any compact subset of B. Since $\{z : \Re z < Q\}$ is the union of all B that satisfies the above assumption, we may infer from the Unique Continuation Theorem 3.17 that there is some sequence $\{d_k\}$ such that

$$f(z) = \lim_{k \to \infty} f(d_k, z)$$

is the analytic solution of (5.119). The proof is complete.

We remark that in the above Theorem, if $\varphi_1(z), ..., \varphi_k(z), \alpha_1(z), ..., \alpha_n(z), \beta_1(z), ..., \beta_s(z)$ and $g(z)$ are rational functions, then letting $a_1, a_2, ..., a_N$ be the totality of all their poles, we may take

$$D > \max\{|a_1|, |a_2|, ... |a_N|\}$$

so that for any negative number b with sufficiently large absolute value, $\varphi_1(z), ..., \varphi_k(z), \alpha_1(z), ..., \alpha_n(z), \beta_1(z), ..., \beta_s(z)$ are analytic for $|z - b| < |b| - D$.

Example 5.11. Consider the equation

$$f'''(z) - \frac{5}{(1+z)^2} f'(z - \tau) = \frac{1}{(1+z)^4}, \quad \tau > 0, \quad (5.140)$$

For any negative b with sufficiently large absolute value, the functions $\alpha(x) = -\frac{5}{(1+z)^2}$ and $g(z) = \frac{1}{(1+z)^4}$ are analytic for $|z - b| < |b| - 2$. The approximating equation of (5.140) is

$$f'''(z) - \frac{5}{(1+z)^2} f'(pz - \tau) = \frac{1}{(1+z)^4}, \quad (5.141)$$

where $p = 1 + \tau b^{-1}$. Furthermore, the equation

$$\varphi'''(z) - \frac{5}{(1+z)^2} f'(z - \tau) = \frac{1}{(1+z)^4}$$

on $|z - b| < |b| - 2$ has the bounded analytic solution $\varphi(z) = z/(1+z)$. In view of Theorem 5.23, there exists $Q < 0$ such that $|Q| > 2$ and equation (5.140) has an analytic solution of the form

$$f(z) = \lim_{b \to -\infty} f(b, z),$$

on $(-\infty, Q)$, where $f(b, z)$ is an analytic solution of equation (5.141).

5.8 Notes

Equation (5.6) and nonhomogeneous differential equations of the form

$$y'' + p(x) y' + q(x) y = h(x)$$

arise in a large number of mathematical models in mechnical vibration theory, quantum mechanics, etc. Therefore there are many results on the analytic solutions of these equations which can be found in standard text books and references [81, 78, 92]. Therefore we have restricted ourselves to the basic Theorem 5.1. However, the following remarks may be of further interest.

First of all, there are equations which do not allow nontrivial analytic solutions. As a simple example, consider the equation
$$x^3 y'' + y = 0.$$
If we seek a solution of the form $y = \widehat{a}(x)$ which is analytic at 0, then substituting y into the above equation, we see that $a_0 = 0$ and
$$a_k = -(k-1)(k-2)a_{k-1}, \; k \in \mathbf{Z}^+.$$
But then $a_k = 0$ for $k \in \mathbf{N}$. This shows that the only solution which is analytic at 0 is the trivial one. As another example, consider the differential equation
$$x^2 y' = y - x. \tag{5.142}$$
If $y = \widehat{a}(x)$ is an analytic solution at 0, then substituting it into the above equation, we obtain
$$\widehat{a}(x) = \sum_{n=0}^{\infty} n! x^{n+1} \tag{5.143}$$
with radius of convergence $\rho(a) = 0$. In other words, equation (5.143) does not allow solutions which are analytic at 0. Note that although (5.143) is not analytic at 0, yet it is a formal solution of (5.142). We can also find equations where no formal solution can exist. For instance, consider the equation
$$xy' = y - x.$$
Substituting $y = \widehat{a}(x)$ will formally yield
$$a_1 = a_1 - 1,$$
which is impossible.

Although nontrivial analytic solutions cannot be found in general, it is possible in some cases to find solutions of the form $x^s f(x)$ where f is analytic in a neighborhood of zero. The method for finding such solutions is called the method of Frobenius. Since this method has been discussed quite extensively in many texts, we will only refer the interested readers to the references [81, 78, 92].

Examples 5.2, 5.3, 5.4 and 5.5 are due to Herron in [77]. Further examples can be found in [116].

The neutral system (5.12) is studied by Cherepennikov in [35]. Theorems 5.2, 5.3 and 5.4 can be found in his work. Related systems and more general systems have also been studied by him in [33, 34, 36–38].

The Cauchy Kowalewski existence theorem has been extended to more general partial differential equations or systems. For instance, we may seek solutions $\widehat{z}, \widehat{v}, \widehat{w}$ of the system
$$\begin{aligned} \widehat{z}_y &= \widehat{F}^1(x, y, \widehat{z}, \widehat{v}, \widehat{w}, \widehat{z}_x, v_x, \widehat{w}_x), \\ \widehat{v}_y &= \widehat{F}^2(x, y, \widehat{z}, \widehat{v}, \widehat{w}, \widehat{z}_x, v_x, \widehat{w}_x), \\ \widehat{w}_y &= \widehat{F}^3(x, y, \widehat{z}, v, \widehat{w}, \widehat{z}_x, \widehat{v}_x, \widehat{w}_x), \end{aligned} \tag{5.144}$$

which are C^1-functions of x and y and which assume prescribed initial various along the x-axis:
$$\hat{z}(x,0) = \hat{z}_0(x), \ \hat{v}(x,0) = \hat{v}_0(x), \ \hat{w}(x,0) = \hat{w}_0(x). \tag{5.145}$$
From (5.145) the values of $\hat{z}_x(x,0), \hat{v}_x(x,0), \hat{w}_x(x,0)$ are known at every point on the x-axis, and from substitution in (5.144) so are the derivatives with respect to y. The Cauchy-Kowalewski theorem now asserts that if the functions z_0, v_0, w_0, are analytic in a neighborhood of $x = 0$, and if the functions F^1, F^2, F^3 are analytic in a neighborhood of
$$x = y = 0, \ z(0,0), \ v(0,0), \ w(0,0), \ z_x(0,0), \ v_x(0,0), \ w_x(0,0),$$
then the initial-value problem (5.145) has precisely one solution (z, v, w), which is analytic in a neighborhood of $x = y = 0$.

By means of such generalized Cauchy-Kowalewski theorems, we may handle the existence theorems for analytic solutions of partial differential equations of the form
$$G(x, y, z, p, q, r, s, t) = 0 \tag{5.146}$$
where
$$p = z_x, \ q = z_y, \ r = z_{xx}, \ s = z_{xy}, \ t = z_{yy}, \tag{5.147}$$
under prescribed initial values on an initial curve $y_0(x)$:
$$z_0(x) = z(x, y_0(x)), \ p_0(x) = p(x, y_0(x)), \ q_0(x) = q(x, y_0(x)). \tag{5.148}$$

Equation (5.44) is studied by Li [118], in which Theorem 5.5 is obtained. The two results related to equation (5.49) are contained in [118].

Neutral differential equations and systems with proportional delays have been studied for some time and found potential important applications in a number of scientific fields. In Carr and Dyson [23] and Kato and McLeod [95], asymptotic behaviors of solutions of the equation
$$y'(x) = ay(\lambda x) + by(x). \tag{5.149}$$
are discussed. In Feldstein and Jackiewicz [64], the 'exponential order' of the solutions of the pantograph equation
$$y'(z) = Ay(z) + B(\lambda z) + Cy'(yz) \tag{5.150}$$
is discussed. Viorica in [216] discussed the analytic solutions of the equation
$$y'(z) = \sum_{j=1}^{p} k_j y(\lambda_j z) + by(z), \ \lambda_j \in (0, 1) \tag{5.151}$$
under the initial condition $y(0) = 1$. In Iserles and Liu [84], the pantograph equation (5.150) is illustrated by means of interesting examples and figures. In [82], Ifantis discussed the following linear functional differential equation
$$\frac{d^k f(z)}{dz^k} + \sum_{i=1}^{k} \varphi_i(z) \frac{d^{k-i} f(z)}{dz^{k-i}} + \sum_{j=1}^{n} \alpha_j(z) f(q_j z) = g(z),$$

where $k \geq 1$ and $|q_j| \leq 1$ for $j = 1, 2, ..., n$. He transforms the problem of the existence of analytic solutions into the problem of finding the null space of an operator defined on a separable Hilbert space. In particular, the equation

$$f'(x) = af(\lambda x) + bf(x), \ a \in \mathbf{C}, b \in \mathbf{R}, \lambda \in \mathbf{C}$$

is discussed in detail.

By methods similar to that in the proof of Theorem 5.12, we may show the following for equation (5.61):

Theorem 5.24. *Suppose $A_1(z), ..., A_l(z), f_1(z), ..., f_l(z)$ and $h(z)$ are analytic for $|x - \xi| < R_1$ and $f_1(\xi) = f_2(\xi) = \cdots = f_l(\xi) = \xi$, $f_i'(\xi) = s_i$ for $i = 1, ..., l$, where $0 < |s_i| < 1$. Suppose further that there exist $R_2 > 0, N_1, ..., N_l > 0$ such that for $|z - \xi| < R_2$, we have $|A_i(z)| \leq N_i$ for $i = 1, ..., l$, and $R_2 (N_1 + \cdots + N_l) < 1$. Then (5.61) has a solution of the form*

$$x(z) = h(\xi) + \sum_{n=1}^{\infty} b_n (z - \xi)^n,$$

which is analytic on a neighborhood of the point ξ.

The above result and Theorem 5.12 can be found in Si and Li [184].

Equation (5.66) and the corresponding Theorems 5.14 and 5.15 can be found in Si [178].

As mentioned before, there are many differential equations with second order unknown derivatives which allow analytic solutions and they can be found in standard text books. There are also many functional differential equations with higher order unknown derivates which allow analytic solutions. We have only presented some simple ones. In particular, equation (5.79) and the corresponding Theorem 5.17 are in Si and Li [184], while equations (5.91) and (5.119) and the corresponding existence theorems are in Si [174].

In Example 5.7, we have demonstrated how analytic solutions can be used to generate approximate solutions of differential equations under various conditions. Similar ideas have been employed in a number of recent studies under the so called 'differential transformation method', see e.g. [237, 24, 26, 25, 88, 89, 10, 6]. However, most of the derivations in these studies are heuristic and error analyses are not provided. Therefore, there is much to be done in this area.

Chapter 6

Functional Equations with Iteration

Recall that $\phi^{[0]}(z) = z$, $\phi^{[1]}(z) = \phi(z)$, $\phi^{[2]}(z) = \phi(\phi(z)), ..., \phi^{[n]}(z) = \phi(\phi^{[n-1]}(z))$ are the zeroth, first, second, ..., and the n-th iterate of the function $\phi(z)$. Functional equations (with or without differentiation) that involve iterates of the unknown functions are called iterative functional equations. Such equations arise naturally in many problems. In this Chapter, we allow functional equations involving more complicated composition of known or unknown functions.

6.1 Equations without Derivatives

For motivation, let x_n be the amount of money saved in a bank during the time period n. Then the amount of money during the time period $n + 1$ is commonly given by

$$x_{n+1} = x_n + rx_n, \ n \in \mathbf{N},$$

where r is the interest rate offered by the bank for one period of time. Let $f(x) = (1 + r)x$. Then the above recurrence relation can also be written as $x_{n+1} = f(x_n)$. Given $x_0 = \lambda$, then $x_1 = f(x_0) = f(\lambda)$, $x_2 = f(x_1) = f(f(x_0))$, ..., and in general $x_n = f^{[n]}(\lambda)$ for $n \in \mathbf{N}$. A natural question is when it is true that $f^{[n]}(\lambda)$ is equal to a prescribed number. A more general question naturally arises as to what kind of function ϕ such that its n-th iterate is equal to a given function Q. Indeed such a question has been considered by Babbage [11, 12]. Another well known recurrence equation is

$$x_{n+1} = \mu x_n(1 - x_n), \ n \in \mathbf{N}.$$

Here $f(x) = \mu x(1-x)$. By asking the question as to when $f^{[n]}(\lambda) = \lambda$ (which corresponds to whether periodic solutions exist, see e.g. Li and Yorke [111], Feigenbaum [60]), there follows a great many number of research works related to 'chaos theory'!

As another example, let L be a curve in the x, y-plane which can be described by a function $y = \phi(x)$. By means of a transformation T on the real plane

$$T(x, y) = (f(x, y), g(x, y)),$$

the curve L is transformed into another curve $\bar{L}$. If $\bar{L}$ can be described by another function $y = \psi(x)$, we call the function ψ the transform of ϕ and we write $\psi = T\phi$. It may happen that L is transformed by T into itself, that is, $\phi = T\phi$. Then L is called an invariant curve under T. This means that $y = \phi(x)$ satisfies

$$\phi(f(x, \phi(x))) = g(x, \phi(x)), \tag{6.1}$$

which is called the equation of invariant curves.

We first take up iterative equations of the form

$$F\left(z, \phi(z), \phi(f^{[1]}(z)), ... \phi(f^{[m]}(z))\right) = 0, \tag{6.2}$$

and

$$F\left(z, \phi^{[1]}(f(z)), ..., \phi^{[m]}(f(z))\right) = 0. \tag{6.3}$$

6.1.1 Babbage Type Equations

We first consider a simple case of (6.3), namely,

$$\phi(\phi(\lambda z)) = g(z), \quad \lambda \neq 0. \tag{6.4}$$

Suppose

$$g(z) = \sum_{n=0}^{\infty} a_n z^n \tag{6.5}$$

where $a_0 = 0$ and $a_1 = g'(0) \neq 0$, is analytic on some disk $B(0; \delta)$, and $|g'(0)/\lambda| < 1$. Then the equation

$$\psi(g(x)) = \frac{a_1}{\lambda} \psi(\lambda x)$$

can be written as a Schröder equation

$$\psi(f(x)) = s\psi(x) \tag{6.6}$$

by letting $s = a_1/\lambda$ and

$$f(x) = g\left(\frac{x}{\lambda}\right) = sx + \sum_{n=2}^{\infty} \frac{a_n}{\lambda^n} x^n.$$

Since $0 < |s| = |a_1/\lambda| = |g'(0)/\lambda| < 1$, in view of Theorem 4.9, (6.6) has a solution $\psi(x)$ which is analytic on some $B(0; \eta)$. From the Analytic Inverse Function Theorem 4.8, ψ^{-1} is analytic on some $B(0; \beta)$. Then the composite function

$$h(x) = \psi^{-1}(b_1 \psi(x)),$$

where

$$b_1^2 = \frac{a_1}{\lambda},$$

is analytic on some $B(0;\gamma)$ and

$$\begin{aligned}h(h(\lambda x)) &= \psi^{-1}(b_1\psi(h(\lambda x))) = \psi^{-1}\left(b_1\psi\left(\psi^{-1}\left(b_1\psi(\lambda x)\right)\right)\right)\\ &= \psi^{-1}\left(b_1^2\psi(\lambda x)\right) = \psi^{-1}\left(\frac{a_1}{\lambda}\psi(\lambda x)\right)\\ &= \psi^{-1}\left(\psi(g(x))\right) = g(x),\end{aligned}$$

that is, h is an analytic solution of (6.4) and $h(0) = \psi^{-1}(b_1\psi(0)) = 0$.

Once the existence of analytic solutions is guaranteed, we may find them by setting

$$\phi(x) = \sum_{n=0}^{\infty} b_n x^n, \tag{6.7}$$

where $b_0 = 0$. Substituting it into (6.4), we see that

$$a = b \circ (b \cdot \underline{\lambda}).$$

Hence

$$a_n = \sum_{i=0}^{n} b_i\, (\underline{\lambda}\cdot b)_n^{\langle i\rangle} = \sum_{i=0}^{n} b_i\left(\underline{\lambda}\cdot b^{\langle i\rangle}\right)_n = \left(\sum_{i=0}^{n} b_i b_n^{\langle i\rangle}\right)\lambda^n,\ n\in\mathbf{N},$$

which yields

$$\lambda b_1^2 = a_1, \tag{6.8}$$

and

$$\lambda^n(b_1 + b_1^n)b_n + P_n(b_1, b_2, ..., b_{n-1}, \lambda) = a_n,\ n\geq 2, \tag{6.9}$$

where P_n is an n-variate polynomial. We may thus determine two sequences $\{0, b_1, b_2, ...\}$ as expected.

We remark that any other existence results for the Schröder equation (such as Theorem 4.10) will yield additional existence results for (6.4).

We remark further that similar principles will lead us to the existence of analytic solutions of the equation

$$\phi^{[n]}(z) = g(z),\ n = 2, 3, ..., \tag{6.10}$$

where g stands for a given nontrivial function analytic on some disk $B(0;\delta)$ and satisfies $g(0) = 0$ and $g'(0) = s$. In case ϕ is a solution of (6.10), it is sometimes called a n-th iterative root of g. Indeed, assume that g is a nontrivial function analytic on a neighborhood of zero and $g(0) = 0$ as well as $g'(0) = s$. If $0 < |s| < 1$, then there are n local analytic solutions of the form

$$\phi(z) = \sigma^{-1}\left(s^{1/n}\sigma(z)\right),\ z\in U,$$

where $s^{1/n}$ stands for any of the n possible values of the complex root of s, U is a neighborhood of zero and $\sigma(z)$ is a local analytic solution of the Schröder equation $\sigma(g(z)) = s\sigma(z)$.

We now look for analytic functions f that satisfies the equation [120]
$$f(p(z) + bf(z)) = h(z), \qquad (6.11)$$
where b is a nonzero complex number, and p, h are given complex functions of a complex variable. The basic conditions that $p(z)$ and $h(z)$ are analytic in a neighborhood of the origin, $p(0) = h(0) = 0$, $p'(0) = r$ and $h'(0) = s \neq 0$ will be assumed throughout the rest of this section.

Let β be a root of the equation $z^2 - rz - bs = 0$ (note that $\beta \neq 0$ since $bs \neq 0$). We will need an auxiliary functional equation in the unknown function g:
$$g(\beta^2 z) - p(g(\beta z)) = bh(g(z)). \qquad (6.12)$$
Once we can show the existence of a solution g of this equation which satisfies $g(0) = 0$, $g'(0) \neq 0$ and is analytic on a neighborhood of the origin, then it is easily seen that f defined by
$$f(z) = \frac{1}{b}\{g(\beta g^{-1}(z)) - p(z)\} \qquad (6.13)$$
is a solution of (6.11) which is also analytic in a neighborhood of the origin. Indeed, by the assumptions on g, we see that g^{-1} and hence f are analytic in a neighborhood of the origin and
$$\begin{aligned}
f(p(z) + bf(z)) &= f\left(p(z) + g(\beta g^{-1}(z)) - p(z)\right) \\
&= f\left(g(\beta g^{-1}(z))\right) \\
&= \frac{1}{b}\{g(\beta g^{-1}(g(\beta g^{-1}(z)))) - p(g(\beta g^{-1}(z)))\} \\
&= \frac{1}{b}\{g(\beta^2 g^{-1}(z)) - p(g(\beta g^{-1}(z)))\} \\
&= \frac{1}{b}bh(g(g^{-1}(z))) \\
&= h(z).
\end{aligned}$$

We will therefore seek analytic solutions of (6.12) which vanish at the origin but not their first derivatives.

Theorem 6.1. *Suppose the equation $z^2 - rz - bs = 0$ has a root β such that $|\beta| \neq 1$ and $|\beta| \neq |bs|^{1/(n+1)}$ for $n = 2, 3, \ldots$. Then for any nonzero complex number η, (6.12) has a solution $g(z)$ which is analytic on a neighborhood of the origin and satisfies the conditions $g(0) = 0$ and $g'(0) = \eta$.*

Proof. Under our assumptions on p and h,
$$p(z) = rz + \sum_{n=2}^{\infty} p_n z^n, \qquad (6.14)$$
$$h(z) = sz + \sum_{n=2}^{\infty} h_n z^n, \qquad (6.15)$$

and the Cauchy's Estimation (Theorem 3.26) asserts that there is $\rho > 0$ such that
$$|p_n|, |h_n| \leq \rho^{n-1}, \ n \geq 2.$$
Introducing new functions
$$P(z) = \rho p\left(\rho^{-1}z\right), \ H(z) = \rho h\left(\rho^{-1}z\right), \ G(z) = \rho g\left(\rho^{-1}z\right),$$
we obtain from $g(0) = 0$ and $g'(0) = \eta$ that $G(0) = 0$ and $G'(0) = \eta$ respectively, and from (6.12) that
$$G\left(\beta^2 z\right) - P\left(g(\beta z)\right) = bH(g(z)),$$
which is again an equation of the form (6.12). Here P and H are of the form
$$P(z) = \sum_{n=1}^{\infty} P_n z^n,$$
and
$$H(z) = \sum_{n=1}^{\infty} H_n z^n$$
respectively, but $|P_n| = \left|p_n \rho^{1-n}\right| \leq 1$ and $|H_n| = \left|h_n \rho^{1-n}\right| \leq 1$ for $n \geq 2$. Consequently, we may assume that
$$|p_n| \leq 1, \ |h_n| \leq 1, \ n \geq 2. \tag{6.16}$$
We seek solutions of (6.12) in the form
$$g(z) = \widehat{g}(z) = \sum_{n=1}^{\infty} g_n z^n. \tag{6.17}$$
By formally substituting g into (6.12), we obtain
$$\underline{\beta^2} \cdot g - p \circ (g \cdot \underline{\beta}) = b(h \circ g).$$
Since $g_0 = 0$, we see that
$$p \circ (g \cdot \underline{\beta}) = \underline{\beta} \cdot (p \circ g),$$
hence
$$(\beta^2 - r\beta - bs)g_1 = 0, \tag{6.18}$$
and
$$(\beta^{2n} - r\beta^n - bs)g_n = \beta^n \sum_{m=2}^{n} p_m g_n^{\langle m \rangle} + b \sum_{m=2}^{n} h_m g_n^{\langle m \rangle}, \ n \geq 2. \tag{6.19}$$
Since β is a nonzero root of $z^2 - rz - bs$, we can choose g_1 as η so that (6.18) is satisfied. Furthermore, we assert that $(\beta^n)^2 - r\beta^n - bs \neq 0$ for $n \geq 2$. Indeed, if the contrary holds, then β^n is a root of the equation $z^2 - rz - bz = 0$. Thus $\beta^n = \beta$ or $\beta^n \neq \beta$. In the former case, $|\beta| = 1$ and in the latter, $\beta^n \beta = -bs$ so that $|\beta| = |bs|^{1/(n+1)}$. Both are contrary to our assumptions on β.

We may now see from (6.19) that the resulting relation defines $g_2, g_3, \ldots$ in a unique manner. We need to show that the subsequent series (6.17) converges in a neighborhood of the origin. To see this, note that

$$\lim_{n \to \infty} \frac{\beta^n}{\beta^{2n} - r\beta^n - bs} = 0,$$

and

$$\lim_{n \to \infty} \frac{b}{\beta^{2n} - r\beta^n - bs} = \begin{cases} 0, & |\beta| > 1, \\ -1/s, & 0 < |\beta| < 1. \end{cases}$$

Thus there is some positive number M such that

$$\left| \frac{\beta^n}{\beta^{2n} - r\beta^n - bs} \right|, \left| \frac{b}{\beta^{2n} - r\beta^n - bs} \right| \leq \frac{M}{2} \quad (6.20)$$

for $n \geq 2$. From (6.19) and (6.20), we have

$$|g_n| \leq M \sum_{m=2}^{n} |g|_n^{\langle m \rangle}, \quad n \geq 2.$$

If we now define a sequence $\{q_n\}_{n \in \mathbf{N}}$ by $q_0 = 0$, $q_1 = |\eta|$ and

$$q_n = M \sum_{m=2}^{n} q_n^{\langle m \rangle}, \quad n \geq 2,$$

then it is easily seen that

$$|g_n| \leq q_n, \quad n \geq 1.$$

In other words, the series $Q(z) = q_1 z + q_2 z^2 + \cdots$ is a majorant series of g. But by Example 4.8, $Q(z)$ is a solution of the implicit relation

$$F(z, Q) \equiv Q - |\eta| z - \frac{MQ^2}{1 - Q} = 0,$$

and is analytic on a neighborhood of the origin. Thus $g(z)$ is analytic there as well. This completes the proof.

In the above result, we assume that $|\beta| \neq 1$. Next, we deal with the case when $|\beta| = 1$.

Theorem 6.2. *Suppose $|bs| < 1$ and the equation $z^2 - rz - bs = 0$ has a nonzero root β which is a Siegel number. Then (6.12) has a solution $g(z)$ which is analytic on a neighborhood of the origin and satisfies the conditions $g(0) = 0$ and $g'(0) = \eta \neq 0$.*

Proof. As in the proof of the previous Theorem 6.1, we may assume that (6.14) and (6.15) hold, and there is a formal solution $g(z)$ of (6.12) given by (6.17) with $g_1 = \eta$ and

$$(\beta^{2n} - r\beta^n - bs)g_n = \beta^n \sum_{m=2}^{n} p_m g_n^{\langle m \rangle} + b \sum_{m=2}^{n} h_m g_n^{\langle m \rangle}, \quad n \geq 2.$$

Since $\beta^2 - r\beta - bs = 0$, we have
$$\beta - r = \frac{bs}{\beta},$$
and
$$\beta^{2n} - r\beta^n - bs = \left(\beta^{2n} - \beta^2\right) - r\left(\beta^n - \beta\right) = \beta\left(\beta^{n-1} - 1\right)\left(\beta^n + \beta - r\right).$$
Thus,
$$\beta\left(\beta^{n-1} - 1\right)\left(\beta^n + \beta - r\right) g_n = \beta^n \sum_{m=2}^{n} p_m g_n^{\langle m \rangle} + b \sum_{m=2}^{n} h_m g_n^{\langle m \rangle}, \ n \geq 2.$$
Note that
$$\frac{1}{|\beta^n + \beta - r|} = \frac{1}{|\beta^n + bs/\beta|} = \frac{1}{|\beta^{n+1} + bs|} \leq \frac{1}{1 - |bs|}.$$
If we now define a sequence $\{w_n\}_{n \in \mathbf{N}}$ by $w_0 = 0$, $w_1 = |\eta|$ and
$$w_n = M\left|\beta^{n-1} - 1\right|^{-1} \sum_{m=2}^{n} w_n^{\langle m \rangle}, \ n \geq 2,$$
where
$$M = \frac{1 + |b|}{1 - |bs|} > 0, \ n \geq 2,$$
then it is not difficult to show by induction that
$$|g_n| \leq w_n, \ n \geq 2.$$
In other words, $W(z) = w_1 z + w_2 z^2 + \cdots$ is a majorant series of g. We now need to show that $W(z)$ has a positive radius of convergence. To see this, note that Example 4.8 asserts that the implicit relation
$$F(z, Q) \equiv Q - |\eta| z - \frac{MQ^2}{1 - Q} = 0, \ M > 0,$$
defines an analytic function
$$Q(z) = \sum_{n=0}^{\infty} q_n z^n,$$
with $q_0 = Q(0) = 0$, $q_1 = Q'(0) = |\eta| > 0$, and
$$q_n = M \sum_{m=2}^{n} q_n^{\langle m \rangle}, \ n \geq 2.$$
Since Cauchy's Estimation (Theorem 3.26) asserts that there is a positive number A such that
$$q_n \leq A^n, \ n \geq 1,$$

we see from Theorem 3.32 that

$$w_n \leq A^n \left(2^{5\delta+1}\right)^{n-1} n^{-2\delta}, \ n \geq 1,$$

which shows that $W(z)$ has a positive radius of convergence. The proof is complete.

We remark that once existence is guaranteed, it may be possible to expand (6.11) in series form and seek the desired solution instead of first finding a solution of (6.12). As an example, consider the functional equation

$$f(z + f(z)) = sz, \ s \neq 0. \tag{6.21}$$

If $z^2 - z - s = 0$ has a nonzero root β, then writing $f(z) = f_1 z + f_2 z^2 + \cdots$, we see that

$$f_1(1 + f_1) = s,$$
$$f_1 f_2 + f_2(1 + f_1)^2 = 0,$$
$$f_1 f_3 + (1 + b f_1) f_3 = 0,$$
$$\ldots = \ldots$$

The conditions in Theorem 6.1 or Theorem 6.2 assure that there is at least one nontrivial solution to the above system of equations. Indeed, one solution of the above system is $\{f_1, f_2, f_3, \ldots\} = \{\beta, 0, 0, \ldots\}$, and the corresponding $f(z) = \beta z$ is an analytic solution of (6.21).

6.1.2 Equations Involving Several Iterates

A natural extension of the problem of the existence of iterative roots of (6.10) is to find a function f such that a linear combination of its iterates is equal to a given function F:

$$\lambda_1 f(z) + \lambda_2 f^{[2]}(z) + \cdots + \lambda_n f^{[n]}(z) = F(z), \tag{6.22}$$

where $\lambda_1, \ldots, \lambda_n$ are complex numbers, not all zero (see [186]).

Theorem 6.3. *Suppose the power series function $F(z) = \widehat{c}(z)$, where $c_0 = 0$ and $c_1 = s$, is analytic on some $B(0; r_1)$. Suppose further that α is a root of the equation*

$$\lambda_1 z + \lambda_2 z^2 + \cdots + \lambda_n z^n = s \tag{6.23}$$

and there is some positive number β such that for $m \geq 2$,

$$\left|\lambda_1 \alpha^m + \lambda_2 \alpha^{2m} + \cdots + \lambda_n \alpha^{nm} - s\right| \geq \beta.$$

Then (6.22) has a solution which is analytic on some disk $B(0; r)$.

Proof. We first look for an analytic solution of the form

$$\phi(z) = \widehat{b}z = \eta z + \sum_{m=1}^{\infty} b_m z^m, \ \eta \neq 0, \tag{6.24}$$

for the equation
$$\lambda_1\phi(\alpha z) + \lambda_2\phi(\alpha^2 z) + \cdots + \lambda_n\phi(\alpha^n z) = F(\phi(z)) \qquad (6.25)$$
in a neighborhood of the origin. In view of the assumptions on F, the Cauchy's Estimation (Theorem 3.26) asserts the existence of some positive number p such that
$$|c_m| \le p^{m-1}, \; m = 2, 3, \ldots .$$
Introducing transformations $\psi(z) = p(z/p)$ and $G(z) = pF(z/p)$, we may obtain from (6.25) that
$$\lambda_1\psi(\alpha z) + \lambda_2\psi(\alpha^2 z) + \cdots + \lambda_n\psi(\alpha^n z) = G(\psi(z)),$$
which is again an equation of the form (6.25). Here G is of the form
$$G(z) = \sum_{m=1}^{\infty} d_m z^m = sz + \sum_{m=2}^{\infty} \frac{c_m}{p^{m-1}} z^m,$$
but
$$|d_m| = \left|\frac{c_m}{p^{m-1}}\right| \le 1, \; m = 2, 3, \ldots .$$
Consequently, we may assume that $|c_m| \le 1$ for $m \ge 2$.

By substituting (6.24) into (6.25), and comparing coefficients, we see that
$$\lambda_1 \underline{\alpha} \cdot b + \lambda_2 \underline{\alpha}^2 \cdot b + \cdots + \lambda_n \underline{\alpha}^n \cdot b = c \circ b.$$
Hence
$$(\lambda_1 \alpha + \lambda_2 \alpha^2 + \cdots + \lambda_n \alpha^n - s) b_1 = 0$$
and
$$(\lambda_1 \alpha^m + \lambda_2 \alpha^{2m} + \cdots + \lambda_n \alpha^{nm} - s) b_m = \sum_{t=2}^{m} c_t b_m^{\langle t \rangle}, \; m \ge 2. \qquad (6.26)$$
Since α is a root of (6.23), we may choose $b_1 = \eta$, and then b_m can be uniquely determined by the recurrence relation (6.26) for $m \ge 2$. This shows that (6.25) has a formal solution of the form (6.24) for any given η which is not zero.

Note that by Example 4.8, the implicit relation
$$W(z, Q) \equiv Q - |\eta| z - \frac{1}{\beta} \frac{Q^2}{1-Q} = 0$$
has a solution
$$Q(z) = \sum_{n=0}^{\infty} q_n z^n$$
which is analytic near 0 and the sequence $q = \{q_n\}_{n \in \mathbf{N}}$ satisfies $q_0 = 0$, $q_1 = |\eta| > 0$ and
$$q_n = \frac{1}{\beta} \sum_{i=2}^{n} q_n^{\langle i \rangle}, \; n \ge 2.$$

Note that $|b_1| = |\eta| \le q_1$. Assume by induction that $|b_k| \le q_k$ for $k = 1, 2, ..., m-1$, then

$$\beta |b_m| \le \left|\lambda_1 \alpha^m + \lambda_2 \alpha^{2m} + \cdots + \lambda_n \alpha^{nm} - s\right| |b_m|$$
$$\le \sum_{t=2}^{m} |c_t| \, |b|_m^{\langle t \rangle} \le \sum_{t=2}^{m} |b|_m^{\langle t \rangle} \le \sum_{t=2}^{m} q_m^{\langle t \rangle} \le \beta q_m.$$

Thus $|b_m| \le q_m$ for $m \ge 2$, which shows that $Q(z)$ is a majorant of $\phi(z)$. We have thus shown that $\phi(z)$ is an analytic solution of (6.25) on some $B(0; r_2)$. Since $\phi(0) = 0$ and $\phi'(0) = \eta \ne 0$, the inverse function ϕ^{-1} is also analytic in a neighborhood $B(0; r)$ where $r \le r_2$. Let

$$f(z) = \phi(\alpha \phi^{-1}(z)), \quad |z| < r.$$

Then f is analytic near 0 and is a solution of (6.22):

$$\lambda_1 f(z) + \lambda_2 f^{[2]}(z) + \cdots + \lambda_n f^{[n]}(z)$$
$$= \lambda_1 \phi(\alpha \phi^{-1}(z)) + \lambda_2 \phi(\alpha^2 \phi^{-1}(z)) + \cdots + \lambda_n \phi(\alpha^n \phi^{-1}(z))$$
$$= F(z).$$

The proof is complete.

As a corollary, we may obtain the following result.

Theorem 6.4. *Suppose $F(0) = 0$, $F'(0) = s$ where $|s| > (1 + \sqrt{5})/2$ and F is analytic in a neighborhood of the origin. Then there is an function f analytic on some $B(0; \delta)$ such that $f^{[n]}(z) = F(z)$ for $n \ge 2$ and $z \in B(0; \delta)$.*

Indeed, if we let α be a root of $\alpha^n = s$, then for $m \ge 2$,

$$|\alpha^{nm} - s| = |s^m - s| \ge |s| \left(|s|^{m-1} - 1\right) = \frac{1+\sqrt{5}}{2} \left(\left(\frac{1+\sqrt{5}}{2}\right)^{m-1} - 1\right) \ge 1.$$

In view of Theorem 6.3, the equation $f^{[n]}(z) = F(z)$ has a solution which is analytic near 0.

Theorem 6.5. *Suppose F is a function analytic on some disk $B(\beta; \delta)$, $F(\beta) = 0$ and $F'(\beta) = \lambda_1 \alpha + \lambda_2 \alpha^2 + \cdots + \lambda_n \alpha^n \ne 0$ where $|\alpha| < 1$. Suppose further that*

$$\sum_{k=1}^{n} \lambda_k = 0 \text{ and } |\alpha| \sum_{k=1}^{n} |\lambda_k| < 1,$$

and there is an integer m such that

$$F'(\beta) = \lambda_1 \alpha^m + \lambda_2 \alpha^{2m} + \cdots + \lambda_n \alpha^{nm}.$$

Then (6.22) has analytic solutions near β.

Proof. Since $F(\beta) = 0$ and $F'(\beta) \neq 0$, the inverse function F^{-1} of F is analytic in a neighborhood of the origin and is of the form

$$F^{-1}(z) = \widehat{c}(z) = \beta + \sum_{m=1}^{\infty} c_m z^m. \tag{6.27}$$

In terms of F^{-1}, we may obtain from (6.25) the equation

$$\phi(z) = F^{-1}\left(\sum_{k=1}^{n} \lambda_k \phi(\alpha^k z)\right). \tag{6.28}$$

We first look for solutions of (6.28) of the form

$$\phi(z) = \widehat{b}(z) = \sum_{m=0}^{\infty} b_m z^m. \tag{6.29}$$

Substituting (6.29) and (6.27) into (6.28), we obtain

$$b = c \circ \left(\sum_{k=1}^{n} \lambda_k b \cdot \underline{\alpha}^k\right) = c \circ d,$$

where

$$d = \left\{\left(\sum_{k=1}^{n} \lambda_k \alpha^{kj}\right) b_j\right\}_{j \in \mathbf{N}}.$$

Since

$$d_0 = \sum_{k=1}^{n} \lambda_k = 0,$$

we see that

$$b = \left\{\sum_{i=0}^{j} c_i d_j^{\langle i \rangle}\right\}_{j \in \mathbf{N}}.$$

These lead us to

$$b_0 = \beta,$$

$$\{1 - c_1(\lambda_1 \alpha + \lambda_2 \alpha^2 + \cdots + \lambda_n \alpha^n)\} b_1 = 0 \tag{6.30}$$

and

$$\{1 - c_1(\lambda_1 \alpha^m + \lambda_2 \alpha^{2m} + \cdots + \lambda_n \alpha^{nm}\} b_m$$
$$= \sum_{l_1+l_2+\cdots+l_t=m; t=2,3,\ldots,m} c_t \prod_{i=1}^{t}\left(\sum_{k=1}^{n} \lambda_k \alpha^{kl_i}\right) b_{l_1} b_{l_2} \cdots b_{l_t} \tag{6.31}$$

for $m \geq 2$. Since

$$c_1 = (F^{-1})'(0) = \frac{1}{F'(F^{-1}(0))} = \frac{1}{F'(\beta)} = \frac{1}{\lambda_1 \alpha + \lambda_2 \alpha^2 + \cdots + \lambda_n \alpha^n},$$

thus
$$1 - c_1(\lambda_1\alpha + \lambda_2\alpha^2 + \cdots + \lambda_n\alpha^n) = 0.$$

In view of (6.30), we may choose $b_1 = \eta \neq 0$. Furthermore, since $|\alpha| < 1$, there exists a positive integer K such that for $m > K$,
$$\left|c_1(\lambda_1\alpha^m + \lambda_2\alpha^{2m} + \cdots + \lambda_n\alpha^{nm})\right| < 1 - \gamma, \ \gamma \in (0,1).$$

Thus, for $m > K$,
$$\left|\{1 - c_1(\lambda_1\alpha^m + \lambda_2\alpha^{2m} + \cdots + \lambda_n\alpha^{nm})\}\right|$$
$$> 1 - \left|\{1 - c_1(\lambda_1\alpha^m + \lambda_2\alpha^{2m} + \cdots + \lambda_n\alpha^{nm})\}\right| > \gamma. \quad (6.32)$$

If $1 - c_1(\lambda_1\alpha^m + \lambda_2\alpha^{2m} + \cdots + \lambda_n\alpha^{nm}) \neq 0$ for $m = 2, 3, ..., K$, then we see that $b_0 = \beta$ and $b_1 = \eta \neq 0$ and that $b_2, b_3, ...$ can be determined from (6.31) in a unique manner to obtain
$$\phi(z) = \beta + \eta z + \sum_{m=2}^{\infty} b_m z^m, \ \eta \neq 0.$$

If $1 - c_1(\lambda_1\alpha^m + \lambda_2\alpha^{2m} + \cdots + \lambda_n\alpha^{nm}) = 0$ for some m, then in view of (6.32), there can only be a finite number of such integers, say, $m_1, m_2, ..., m_r \in \{2, 3, ..., K\}$. Thus, we may choose $b_{m_i} = \eta_{m_i}$ for $i = 1, ..., r$, and $b_m = \eta_m$ for $m \in \{2, 3, ..., K\} \setminus \{m_1, m_2, ..., m_r\}$ so as to satisfy the equation
$$\sum_{l_1+l_2+\cdots+l_t=m; t=2,3,...,m} \frac{(F^{-1})^{(t)}(0)}{t!} \prod_{i=1}^{t}\left(\sum_{k=1}^{n}\lambda_k\alpha^{kl_i}\right)\eta_{l_1}\eta_{l_2}\cdots\eta_{l_t} = 0, \quad (6.33)$$

and then determine $b_{K+1}, b_{K+2}, ...$ by (6.31) in a unique manner to obtain
$$\phi(z) = \beta + \eta z + \sum_{j=1}^{r}\eta_{m_j}z^{m_j} + \sum_{m=2; m\neq m_1, m_2, ..., m_r}^{K}\eta_m z^m + \sum_{m=K+1}^{\infty}b_m z^m,$$

where $\eta \neq 0$ and $\eta_{m_1}, \eta_{m_2}, ..., \eta_{m_r}$ are arbitrary complex numbers.

We now need to prove that (6.29) is convergent in a neighborhood of the origin. First in view of (6.27), we see from Cauchy's Estimation (Theorem 3.26)) that there is some number $p > 0$ such that
$$|c_m| \leq p^{m-1}, \ m = 2, 3, ... \ .$$

Introducing transformations $\psi(z) = p^{-1}\phi(z/p)$ and $G(z) = p^{-1}F(z/p)$, we may then see from (6.28) that
$$\psi(z) = G^{-1}\left(\sum_{k=1}^{n}\lambda_k\psi(\alpha^k z)\right),$$

which is of the same form as (6.28) but
$$G^{-1}(z) = pF^{-1}(z/p) = \sum_{m=0}^{\infty}d_m z^m = p\beta + \frac{1}{F'(\beta)}z + \sum_{m=2}^{\infty}\frac{c_m}{p^{m-1}}z^m$$

and
$$|d_m| = \left|\frac{c_m}{p^{m-1}}\right| \leq 1, \ m = 2, 3, \dots .$$

Therefore, we may assume without loss of generality that $|c_m| \leq 1$ for $m \geq 2$.

Consider the sequence $u = \{u_m\}_{m \in \mathbf{N}}$ defined by $u_0 = 0$,
$$u_m = |\eta|, \ m = 1,$$
$$u_m = |\eta_m|, \ 2 \leq m \leq K,$$

and
$$u_m = \frac{1}{\gamma} \sum_{t=2}^{m} u_m^{\langle t \rangle}, \ m \geq K+1.$$

Clearly, $|b_m| = u_m$ for $m = 1, 2, \dots, K$. Furthermore, in view of (6.31), (6.32) and our assumptions on $\lambda_1, \dots, \lambda_n$,

$$\gamma |b_{K+1}| \leq \left|1 - c\left(\lambda_1 \alpha^{K+1} + \lambda_2 \alpha^{2(N+1)} + \cdots + \lambda_n \alpha^{n(N+1)}\right)\right| |b_{K+1}|$$
$$\leq \sum_{l_1+l_2+\cdots+l_t=m; t=2,3,\dots,m} |c_t| \prod_{i=1}^{t} \left(\sum_{k=1}^{n} |\lambda_k| |\alpha|^{kl_i}\right) |b_{l_1} b_{l_2} \cdots b_{l_t}|$$
$$\leq \sum_{t=2}^{m} \left(|\alpha| \sum_{k=1}^{t} |\lambda_k|\right)^t |b|_m^{\langle t \rangle}$$
$$\leq \sum_{t=2}^{m} u_m^{\langle t \rangle}$$
$$= \gamma u_{K+1}.$$

By induction, we may then easily show that $|b_m| \leq u_m$ for $m \geq 1$. In other words, the power series function $u(z) = \sum_{m=0}^{\infty} u_m z^m$ is a majorant of the power series function $\sum_{m=1}^{\infty} b_m z^m$. We assert further that $u(z)$ is convergent in a neighborhood of the origin. To see this, note that Example 4.8 asserts that $u(z)$ is a solution of the implicit relation

$$W(z, u) \equiv u - |\eta| z - \sum_{m=2}^{K} |\eta_m| z^m - \frac{1}{\gamma} \frac{u^{K+1}}{1-u} = 0,$$

and is analytic near 0. Thus we may conclude that $\phi(z) = \beta + \sum_{m=1}^{\infty} b_m z^m$ is analytic on a neighborhood of the origin. Since $\phi(0) = \beta$ and $\phi'(0) = \eta \neq 0$, its inverse $\phi^{-1}(z)$ is analytic in a neighborhood of β. If we let

$$f(z) = \phi(\alpha \phi^{-1}(z)),$$

then we may easily verified that f is an analytic solution of (6.22) in a neighborhood of β. The proof is complete.

Example 6.1. Consider the equation

$$f^{[2]}(z) - f(z) - z = 0. \tag{6.34}$$

Let $F(z) = z$. Then F is analytic in C, $F(0) = 0$ and $F'(0) = 1$. Since $\alpha = (1+\sqrt{5})/2$ is a root of the equation $z^2 - z - 1 = 0$ and for $m \geq 2$,

$$\left| \left(\frac{1+\sqrt{5}}{2}\right)^{2m} - \left(\frac{1+\sqrt{5}}{2}\right)^m - 1 \right| \geq \left(\frac{1+\sqrt{5}}{2}\right)^m \left| \left(\frac{1+\sqrt{5}}{2}\right)^m - 1 \right| - 1$$

$$\geq \left(\frac{1+\sqrt{5}}{2}\right)^m - 1$$

$$\geq 1,$$

if we take $\beta = 1$, then the conditions in Theorem 6.5 are satisfied. Thus (6.34) has an analytic solution of the form

$$f(z) = \phi\left(\frac{1+\sqrt{5}}{2}\phi^{-1}(z)\right),$$

where $\phi(z)$ is an analytic solution of

$$\phi\left(\left(\frac{1+\sqrt{5}}{2}\right)^2 z\right) - \phi\left(\frac{1+\sqrt{5}}{2}z\right) = \phi(z). \tag{6.35}$$

Let

$$\phi(z) = \sum_{m=1}^{\infty} b_m z^m. \tag{6.36}$$

After substituting into (6.35), we obtain

$$\sum_{m=1}^{\infty} b_m \left(\frac{1+\sqrt{5}}{2}\right)^{2m} z^m - \sum_{m=1}^{\infty} b_m \left(\frac{1+\sqrt{5}}{2}\right)^m z^m = \sum_{m=1}^{\infty} b_m z^m,$$

and

$$\left\{ \left(\frac{1+\sqrt{5}}{2}\right)^{2m} - \left(\frac{1+\sqrt{5}}{2}\right)^m - 1 \right\} b_m = 0, \; m \in \mathbf{Z}^+.$$

If we take $b_1 = \eta \neq 0$, then $b_m = 0$ for $m \geq 2$, so that $\phi(z) = \eta z$. Since $\phi^{-1}(z) = z/\eta$, we see that

$$f(z) = \frac{1+\sqrt{5}}{2} z.$$

Example 6.2. Consider the equation
$$f^{[2]}(z) = \lambda f(z) + (1-\lambda)(z-\beta), \ 0 < \lambda < 1, \ \beta \in C, \quad (6.37)$$
which can be written as
$$\frac{1}{1-\lambda}f^2(z) - \frac{\lambda}{1-\lambda}f(z) = z - \beta. \quad (6.38)$$
Let $\lambda_1 = -\lambda/(1-\lambda)$, $\lambda_2 = 1/(1-\lambda)$ and $F(z) = z - \beta$. Then $\lambda_1 + \lambda_2 = 0$, $F(\beta) = 0$, $F'(\beta) = 1$ and $\alpha = \left(\lambda - \sqrt{\lambda^2 + 4(1-\lambda)}\right)/2$ is a root of the equation
$$\frac{1}{1-\lambda}z^2 - \frac{\lambda}{1-\lambda}z - 1 = 0.$$
Furthermore,
$$|\alpha| = \frac{\sqrt{\lambda^2 + 4(1-\lambda)} - \lambda}{2} < \frac{\lambda + 2(1-\lambda) - \lambda}{2} = 1 - \lambda < 1,$$
and
$$|\alpha|(|\lambda_1| + |\lambda_2|) = \frac{\sqrt{\lambda^2 + 4(1-\lambda)} - \lambda}{2}\left(\frac{\lambda}{1-\lambda} + \frac{1}{1-\lambda}\right)$$
$$= \frac{\sqrt{\lambda^2 + 4(1-\lambda)} - \lambda}{2}\frac{1+\lambda}{1-\lambda}$$
$$< \frac{1-\lambda}{2}\frac{1+\lambda}{1-\lambda} = \frac{1+\lambda}{2} < 1.$$

Let
$$\phi(z) = \beta + \sum_{m=1}^{\infty} b_m z^m$$
be an analytic solution of
$$\frac{1}{1-\lambda}\phi(\alpha^2 z) - \frac{\lambda}{1-\lambda}\phi(\alpha z) = \phi(z) - \beta.$$
Then
$$\left\{\frac{1}{1-\lambda}\alpha^{2m} - \frac{\lambda}{1-\lambda}\alpha^m - 1\right\}b_m = 0, \ m \in \mathbf{Z}^+.$$
Since
$$\frac{1}{1-\lambda}\alpha^2 - \frac{\lambda}{1-\lambda}\alpha - 1 = 0,$$
and
$$\frac{1}{1-\lambda}\alpha^{2m} - \frac{\lambda}{1-\lambda}\alpha^m - 1 \neq 0$$
for $m \geq 2$, if we take $b_1 = \eta \neq 0$, then $b_m = 0$ for $m \geq 2$. Thus $\phi(z) = \beta + \eta z$ and $\phi^{-1}(z) = (z-\beta)/\eta$. This shows that
$$f(z) = \phi\left(\frac{\lambda - \sqrt{\lambda^2 + 4(\lambda-1)}}{2}\phi^{-1}(z)\right) = \beta + \frac{\lambda - \sqrt{\lambda^2 + 4(\lambda-1)}}{2}(z-\beta)$$
is an analytic solution of (6.38). In particular, when $\beta = 0$, we see that
$$f(z) = \frac{\lambda - \sqrt{\lambda^2 + 4(\lambda-1)}}{2}z$$
is an analytic solution of
$$f^{[2]}(z) = \lambda f(z) + (1-\lambda)f(z), \ \lambda \in (0,1).$$

6.1.3 Equations of Invariant Curves

If we take $f(x,y) = x + y$ and $g(x,y) = y$ in (6.1), then it reduces to the Euler equation

$$\phi(x + \phi(x)) = \phi(x).$$

If we take $f(x,y) = x + y$ and $g(x,y) = \psi(y)$, then (6.1) reduces to the functional equation

$$\phi(x + \phi(x)) = \psi(\phi(x)). \tag{6.39}$$

If we take $f(x,y) = y$ and

$$g(x,y) = 2y - x - \frac{1}{2}(h(x) + h(y)),$$

then (6.1) reduces to the functional equation

$$\phi(\phi(x)) = 2h(x) - x - \frac{1}{2}(h(\phi(x)) + h(x)). \tag{6.40}$$

6.1.3.1 Equation I

In this section, we prove a theorem concerning the existence of analytic solutions of equation (6.39) in the complex field. More specifically, we consider the equation [176]

$$\phi(z + \phi(z)) = \psi(\phi(x)), \tag{6.41}$$

where $\phi(z)$ is the unknown function and $\psi(z)$ is a given complex-valued function of a complex variable.

Suppose $\psi(z)$ is analytic on a neighborhood of zero, $\psi(0) = 0$ and $\psi'(0) = \alpha$. Consider the following three cases (i) $|\alpha| \geq \frac{1+\sqrt{5}}{2}$; (ii) $0 < |\alpha| < 1$; and (iii) α is a Siegel number.

Theorem 6.6. *Assume that one of the condition (i)-(iii) is fulfilled. Then equation (6.41) has a solution which is analytic on a neighborhood of zero.*

Observe that, if $f(z)$ is an analytic solution of the equation

$$f(\alpha^2 z) - f(\alpha z) = \psi(f(\alpha z) - f(z)) \tag{6.42}$$

and $f'(0) \neq 0$, then the formula

$$\phi(z) = f(\alpha f^{-1}(z)) - z$$

defines an analytic function satisfying equation (6.41) on a neighborhood of the origin. Thus our Theorem follows immediately from the following.

Theorem 6.7. *Assume that one of the conditions (i)-(iii) is fulfilled. For any $\eta \in \mathbf{C}$ in the cases (i) and (ii), and for $\eta = 1$ in the case (iii), equation (6.42) has a solution $f(z)$ which is analytic on a neighborhood of zero and $f(0) = 0$ as well as $f'(0) = \eta$.*

Proof. Fix an $\eta \in \mathbf{C}$. If $\eta = 0$ then the zero function satisfies the assertion. So assume that $\eta \neq 0$ and, in addition, $\eta = 1$ in the case (iii). Let

$$\psi(z) = \widehat{c}(z) = \sum_{n=0}^{\infty} c_n z^n, \tag{6.43}$$

where $c_0 = 0$ and $c_1 = \alpha$. Since $\psi(z)$ is analytic on a neighborhood of zero, by Cauchy's Estimation (Theorem 3.26), there exists a positive number β such that $|c_n| \leq \beta^{n-1}$ for $n \geq 2$. Observe that (6.42) is invariant with respect to the transformations $f(z) = \widetilde{f}(\beta z)/\beta$ and $\psi(z) = \widetilde{\psi}(\beta z)/\beta$. Consequently, in the sequel we may assume that

$$|c_n| \leq 1, \ n \in \mathbf{Z}^+. \tag{6.44}$$

Let

$$f(z) = \widehat{b}(z) = \sum_{n=0}^{\infty} b_n z^n, \ b_0 = 0, \tag{6.45}$$

be the expansion of a formal solution $f(z)$ of equation (6.42). Inserting (6.43) and (6.45) into (6.42), we see that

$$\underline{\alpha^2} \cdot b - \underline{\alpha} \cdot b = c \circ (\underline{\alpha} \cdot b - b).$$

Thus

$$\left(\alpha^2 - (1 + c_1)\alpha + c_1\right) b_1 = 0, \tag{6.46}$$

and

$$\left(\alpha^{2n} - (1 + c_1)\alpha^n + c_1\right) b_n = \sum_{l_1 + \cdots + l_t = n; t = 2, 3, \ldots, n} c_t \prod_{k=1}^{t} \left(\alpha^{l_k} - 1\right) b_{l_k}, \tag{6.47}$$

for $n \geq 2$.

Since $c_1 = \alpha$, $\alpha^2 - (1 + c_1)\alpha + c_1 = 0$ so that we may choose $b_1 = \eta$ in (6.46). Let $a_1 = b_1$ and let

$$(\alpha^n - \alpha) a_n = \sum_{t=2}^{n} c_t a_n^{\langle t \rangle}, \ n \geq 2.$$

Then in view of (6.47), $a_n = (\alpha^n - 1) b_n$ for $n \geq 2$.

Clearly, the sequence $\{a_n\}_{n \in \mathbf{Z}^+}$ is uniquely determined. We shall prove the convergence of the series $\sum_{n=1}^{\infty} a_n z^n$ for each z in a neighborhood of the origin.

In case (ii) where $0 < |\alpha| < 1$, for any $\gamma \in (0, |\alpha|)$, it is easy to find a positive integer q such that $|\alpha|^n \leq |\alpha| - \gamma$ for every $n \geq q$. In case (i) or (iii), we choose $q = \gamma = 1$. Note that Example 4.8 asserts that a solution $\omega(z)$ of the implicit relation

$$R(z, \omega) \equiv \omega - |\eta| z - \sum_{n=2}^{q} |a_n| z^n - \frac{1}{\gamma} \frac{\omega^{q+1}}{1 - \omega} = 0$$

exists which is analytic on a neighborhood of the origin and

$$\omega(z) = \sum_{n=0}^{\infty} u_n z^n$$

where $u = \{u_n\}_{n \in \mathbb{N}}$ is determined by $u_0 = 0$, $u_1 = |\eta|$, $u_k = |a_k|$ for $k = 2, ..., q$, and

$$u_n = \frac{1}{\gamma} \sum_{k=q+1}^{n} u_n^{\langle k \rangle}, \quad n \geq q+1. \tag{6.48}$$

If case (i) holds, then $|\alpha^n - \alpha| \geq |\alpha|^n - |\alpha| \geq |\alpha|^2 - |\alpha| \geq 1$ for $n \geq 2$. If case (ii) holds, then $|\alpha^2 - \alpha| \geq |\alpha| - |\alpha|^2 > \gamma$ for $n \geq q+1$. Hence, using induction and the inequality (6.44), we infer that $|a_n| \leq u_n$ for $n \in \mathbb{Z}^+$. Thus the series $\sum_{n=1}^{\infty} a_n z^n$ and, consequently, the series $\sum_{n=1}^{\infty} b_n z^n$ is convergent in a neighborhood of zero.

Now consider the case (iii). Since the series $\sum_{n=1}^{\infty} u_n z^n$ converges in a neighborhood of the origin, by Cauchy's Estimation (Theorem 3.26), there is a positive A such that $u_n \leq A^n$ for $n \in \mathbb{Z}^+$. By Theorem 3.32, we see that there are positive numbers δ and N such that $|a_n| \leq M A^n N^{n-1} n^{-2\delta}$ for $n \in \mathbb{Z}^+$, and hence

$$|b_n| = |\alpha^n - \alpha|^{-1} |a_n| \leq (2(n-1))^{\delta} A^n N^{n-1} n^{-2\delta}, \quad n \in \mathbb{Z}^+.$$

Thus the series $\sum_{n=1}^{\infty} b_n z^n$ converges for each z in a neighborhood of the origin. This completes the proof.

6.1.3.2 Equation II

In this section, we consider the equation (6.40) in the complex domain, that is,

$$\psi(\psi(z)) = 2\psi(z) - z - \frac{1}{2}(h(\psi(z)) + h(z)), \quad z \in \mathbb{C}, \tag{6.49}$$

where ψ is unknown and h is a given function which is analytic in a neighborhood of $0 \in \mathbb{C}$ such that $h(0) = 0$ and its derivative $h'(0) = \xi \neq 0$ (see [195]).

The existence of analytic solutions for (6.49) is accomplished by transforming the equation to another functional equation without iteration

$$\phi(\lambda^2 z) = 2\phi(\lambda z) - \phi(z) - \frac{1}{2}(h(\phi(\lambda z)) + h(\phi(z))), \quad z \in \mathbb{C}, \tag{6.50}$$

called the auxiliary equation of (6.49), where $\lambda \neq 0$ satisfies the algebraic equation

$$2\lambda^2 - (4 - \xi)\lambda + 2 + \xi = 0, \tag{6.51}$$

and by constructing analytic solutions for (6.50).

Theorem 6.8. *Assume that $0 < |\lambda| \neq 1$. Then for any $\tau \in \mathbb{C}$, the auxiliary equation (6.50) has a solution $\phi(z)$ which is analytic on a neighborhood of the origin and $\phi(0) = 0$ as well as $\phi'(0) = \tau$.*

Proof. Clearly, if $\tau = 0$, (6.50) has a trivial solution. Assume $\tau \neq 0$. By our assumption on h, we may let

$$h(z) = \widehat{a}(z) = \sum_{n=0}^{\infty} a_n z^n, \qquad (6.52)$$

where $a_0 = 0$ and $a_1 = \xi$. Since h is analytic in a neighborhood of the origin, by Cauchy's Estimation (Theorem 3.26), there exists a constant $\rho > 0$ such that $|a_n| \leq \rho^{n-1}$ for $n \geq 2$. By means of the transformations $\tilde{\phi}(z) = \rho \phi(\rho^{-1} z)$ and $\tilde{h}(z) = \rho h(\rho^{-1} z)$, in view of (6.50), we see $\tilde{\phi}(z)$ satisfies

$$\tilde{\phi}(\lambda^2 z) = 2\tilde{\phi}(\lambda z) - \tilde{\phi}(z) - \frac{1}{2}\left(\tilde{h}\left(\tilde{\phi}(\lambda z)\right) + \tilde{h}\left(\tilde{\phi}(z)\right)\right), \quad z \in C,$$

which is of the same form as (6.50) but

$$\tilde{g}(z) = \rho h(\rho^{-1} z) = \xi z + \sum_{n=2}^{\infty} a_n \rho^{1-n} z^n,$$

where obviously the coefficient $|a_n \rho^{1-n}| \leq 1$ for $n \geq 2$. Thus, we may assume without loss of generality that

$$|a_n| \leq 1, \quad n \geq 2. \qquad (6.53)$$

Let

$$\phi(z) = \widehat{b}(z) = \sum_{n=0}^{\infty} b_n z^n, \qquad (6.54)$$

where $b_0 = 0$, be a formal solution of (6.50). Substituting it into (6.50), we have

$$\lambda^2 \cdot b = 2\lambda \cdot b - b - \frac{1}{2} a \circ (\lambda \cdot b) - \frac{1}{2} a \circ b. \qquad (6.55)$$

Hence

$$\left(\lambda^{2n} - 2\lambda^n + 1\right) b_n = -\frac{1}{2}\left(\lambda^n + 1\right) \sum_{k=1}^{n} a_k b_n^{\langle k \rangle}, \quad n \in \mathbf{Z}^+. \qquad (6.56)$$

In other words,

$$\left(\lambda^2 - 2\lambda + 1 + \frac{1}{2}(\lambda + 1)\xi\right) b_1 = 0,$$

and

$$\left(\lambda^{2n} - 2\lambda^n + 1 + \frac{1}{2}(\lambda^n + 1)\xi\right) b_n = -\frac{1}{2}(\lambda^n + 1) \sum_{k=2}^{n} a_k b_n^{\langle k \rangle}, \quad n \geq 2. \qquad (6.57)$$

In view of (6.51), $\xi = -2(\lambda - 1)^2 / (\lambda + 1)$. Hence the coefficient of b_1 is zero, and (6.57) is reduced to

$$b_n = -\frac{(\lambda + 1)(\lambda^n + 1)}{2(\lambda^n - \lambda)(\lambda^{n+1} + \lambda^n + \lambda - 3)} \sum_{k=2}^{n} a_k b_n^{\langle k \rangle}, \quad n \geq 2. \qquad (6.58)$$

Consequently, we may choose $b_1 = \tau \neq 0$ and then determine the sequence $\{b_n\}_{n=2}^\infty$ by (6.58) recursively.

In what follows we prove the convergence of series (6.54) for each z in a neighborhood of the origin. In view of our assumption that $0 < |\lambda| \neq 1$,

$$\lim_{n \to \infty} \frac{(\lambda+1)(\lambda^n+1)}{2(\lambda^n-\lambda)(\lambda^{n+1}+\lambda^n+\lambda-3)} = \begin{cases} \frac{\lambda+1}{\lambda(3-\lambda)}, & 0 < |\lambda| < 1, \\ 0, & |\lambda| > 1. \end{cases}$$

Hence there exists $M > 0$ such that

$$\left| \frac{(\lambda+1)(\lambda^n+1)}{2(\lambda^n-\lambda)(\lambda^{n+1}+\lambda^n+\lambda-3)} \right| \leq M, \quad n \geq 2. \tag{6.59}$$

From (6.58) and (6.53) we see that

$$|b_n| \leq M \sum_{k=2}^n |b|_n^{\langle k \rangle}, \quad n \geq 2. \tag{6.60}$$

Next, note that Example 4.8 asserts that

$$W - |\tau| z - M \frac{W^2}{1-W} = 0 \tag{6.61}$$

has a solution $W(z)$ which is analytic on a neighborhood of the origin and

$$W(z) = \sum_{n=0}^\infty B_n z^n, \tag{6.62}$$

where $B_0 = 0$, $B_1 = |\tau|$ and

$$B_n = M \sum_{k=2}^n B_n^{\langle k \rangle}, \quad n \geq 2.$$

Furthermore,

$$|b_n| \leq B_n, \quad n \in \mathbf{Z}^+. \tag{6.63}$$

In fact $|b_1| = |\tau| = B_1$. For inductive proof we assume that $|b_j| \leq B_j$ for $j \leq n-1$. Then

$$|b_n| \leq M \sum_{k=2}^n |b|_n^{\langle k \rangle} \leq M \sum_{k=2}^n B_n^{\langle k \rangle} = B_n$$

as required. In other words, $\phi(z)$ is majorized by the analytic function $W(z)$. The proof is complete.

Theorem 6.9. *Assume that λ is a Siegel number. Then for any $\tau \in C$ with $0 < |\tau| \leq 1$, the auxiliary equation (6.50) has an analytic solution $\phi(z)$ in a neighborhood of the origin such that $\phi(0) = 0$ and $\phi'(0) = \tau$.*

Proof. As in the proof of the previous Theorem, we seek a power series solution of (6.50) of the form (6.54). Choosing $b_1 = \tau$ as before and using the same arguments as above we can uniquely determine the sequence $b = \{b_n\}_{n \in \mathbf{N}}$. Note that the Siegel number λ is equal to $\exp(2\pi i\theta)$ for some irrational number θ. Thus

$$|\lambda - 3| = |\cos(2\pi\theta) + i\sin(2\pi\theta) - 3| \geq 3 - \cos(2\pi\theta) \geq 2,$$

that is, $N := |\lambda - 3| - 2 > 0$. From (6.58),

$$|b_n| \leq \frac{(|\lambda| + 1)(|\lambda|^n + 1)}{2|\lambda||\lambda^{n-1} - 1|\left(|\lambda - 3| - |\lambda|^{n+1} - |\lambda|^n\right)} \times \sum_{k=2}^{n} |b|_n^{\langle k \rangle}$$

$$\leq \frac{2}{|\lambda^{n-1} - 1|(|\lambda - 3| - 2)} \sum_{k=2}^{n} |b|_n^{\langle k \rangle}$$

$$\leq \frac{2}{N}\left|\lambda^{n-1} - 1\right|^{-1} \sum_{k=2}^{n} |b|_n^{\langle k \rangle} \qquad (6.64)$$

for $n \geq 2$. Clearly, the sequence b is majorized by the sequence $u = \{u_n\}_{n \in \mathbf{N}}$ defined by $u_0 = 0$, $u_1 = |\tau|$ and

$$u_n = \frac{2}{N}\left|\lambda^{n-1} - 1\right|^{-1} \sum_{k=2}^{n} u_n^{\langle k \rangle}, \ n \geq 2.$$

Therefore it suffices now to show that u has a positive radius of convergence. To this end, note that Example 4.8 asserts that the equation

$$V - z - \frac{2}{N}\frac{V^2}{1-V} = 0 \qquad (6.65)$$

has a solution $V(z)$ which is analytic on a neighborhood of the origin and

$$V(z) = \sum_{n=0}^{\infty} C_n z^n,$$

where $C_0 = 0$, $C_1 = 1$ and

$$C_n = \frac{2}{N}\sum_{k=2}^{n} C_n^{\langle k \rangle}, \ n \geq 2.$$

Furthermore, the Cauchy's Estimation (Theorem 3.26) shows that there is some $T > 0$ such that $|V_n| \leq T^n$ for $n \in \mathbf{N}$. Thus by Theorem 3.32, we see that the radius of convergence of u is positive. The proof is complete.

Theorem 6.10. *Assume that $0 < |\lambda| \neq 1$ or λ is a Siegel number. Then Eq. (6.49) has an analytic solution of the form $\psi(x) = \phi(\lambda\phi^{-1}(z))$ in a neighborhood of the origin, where $\phi(z)$ is an analytic solution of the auxiliary equation (6.50).*

Proof. By Theorems 6.8 and 6.9, we can find an analytic solution $\phi(z)$ of the auxiliary equation (6.50) in the form of (6.54) such that $\phi(0) = 0$ and $\phi'(0) = \tau \neq 0$. Clearly the inverse $\phi^{-1}(z)$ exists and is analytic in a neighborhood of the origin. Let

$$\psi(x) = \phi(\lambda \phi^{-1}(z)), \qquad (6.66)$$

which is also analytic on a neighborhood of the origin. From (6.50), it is easy to see that

$$\psi(\psi(z)) = \phi(\lambda \phi^{-1}(\phi(\lambda \phi^{-1}(z)))) = \phi(\lambda^2 \phi^{-1}(z))$$
$$= 2\phi(\lambda \phi^{-1}(z)) - \phi(\phi^{-1}(z)) - \frac{1}{2}(h(\phi(\lambda \phi^{-1}(z))) + h(\phi(\phi^{-1}(z))))$$
$$= 2\psi(z) - z - \frac{1}{2}(h(\psi(z))h(z)),$$

that is, the function ψ in (6.66), defined on a neighborhood of the origin, satisfies (6.49).

Example 6.3. Let

$$h(z) = 2(1 - e^z) = -\sum_{n=1}^{\infty} \frac{2}{n!} z^n. \qquad (6.67)$$

The algebraic equation corresponding to (6.51) is

$$2\lambda(\lambda - 3) = 0, \qquad (6.68)$$

which has a nonzero root $\lambda_1 = 3$. By Theorem 6.8, the auxiliary equation

$$\phi(9z) = 2\phi(3z) - \phi(z) - \frac{1}{2}(h(\phi(3z)) + h(\phi(z))) \qquad (6.69)$$

has a solution of the form (6.54) where $b_1 = \tau \neq 0$ is arbitrary and $b_2, b_3, \ldots$ are determined by (6.58) recursively, i.e.,

$$b_n = \frac{3^n + 1}{(3^{n-1} - 1) 3^{n+1}} \sum_{k=2}^{n} \left(\frac{1}{k!}\right) b_n^{\langle k \rangle}, \; n \geq 2. \qquad (6.70)$$

In particular,

$$b_2 = \frac{\phi''(0)}{2!} = \frac{5}{54}\tau^2,$$
$$b_3 = \frac{\phi'''(0)}{3!} = \frac{3^3 + 1}{(3^2 - 1) 3^4}\left(b_1 b_2 + \frac{b_1^3}{6}\right) = \frac{7^2}{2 \cdot 3^7}\tau^3,$$

etc. Since $\phi(0) = 0$, $\phi'(0) = \tau \neq 0$ and the inverse $\phi^{-1}(z)$ is analytic near the origin, we can calculate

$$(\phi^{-1})'(0) = \frac{1}{\phi'(\phi^{-1}(0))} = \frac{1}{\phi'(0)} = \frac{1}{\tau},$$

$$\left(\phi^{-1}\right)''(0) = -\frac{\phi''\left(\phi^{-1}(0)\right)\left(\phi^{-1}\right)'(0)}{\left(\phi'\left(\phi^{-1}(0)\right)\right)^2} = -\frac{\phi''(0)\left(\phi^{-1}\right)'(0)}{\left(\phi'(0)\right)^2} = -\frac{5}{27\tau},$$

$$\left(\phi^{-1}\right)'''(0)$$
$$= -\frac{\left\{\phi'''\left(\phi^{-1}(0)\right)\left(\left(\phi^{-1}\right)'(0)\right)^2 + \phi''\left(\phi^{-1}(0)\,\phi^{-1}(0)\right)\left(\phi^{-1}\right)''(0)\right\}\left(\phi'\left(\phi^{-1}(0)\right)\right)^2}{\left(\phi^{-1}\left(\phi^{-1}(0)\right)\right)^4}$$
$$+ \frac{\phi''\left(\phi^{-1}(0)\right)\left(\phi^{-1}\right)'(0)\cdot 2\phi'\left(\phi^{-1}(0)\right)\phi''\left(\phi^{-1}(0)\right)\left(\phi^{-1}\right)'(0)}{\left(\phi'\left(\phi^{-1}(0)\right)\right)^4}$$
$$= \frac{26}{3^6\tau},$$

etc. Furthermore,

$$\psi(0) = \phi\left(3\phi^{-1}(0)\right) = \phi(0) = 0,$$

$$\psi'(0) = \phi'\left(3\phi^{-1}(0)\right)\cdot 3\left(\phi^{-1}\right)'(0) = 3\phi'(0)\left(\phi^{-1}\right)'(0) = 3\tau\frac{1}{\tau} = 3,$$

$$\psi''(0) = 9\phi''\left(3\phi^{-1}(0)\right)\left(\left(\phi^{-1}\right)'(0)\right)^2 + 3\phi'\left(3\phi^{-1}(0)\right)\left(\phi^{-1}\right)''(0) = \frac{10}{9},$$

$$\psi'''(0) = 27\phi'''\left(3\phi^{-1}(0)\right)\left(\left(\phi^{-1}\right)'(0)\right)^3 + 18\phi''\left(3\phi^{-1}(0)\right)\left(\phi^{-1}\right)'(0)\left(\phi^{-1}\right)''(0)$$
$$+ 9\phi''\left(3\phi^{-1}(0)\right)\left(\phi^{-1}\right)'(0)\left(\phi^{-1}\right)''(0) + 3\phi'\left(3\phi^{-1}(0)\right)\left(\phi^{-1}\right)'''(0)$$
$$= \frac{28}{27},$$

etc. Thus near 0, Eq. (6.49) with h in (6.67) has an analytic solution of the form

$$\psi(z) = 3z + \frac{5}{9}z^2 + \frac{17}{81}z^3 + \cdots. \tag{6.71}$$

We remark that if $h(x)$ in (6.49) is an analytic function near 0 with real coefficients, and if $a_1 = \xi$ satisfies $\xi < 0$ or $\xi \geq 16$, then by Theorem 6.8, Eq. (6.49) has an analytic real solution. Indeed, the equation $2\lambda^2 - (4-\xi)\lambda + 2 + \xi = 0$ now has real roots λ_1 and λ_2. Clearly by (6.58) where $\lambda = \lambda_1$ or λ_2, we can define a real sequence $\{b_n\}_{n=2}^\infty$ and obtain a solution $\phi(z)$ of (6.50) with real coefficients. Since ϕ and its inverse are real valued functions, the function $\psi(x) = \phi\left(\lambda_j\phi^{-1}(z)\right)$, where $j = 1$ or 2, is also a real function and Theorem 6.8 implies its analyticity.

6.2 Equations with First Order Derivatives

The existence of solutions of differential equations of the form

$$x'(t) = f(t, x(t))$$

is of fundamental importance in the theory of ordinary differential equations. There are, however, plenty of differential equations that are useful in modeling natural processes, but cannot be written in the above form. For instance, Cooke in [41] describes a model of infection, a prey-predator model, a respiration model and a model in electrodynamics, all of them call for equations of the form

$$x'(t) = x(t - x(t))$$

or

$$x'(t) + ax(t - rx(t))) = 0.$$

Earlier Driver in [50] investigated an equation of the form

$$y'(t) = f(t, y(t), y(g(t, y(t))), y'(g(t, y(t))))$$

which is related to the Dirac equation of classical electrodynamics without radiation effect.

As a further example, consider a sequence of curves in the plane that can be described by a sequence of functions $x_0(t), x_1(t), x_2(t), \ldots$. Suppose the slope of each function $x_k(t)$ is related to its value at some u, that is

$$x'_k(t) = Lx_k(u),$$

for some real number L and u is calculated at $x_{k-1}(t)$, that is,

$$u = x_{k-1}(t).$$

The question then arise as what are $\{x_0(t), x_1(t), x_2(t), \ldots\}$. Such a family of functions is naturally called a solution of the above relations. Such a problem can be quite difficult, but one approach is to find a family of functions such that $x_k(t) = x_{k+1}(t)$ for all large k (usually called a stationary solution). Then we are led to

$$x'(t) = Lx(x(t)).$$

6.2.1 Equation I

Consider the following differential equation [185]

$$x'(z) = x^{[m]}(z) \qquad (6.72)$$

where m is a positive integer greater than or equal to 2, and $x^{[m]}(z)$, as defined before, denotes the m-th iterate of the function $x(z)$. We will find its solutions which are analytic over a neighborhood of a complex number α which is a Siegel number or satisfies $0 < |\alpha| < 1$.

To this end, we first seek a formal power series solution for the following equation

$$y'(\alpha z) = \frac{1}{\alpha} y'(z) y(\alpha^m z), \qquad (6.73)$$

subject to the initial condition
$$y(0) = \alpha \tag{6.74}$$
Then we show that such a power series solution is majorized by a convergent power series. Finally, we show that
$$x(z) = y\left(\alpha y^{-1}(z)\right) \tag{6.75}$$
is an analytic solution of (6.72) in a neighborhood of α.

Theorem 6.11. *Suppose $0 < |\alpha| < 1$. Then for each complex number $\eta \neq 0$, equation (6.73) has a solution of the form*
$$y(z) = \sum_{n=0}^{\infty} b_n z^n \tag{6.76}$$
which is analytic on a neighborhood of the origin and satisfies $b_0 = \alpha$ and $b_1 = \eta$.

Proof. Assume (6.73) has an analytic solution of the form $y(z) = \widehat{b}(z)$, where $b = \{b_n\} \in l^{\mathbf{N}}$ satisfies $b_0 = \alpha$ and $b_1 = \eta$. Substituting (6.76) into (6.73), we see that
$$\underline{\alpha} \cdot Db = \frac{1}{\alpha}(Db) * (\underline{\alpha}^m \cdot b).$$
Hence the sequence b can be determined by
$$\left(\alpha^{n+1} - \alpha\right)(n+1)b_{n+1} = \sum_{k=0}^{n-1}(k+1)\alpha^{m(n-k)}b_{k+1}b_{n-k}, \quad n \in \mathbf{Z}^+, \tag{6.77}$$
in a unique manner. Furthermore, there is some $M > 0$ such that
$$\left|\frac{(k+1)\alpha^{m(n-k)}}{(n+1)(\alpha^{n+1} - \alpha)}\right| \leq \frac{1}{|\alpha^n - 1|} \leq \frac{1}{M}, \quad n \geq 2, 0 \leq k \leq n-1.$$
Thus if we define a sequence $\{B_n\}_{n=0}^{\infty}$ by $B_0 = \alpha$, $B_1 = |\eta|$ and $B_{n+1} = M^{-1}B_{n+1}B_{n-k}$ for $n \in \mathbf{Z}^+$, then in view of (6.77),
$$|b_n| \leq B_n, \quad n \in \mathbf{Z}^+,$$
that is, b is majorized by the sequence $\{B_n\}_{n \in \mathbf{N}}$. Therefore our proof will be complete if we can show that the radius of convergence of $\{B_n\}_{n \in \mathbf{N}}$ is positive. To this end, note that the equation
$$G^2(z) - MG(z) - M|\eta|z = 0,$$
being a special case of (4.10), has a solution
$$G(z) = \sum_{n=0}^{\infty} g_n z^n$$
which is analytic on a neighborhood of the origin and the sequence $g = \{g_n\}_{n \in \mathbf{N}}$ is given by $g_0 = 0$, $g_1 = |\eta|$ and
$$g_{n+1} = M^{-1}g_{n+1}g_{n-k}, \quad n \geq 1.$$

Since $B_0 = g_0$ and $B_1 = g_1$, it is clear that $\{B_n\}_{n \in \mathbf{N}} = g$ so that $\{B_n\}_{n \in \mathbf{N}}$ has a positive radius of convergence. The proof is complete.

Theorem 6.12. *Suppose α is a Siegel number. If $\eta = 1$, then equation (6.73) has a solution of the form (6.76) which is analytic on a neighborhood of the origin and satisfies $b_0 = \alpha$ and $b_1 = 1$.*

Proof. As in the previous proof, assume the existence of an analytic solution of the form (6.76) with $b_0 = \alpha$ and $b_1 = 1$. Then (6.77) holds again, so that

$$|b_{n+1}| \leq \frac{1}{|\alpha^n - 1|} \sum_{k=0}^{n-1} |b_{k+1}| |b_{n-k}|, \ n \in \mathbf{Z}^+. \tag{6.78}$$

Note that the equation

$$G^2(z) - G(z) + z = 0$$

being a special case of (4.10), has a solution $G(z)$ which is analytic on a neighborhood of the origin and

$$G(z) = \sum_{n=0}^{\infty} C_n z^n$$

where the sequence $C = \{C_n\}_{n \in \mathbf{N}}$ is defined by $C_0 = 0$, $C_1 = 1$ and

$$C_n = C_n^{(2)}, \ n \geq 2.$$

As in the proof of Theorem 3.32, we may apply Siegel's Lemma (Theorem 3.31) to conclude that the sequence b has a positive radius of convergence. The proof is complete.

We now state and prove our main result in this section.

Theorem 6.13. *Suppose $0 < |\alpha| < 1$ or α is a Siegel number. Then equation (6.72) has a solution $x(z)$ which is analytic on a neighborhood of α. Furthermore, $x(z)$ is of the form*

$$x(z) = y\left(\alpha y^{-1}(z)\right),$$

where $y(z)$ is a solution of (6.73) that is analytic at 0 and satisfies $y(0) = \alpha$.

Proof. It suffices to show that the power series function $y(z)$ generated by the sequence b defined by $b_0 = \alpha$, $b_1 = \eta \neq 0$, and (6.77) satisfies (6.72). Indeed, since $y'(0) = \eta \neq 0$, the function $y^{-1}(z)$ is analytic in a neighborhood of the point $y(0) = \alpha$. If we now define $x(z)$ by means of (6.75), then

$$x'(z) = \alpha y'\left(\alpha y^{-1}(z)\right)\left(y^{-1}\right)'(z)$$

$$= \alpha y'\left(\alpha y^{-1}(z)\right) \frac{1}{y'(y^{-1}(z))}$$

$$= y\left(\alpha^m y^{-1}(z)\right) = x^{[m]}(z),$$

as required. The proof is complete.

We remark that in the above proof, since
$$y\left(\alpha y^{-1}(\alpha)\right) = y(\alpha \cdot 0) = \alpha,$$
α is a fixed point of the solution $x(z)$.

We now show how to explicitly construct an analytic solution of (6.72) by means of (6.77). Let α be a complex number which either satisfies $0 < |\alpha| < 1$ or is a Siegel number. By means of the previous Theorem, equation (6.73) has an analytic solution of the form $y(z) = \hat{b}(z)$, where $b_0 = \alpha, b_1 = \eta \neq 0$, and $\{b_n\}_{n=2}^{\infty}$ is determined by (6.77). It is not difficult to calculate the coefficients b_n by means of (6.77), indeed the first few terms are as follows:

$$b_2 = \frac{y''(0)}{2} = \frac{\alpha^{m-1}\eta^2}{2(\alpha-1)},$$

$$b_3 = \frac{y'''(0)}{3!} = \frac{\alpha^{2(m-1)}(\alpha^m + 2)\eta^3}{3!(\alpha^2 - 1)(\alpha - 1)},$$

$$b_4 = \frac{y^{(4)}(0)}{4!} = \frac{\alpha^{3m-3}\left[(\alpha^{2m} + 3)(\alpha^m + 2)(\alpha - 1) + 3\alpha^m(\alpha^2 - 1)\right]\eta^4}{4!(\alpha^3 - 1)(\alpha^2 - 1)(\alpha - 1)^2}.$$

Furthermroe, since $y^{-1}(z)$ is analytic in a neighborhood of the point $y(0) = \alpha$, it can also be determined once its derivatives at α have been determined

$$\left(y^{-1}\right)'(\alpha) = \frac{1}{y'\left(y^{-1}(\alpha)\right)} = \frac{1}{y'(0)} = \frac{1}{\eta},$$

$$\left(y^{-1}\right)''(\alpha) = -\frac{y''\left(y^{-1}(\alpha)\right)\left(y^{-1}\right)'(\alpha)}{\left(y'\left(y^{-1}(\alpha)\right)\right)^2} = -\frac{y''(0)\left(y^{-1}\right)'(\alpha)}{\left(y'(0)\right)^2} = -\frac{\alpha^{m-1}}{(\alpha-1)\eta},$$

$$\left(y^{-1}\right)'''(\alpha)$$
$$= -\frac{\left\{y'''\left(y^{-1}(\alpha)\right)\left[\left(y^{-1}\right)'(\alpha)\right]^2 + y''\left(y^{-1}(\alpha)\right)\left(y^{-1}\right)''(\alpha)\right\}\left[y'\left(y^{-1}(\alpha)\right)\right]^2}{\left[y'\left(y^{-1}(\alpha)\right)\right]^4}$$
$$+ \frac{y''\left(y^{-1}(\alpha)\right)\left(y^{-1}\right)'(\alpha) \cdot 2 \cdot y'\left(y^{-1}(\alpha)\right)y''\left(y^{-1}(\alpha)\right)\left(y^{-1}\right)'(\alpha)}{\left[y'\left(y^{-1}(\alpha)\right)\right]^4}$$
$$= -\frac{\left[y'''(0)\eta^{-2} - \eta''(0)\alpha^{m-1}/(\alpha-1)\eta\right]\eta^2 - \eta''(0)\eta^{-1} \cdot 2 \cdot \eta y''(0)\eta^{-1}}{\left[y'(0)\right]^4}$$
$$= \frac{\alpha^{2(m-1)}(3\alpha - \alpha^m + 1)}{(\alpha-1)^2(\alpha+1)\eta},$$

etc. Finally, we determine a solution $x(z)$ of (6.72) by finding its derivatives at α:

$$x(\alpha) = y\left(\alpha y^{-1}(\alpha)\right) = y(\alpha \cdot 0) = \alpha,$$

$$x'(\alpha) = y'\left(\alpha y^{-1}(\alpha)\right) \cdot \alpha\left(y^{-1}\right)'(\alpha) = \alpha y'(0)\left(y^{-1}\right)'(\alpha) = \alpha\eta \cdot \frac{1}{\eta} = \alpha,$$

$$x''(\alpha) = \alpha^2 y''(\alpha y^{-1}(\alpha)) \left[(y^{-1})'(\alpha)\right]^2 + \alpha y'(\alpha y^{-1}(\alpha))(y^{-1})''(\alpha) = \alpha^m,$$

$$x'''(\alpha) = y'''(\alpha y^{-1}(\alpha)) \left[\alpha(y^{-1})'(\alpha)\right]^3 + y''(\alpha y^{-1}(\alpha)) \cdot 2\alpha(y^{-1})'(\alpha) \cdot 2(y^{-1})''(\alpha)$$
$$+ y''(\alpha y^{-1}(\alpha)) \cdot \alpha(y^{-1})'(\alpha) \cdot \alpha(y^{-1})''(\alpha) + y'(\alpha y^{-1}(\alpha)) \cdot \alpha(y^{-1})'''(\alpha)$$
$$= \frac{\alpha^{2m-1}(\alpha^m - 1)}{\alpha - 1} = \alpha^{2m-1}(\alpha^{m-1} + \cdots + \alpha + 1),$$

etc. Thus, the desired solution of (6.72) is

$$x(z) = \alpha + \alpha(z - \alpha) + \frac{\alpha^m}{2!}(z-\alpha)^2 + \frac{\alpha^{2m-1}(\alpha^m - 1)}{3!(\alpha - 1)}(z-\alpha)^3 + \cdots \quad (6.79)$$

We remark that a simple program can be used to generate other terms in (6.79). For instance, by means of the following Mathematica program:

num = 4;
$b[0] = \alpha$;
$b[1] = \eta$;
$Do[b[n+1] = \text{Sum}[(k+1) * \alpha^{\wedge}(m * (n-k)) * b[k+1] * b[n-k], \{k, 0, n-1\}] /$
$((\alpha^{\wedge}(n+1) - \alpha) * (n+1), \{n, 1, \text{num}\}]$
$y(0) = \alpha$;
$iy[x\_] := \text{InverseFunction}[y][x]$
$Do[\text{Derivative}[n][y][0] = b[n] * n!, \{n, 1, \text{num}\}]$
$Do[a[n] = \text{Simplify}[Dy[\alpha * iy[x]], \{x, n\}] / .\text{InverseFunction}[y][x] \to 0], \{n, 1, \text{num}\}]$;
$Do[\text{Print}["a[", i, "] =", a[i]], \{i, 1, \text{num}\}]$;

we may obtain

$$a[1] = \alpha$$
$$a[2] = \alpha^m$$
$$a[3] = \frac{\alpha^{-1+2m}(-1 + \alpha^m)}{-1 + \alpha}$$
$$a[4] = \frac{\alpha^{-2+3m}(-1+\alpha^m)(-1-3\alpha+3\alpha^m+\alpha^{2m})}{(-1+\alpha)^2(1+\alpha)}$$

so that $a[i] = x^{(i)}(\alpha)$. By changing the value of num, we may obtain any $a[\text{num}]$ as desired.

6.2.2 Equation II

Next, we will be concerned with a more general class of equation of the form [181]

$$x'(z) = c_1 x(z) + c_2 x^{[2]}(z) + \cdots + c_m x^{[m]}(z), \quad c_1 + \cdots + c_m \neq 0. \quad (6.80)$$

where $c_1, ..., c_m$ are complex numbers, and $x^{[k]}(z)$ denotes the k-th iterate of the function $x(z)$. We will construct analytic solutions for our equation in a neighborhood of a complex number of the form $\alpha/(c_1 + \cdots + c_m)$ where α either satisfies $0 < |\alpha| < 1$ or is a Siegel number.

We first seek a formal power series solution for the following initial value problem

$$y'(\alpha z) = \frac{1}{\alpha} y'(z) \sum_{i=1}^{m} c_i y(\alpha^i z), \qquad (6.81)$$

$$y(0) = \frac{\alpha}{c_1 + \cdots + c_m}. \qquad (6.82)$$

For the sake of convenience, we will set

$$C = c_1 + \cdots + c_m.$$

Theorem 6.14. *Suppose* $0 < |\alpha| < 1$. *Then for any complex number* $\eta \neq 0$, *equation (6.81) has a solution of the form*

$$y(z) = \frac{\alpha}{C} + \eta z + \sum_{n=2}^{\infty} b_n z^n \qquad (6.83)$$

which is analytic on a neighborhood of the origin.

Proof. We seek a solution of (6.81) in a power series of the form $y(z) = \widehat{b}(z)$ where $b_0 = \alpha/C$ and $b_1 = \eta$. Substituting (6.83) into (6.81), we see that

$$\underline{\alpha} \cdot Db = \frac{1}{\alpha}(Db) * \left(\sum_{i=1}^{m} c_i \underline{\alpha}^i \cdot b \right).$$

Hence the sequence $\{b_n\}_{n=2}^{\infty}$ is successively determined by the condition

$$(\alpha^{n+1} - \alpha)(n+1)b_{n+1} = \sum_{k=0}^{n-1}(k+1)\sum_{i=1}^{m} c_i \alpha^{i(n-k)} b_{k+1} b_{n-k}, n \in \mathbf{Z}^+. \qquad (6.84)$$

in a unique manner. Since $0 < |\alpha| < 1$, there exists a positive number N such that for $n > N$,

$$|\alpha|^{n+1} < |\alpha| - \gamma$$

for some γ satisfying $0 < \gamma < |\alpha|$. From (6.76), we see that

$$\gamma |b_{n+1}| \leq \left(|\alpha| - |\alpha|^{n+1} \right) |b_{n+1}| \leq \left| \sum_{i=1}^{m} c_i \right| \sum_{k=0}^{n-1} |b_{k+1}| |b_{n-k}|.$$

Note that Example 4.14 asserts that the polynomial equation

$$G^2 - \frac{\gamma}{|C|} G + \frac{\gamma}{|C|} |\eta| z + \sum_{n=2}^{N} \left(\frac{\gamma}{|C|} |\eta_n| - \sum_{k=1}^{n-1} |\eta_k| |\eta_{n-k}| \right) z^n = 0$$

has a solution $G(z)$ which is analytic on a neighborhood of the origin and

$$G(z) = \sum_{n=0}^{\infty} B_n z^n$$

where $\{B_n\}_{n \in \mathbf{N}}$ is defined by $B_0 = 0$, $B_1 = |\eta|$,
$$B_n = |\eta_n| \equiv |b_n|, n = 2, ..., N,$$
and
$$B_{n+1} = \gamma^{-1} \left| \sum_{i=1}^{m} c_i \right| \sum_{k=0}^{n-1} B_{k+1} B_{n-k}, n \geq N.$$

It is easy to show by induction that $b_n \leq B_n$ for $n \geq 1$, and hence the radius of convergence of b is positive. The proof is complete.

Theorem 6.15. *Suppose α is a Siegel number. If $\eta = 1$, equation (6.81) has an analytic solution of the form (6.83) in a neighborhood of the origin.*

The proof is similar to the that of Theorem 6.12 and hence will be sketched as follows. We first seek a power series solution of the form (6.83) where $b_0 = \alpha/(c_1 + \cdots + c_m)$ and $b_1 = 1$. This leads to (6.84) again so that
$$|b_{n+1}| = \frac{1}{|\alpha^n - 1|} \sum_{k=0}^{n-1} |b_{k+1}| |b_{n-k}|, \; n \geq 1.$$

To show that the formal solution converges in a neighborhood of the origin, note that the polynomial equation
$$|C| G^2(z) + z = G(z),$$
being a special case of (4.10), has a solution $G(z)$ which is analytic on a neighborhood of the origin and $G(z) = \hat{v}(z)$, where the sequence $v = \{v_n\}_{n \in \mathbf{N}}$ is defined by $v_0 = 0$, $v_1 = 1$ and
$$v_{n+1} = |C| v_n^{(2)}, \; n \geq 2.$$

As in the proof of Theorem 3.32, we may apply Siegel's Lemma (Theorem 3.31) to conclude that the radius of convergence of b is positive.

Theorem 6.16. *Suppose $0 < |\alpha| < 1$ or α is a Siegel number. Then equation (6.80) has an analytic solution of the form*
$$x(z) = \frac{\alpha}{C} + \alpha \left(z - \frac{\alpha}{C} \right) + \frac{1}{2!} \left(\sum_{i=1}^{m} c_i \alpha^i \right) \left(z - \frac{\alpha}{C} \right)^2$$
$$+ \frac{1}{3!} \left(\sum_{i=1}^{m} c_i \alpha^{i-1} \right) \left(\sum_{i=1}^{m} c_i \alpha^i \left(\alpha^{i-1} + \alpha^{i-2} + \cdots + 1 \right) \right) \left(z - \frac{\alpha}{C} \right)^3$$
$$+ \sum_{n=4}^{\infty} \frac{1}{n!} \lambda_n \left(z - \frac{\alpha}{C} \right)^n \tag{6.85}$$

in a neighborhood of α/C, where $\lambda_4, \lambda_5, ...$ are constants.

Proof. By the previous two results, (6.81) has an analytic solution of the form (6.83) in a neighborhood of the origin. If we pick $b_1 = \eta \neq 0$, then since $y'(0) \neq 0$, the function $y^{-1}(z)$ is analytic in a neighborhood of the point $b_0 = y(0) = \alpha/C$. If we now define $x(z)$ by means of

$$x(z) = y\left(\alpha y^{-1}(z)\right),$$

then

$$x\left(\frac{\alpha}{C}\right) = y(0) = \frac{\alpha}{C},$$

and

$$x'(z) = \alpha y'\left(\alpha y^{-1}(z)\right) \cdot \left(y^{-1}\right)'(z)$$

$$= \alpha y'\left(\alpha y^{-1}(z)\right) \cdot \frac{1}{y'(y^{-1}(z))}$$

$$= \sum_{i=1}^{m} c_i y\left(\alpha^i y^{-1}(z)\right) = \sum_{i=1}^{m} c_i x^{[i]}(z).$$

This shows that (6.80) has an analytic solution of the form

$$x(z) = \frac{\alpha}{C} + \sum_{n=1}^{\infty} \lambda_n \left(z - \frac{\alpha}{C}\right)^n$$

in a neighborhood of the number α/C.

To find out the first few terms of the coefficient sequence $\{\lambda_n\}_{n=1}^{\infty}$, we follow the above approach. First we calculate from (6.84) that

$$b_2 = \frac{y''(0)}{2!} = \frac{\eta^2}{2!(\alpha - 1)} \sum_{i=1}^{m} c_i \alpha^{i-1},$$

$$b_3 = \frac{y'''(0)}{3!} = \frac{\eta^3}{3!(\alpha^2 - 1)(\alpha - 1)} \left(\sum_{i=1}^{m} c_i \alpha^{i-1}\right) \left(\sum_{i=1}^{m} c_i \left(\alpha^{2i-1} + 2\alpha^{i-1}\right)\right),$$

etc. Next, we determine $y^{-1}(z)$ by calculating the first few terms of the its derivatives at b_0:

$$\left(y^{-1}\right)'(b_0) = \frac{1}{\eta},$$

$$\left(y^{-1}\right)''(b_0) = -\frac{1}{(\alpha - 1)\eta} \sum_{i=1}^{m} c_i \alpha^{i-1},$$

$$\left(y^{-1}\right)'''(b_0) = \frac{1}{(\alpha - 1)^2(\alpha + 1)\eta} \left(\sum_{i=1}^{m} c_i \alpha^{i-1}\right) \left(\sum_{i=1}^{m} c_i \alpha^{i-1}\left(3\alpha - \alpha^i + 1\right)\right),$$

etc. Finally, we determine the first few derivatives of $x(z)$ at b_0:

$$x(b_0) = y\left(\alpha y^{-1}(b_0)\right) = y(\alpha \cdot 0) = b_0,$$

$$x'(b_0) = \alpha,$$

$$x''(b_0) = \sum_{i=1}^{m} c_i \alpha^i,$$

$$x'''(b_0) = \left(\sum_{i=1}^{m} c_i \alpha^{i-1}\right) \left(\sum_{i=1}^{m} c_i \alpha^i \left(\alpha^{i-1} + \alpha^{i-2} + \cdots + 1\right)\right),$$

etc. The proof is complete.

As a final check of our derivation, note that the unique solution of

$$x'(z) = cx(z), \ c \neq 0,$$

$$x\left(\frac{\alpha}{c}\right) = \frac{\alpha}{c},$$

where α is arbitrary, is

$$x(z) = \frac{\alpha}{c} e^{c\left(z - \frac{\alpha}{c}\right)} = \frac{\alpha}{c} + \alpha \left(z - \frac{\alpha}{c}\right) + \frac{c\alpha}{2!}\left(z - \frac{\alpha}{c}\right)^2 + \frac{c^2\alpha}{3!}\left(z - \frac{\alpha}{c}\right)^3 + \cdots$$

which coincides with the formula (6.85) when $c_2 = c_3 = \cdots = c_m = 0$.

6.2.3 Equation III

We consider a class of functional differential equation of the form [180]

$$x'(z) = x(az + bx(z)). \tag{6.86}$$

When $a = 0$ and $b = 1$, equation (6.86) reduces to the iterative functional differential equation $x'(z) = x(x(z))$. When $b = 0$ and $|a| \leq 1$, equation (6.86) reduces to the functional differential equation $x'(z) = x(az)$.

When $a \neq 1$ and $b \neq 0$, we will construct analytic solutions for our equations in a neighborhood of the complex number $(\beta - a)/(1 - a)$, where β either satisfies $0 < |\beta| < 1$ or is a Siegel number.

We first seek a formal power series solution for the following initial value problem

$$y'(\beta z) = \frac{1}{\beta} y'(z) \left\{ y\left(\beta^2 z\right) - ay(\beta z) + a \right\}, \tag{6.87}$$

$$y(0) = \frac{\beta - a}{1 - a}. \tag{6.88}$$

Then we show that such a power series solution is majorized by a convergent power series. Then we show that

$$x(z) = \frac{1}{b} y(\beta y^{-1}(z)) - \frac{a}{b} z \tag{6.89}$$

is an analytic solution of (6.86) in a neighborhood of $(\beta - a)/(1 - a)$. Finally, we make use of a partial difference equation to show how to explicitly construct such a solution.

Theorem 6.17. *Suppose $0 < |\beta| < 1$ holds. Then for any nontrivial complex number η, equation (6.87) has a solution of the form*

$$y(z) = \frac{\beta - a}{1 - a} + \eta z + \sum_{n=2}^{\infty} b_n z^n \qquad (6.90)$$

which is analytic on a neighborhood of the origin. Furthermore, there exists a positive constant M such that for z in this neighborhood,

$$|y(z)| \leq \left|\frac{\beta - a}{1 - a}\right| + \frac{1}{2M}.$$

Proof. We seek a solution of (6.87) in a power series of the form $y(z) = \widehat{b}(z)$ where $b_0 = (\beta - a)/(1 - a)$ and $b_1 = \eta$. Substituting (6.90) into (6.87), we see that

$$\underline{\beta} \cdot Db = \frac{1}{\beta}(Db) * [\underline{\beta}^2 \cdot b - a\underline{\beta} \cdot b + \overline{a}].$$

Hence the sequence $\{b_n\}_{n=2}^{\infty}$ is successively determined by the condition

$$(\beta^{n+1} - \beta)(n+1)b_{n+1} = \sum_{k=0}^{n-1}(k+1)\left(\beta^{2(n-k)} - a\beta^{n-k}\right) b_{k+1} b_{n-k}, \ n \in \mathbf{Z}^+, \quad (6.91)$$

in a unique manner. Furthermore, since $0 \leq k \leq n-1$, we see that

$$\left|\frac{\beta^{2(n-k)} - a\beta^{n-k}}{\beta^{n+1} - \beta}\right| \leq \frac{1 + |a|}{|\beta^n - 1|} \leq M, \ n \geq 2 \qquad (6.92)$$

for some positive number M.

Note that the equation

$$G^2(z) - \frac{1}{M}G(z) + \frac{1}{M}|\eta|z = 0,$$

being a special case of (4.10), has a solution

$$G(z) = \sum_{n=0}^{\infty} B_n z^n$$

which is analytic on a neighborhood of the origin and $b_0 = 0$, $b_1 = |\eta|$, and

$$B_{n+1} = M \sum_{k=0}^{n-1} B_{k+1} B_{n-k}, \ n \in \mathbf{Z}^+.$$

Since it is easily checked that

$$|b_n| \leq B_n$$

for $n \in \mathbf{Z}^+$. We see that the sequence b has a positive radius of convergence.

Next, recall that the solution $G(z)$ can be written as

$$G(z) = \frac{1}{2M}\left\{1 - \sqrt{1 - 4M|\eta|z}\right\}$$

which converges for $|z| \leq 1/(4M|\eta|)$. Since for $|z| \leq 1/(4M|\eta|)$,

$$\frac{1}{G(|z|)} = \frac{2M}{1 - \sqrt{1 - 4M|\eta||z|}} = \frac{1 + \sqrt{1 - 4M|\eta||z|}}{2|\eta||z|} \geq \frac{1}{2|\eta||z|},$$

or

$$G(|z|) \leq 2|\eta||z| \leq 2|\eta|\frac{1}{4M|\eta|} = \frac{1}{2M},$$

thus

$$|y(z)| \leq \left|\frac{\beta - a}{1 - a}\right| + \sum_{n=1}^{\infty} |b_n||z|^n \leq \left|\frac{\beta - a}{1 - a}\right| + \sum_{n=1}^{\infty} B_n |z|^n$$

$$= \left|\frac{\beta - a}{1 - a}\right| + G(|z|) \leq \left|\frac{\beta - a}{1 - a}\right| + \frac{1}{2M}$$

as required. The proof is complete.

Theorem 6.18. *Suppose β is a Siegel number. Then equation (6.87) has an analytic solution of the form*

$$y(z) = \frac{\beta - a}{1 - a} + z + \sum_{n=2}^{\infty} b_n z^n \quad (6.93)$$

in a neighborhood of the origin, and there exists a positive constant δ such that

$$|y(z)| \leq \left|\frac{\beta - a}{1 - a}\right| + \frac{1}{2^{5\delta+1}} \sum_{n=1}^{\infty} \frac{1}{n^{2\delta}}.$$

Proof. As in the previous proof, we seek a power series solution of the form (6.93). Then defining $b_0 = (\beta - a)/(1 - a)$ and $b_1 = 1$, (6.91) and (6.92) again hold so that

$$|b_{n+1}| \leq \frac{1 + |a|}{|\beta^n - 1|} \sum_{k=0}^{n-1} |b_{k+1}||b_{n-k}| \quad (6.94)$$

for $n \in \mathbf{Z}^+$. Note that the equation

$$(1 + |a|)G^2(z) + z = G(z),$$

being a special case of (4.10), has a solution

$$G(z) = \sum_{n=0}^{\infty} C_n z^n = \frac{1}{2(1 + |a|)} \left\{1 - \sqrt{1 - 4(1 + |a|)z}\right\}$$

which is analytic on $B(0; 1/(4(1 + |a|)))$ and $C_0 = 0$, $C_1 = 1$ and

$$C_{n+1} = (1 + |a|) \sum_{k=0}^{n-1} C_{k+1} C_{n-k}, \quad n \in \mathbf{Z}^+.$$

In view of the Cauchy Estimation (Theorem 3.26), there is some $r > 0$ such that $|C_n| \leq r^n$ for $n \geq 1$. As in the proof of Theorem 3.32, we may apply Siegel's Lemma (Theorem 3.31) to conclude that there is $\delta > 0$ such that

$$|b_n| \leq r^n \left(2^{5\delta+1}\right)^{n-1} n^{-2\delta}, \ n \in \mathbf{Z}^+,$$

which shows that the series (6.90) converges for $|z| < \left(r2^{5\delta+1}\right)^{-1}$.

Finally, when $|z| \leq (r2^{5\delta+1})^{-1}$, we have

$$|y(z)| \leq \left|\frac{\beta - a}{1 - a}\right| + \sum_{n=1}^{\infty} |b_n| |z|^n \leq \left|\frac{\beta - a}{1 - a}\right| + \sum_{n=1}^{\infty} C_n d_n |z|^n$$

$$\leq \left|\frac{\beta - a}{1 - a}\right| + \sum_{n=1}^{\infty} r^n \left(2^{5\delta+1}\right)^{n-1} n^{-2\delta} |z|^n$$

$$\leq \left|\frac{\beta - a}{1 - a}\right| + \sum_{n=1}^{\infty} r^n \left(2^{5\delta+1}\right)^{n-1} n^{-2\delta} (r2^{5\delta+1})^{-n}$$

$$= \left|\frac{\beta - a}{1 - a}\right| + \frac{1}{2^{5\delta+1}} \sum_{n=1}^{\infty} \frac{1}{n^{2\delta}}$$

as required. The proof is complete.

Theorem 6.19. *Suppose $0 < |\beta| < 1$ or β is a Siegel number. Then equation (6.86) has an analytic solution $x(z)$ of the form (6.89) in a neighborhood of $(\beta - a)/(1 - a)$, where $y(z)$ is an analytic solution of equation (6.87). Furthermore, when $0 < |\beta| < 1$ holds, there is a positive constant M such that*

$$|x(z)| \leq \frac{1}{|b|} \left(\left|\frac{\beta - a}{1 - a}\right| + \frac{1}{2M} \right) + \left|\frac{a}{b}\right| |z|$$

in a neighborhood of $(\beta - a)/(1 - a)$; and when β is a Siegel number, there is a positive number δ such that

$$|x(z)| \leq \frac{1}{|b|} \left(\left|\frac{\beta - a}{1 - a}\right| + \frac{1}{Q} \sum_{n=1}^{\infty} \frac{1}{n^{2\delta}} \right) + \left|\frac{a}{b}\right| |z|, \ Q = 2^{5\delta+1},$$

in a neighborhood of $(\beta - a)/(1 - a)$.

Proof. In view of the previous two results, we may find a sequence $\{b_n\}_{n=2}^{\infty}$ such that the function $y(z)$ of the form by (6.93) is an analytic solution of (6.87) in a neighborhood of the origin. Since $y'(0) = 1$, the function $y^{-1}(z)$ is analytic in a neighborhood of the point $y(0) = (\beta - a)/(1 - a)$. If we now define $x(z)$ by means of (6.89), then

$$x'(z) = \frac{1}{b} \cdot \beta y' \left(\beta y^{-1}(z)\right) \cdot \left(y^{-1}\right)'(z) - \frac{a}{b} = \frac{\beta}{b} y' \left(\beta y^{-1}(z)\right) \cdot \frac{1}{y'\left(y^{-1}(z)\right)} - \frac{a}{b}$$

$$= \frac{1}{b} \left\{ y \left(\beta^2 y^{-1}(z)\right) - a y \left(\beta y^{-1}(z)\right) + a \right\} - \frac{a}{b}$$

$$= \frac{1}{b} \left\{ y \left(\beta^2 y^{-1}(z)\right) - a y \left(\beta y^{-1}(z)\right) \right\},$$

and
$$x\left(az + bx(z)\right) = x\left(az + b\left[\frac{1}{b}y\left(\beta y^{-1}(z)\right) - \frac{a}{b}z\right]\right) = x\left(y\left(\beta y^{-1}(z)\right)\right)$$
$$= \frac{1}{b}y\left(\beta y^{-1}\left(y\left(\beta y^{-1}(z)\right)\right)\right) - \frac{a}{b}y\left(\beta y^{-1}(z)\right)$$
$$= \frac{1}{b}\left\{y\left(\beta^2 y^{-1}(z)\right) - ay\left(\beta y^{-1}(z)\right)\right\}$$

as required.

Next, if $0 < |\beta| < 1$, then
$$|x(z)| = \frac{1}{|b|}\left|y\left(\beta y^{-1}(z)\right) - az\right| \leq \frac{1}{|b|}\left(\left|y\left(\beta y^{-1}(z)\right)\right| + |a||z|\right)$$
$$\leq \frac{1}{|b|}\left(\left|\frac{\beta - a}{1 - a}\right| + \frac{1}{2M}\right) + \left|\frac{a}{b}\right||z|;$$

and if β is a Siegel number, then
$$|x(z)| = \frac{1}{b}\left|y\left(\beta y^{-1}(z)\right) - az\right| \leq \frac{1}{|b|}\left(\left|y\left(\beta y^{-1}(z)\right)\right| + |a||z|\right)$$
$$\leq \frac{1}{|b|}\left(\left|\frac{\beta - a}{1 - a}\right| + \frac{1}{Q}\sum_{n=1}^{\infty}\frac{1}{n^{2\delta}}\right) + \left|\frac{a}{b}\right||z|.$$

The proof is complete.

We now show how to explicitly construct an analytic solution of (6.86) by means of (6.89). Since
$$x(z) = \frac{1}{b}y\left(\beta y^{-1}(z)\right) - \frac{a}{b}z,$$
thus
$$x\left(\frac{\beta - a}{1 - a}\right) = \frac{1}{b}y(0) - \frac{a}{b}\frac{\beta - a}{1 - a} = \frac{1}{b}\frac{\beta - a}{1 - a} - \frac{a}{b}\frac{\beta - a}{1 - a} = \frac{\beta - a}{b}.$$

Furthermore,
$$x'\left(\frac{\beta - a}{1 - a}\right) = x\left(a \cdot \frac{\beta - a}{1 - a} + bx\left(\frac{\beta - a}{1 - a}\right)\right)$$
$$= x\left(a \cdot \frac{\beta - a}{1 - a} + b \cdot \frac{\beta - a}{b}\right) = x\left(\frac{\beta - a}{1 - a}\right) = \frac{\beta - a}{b}.$$

By calculating the derivatives of both sides of (6.86), we obtain successively
$$x''(z) = x'(az + bx(z))(a + bx'(z)),$$
$$x'''(z) = x''(az + bx(z))(a + bx'(z))^2 + x'(az + bx(z))(bx''(z)),$$
so that
$$x''\left(\frac{\beta - a}{1 - a}\right) = x'\left(a \cdot \frac{\beta - a}{1 - a} + bx\left(\frac{\beta - a}{1 - a}\right)\right)\left(a + bx'\left(\frac{\beta - a}{1 - a}\right)\right)$$
$$= \beta x'\left(\frac{\beta - a}{1 - a}\right) = \frac{\beta(\beta - a)}{b},$$

$$x'''\left(\frac{\beta-a}{1-a}\right) = x''\left(\frac{\beta-a}{1-a}\right)\beta^2 + x'\left(\frac{\beta-a}{1-a}\right) \cdot bx''\left(\frac{\beta-a}{1-a}\right)$$
$$= \frac{1}{b}\left[\beta(\beta-a)\left(\beta^2+\beta-a\right)\right].$$

It seems from the above calculations that the higher derivatives $x^{(m)}(z)$ at $z = \xi \equiv (\beta-a)/(1-a)$ can be determined uniquely in similar manners. To see this, let us denote the derivative $\left(x^{(i)}(az+bx(z))\right)^{(j)}$ at $z=\xi$ by λ_{ij}, where $i,j \geq 0$. Note that the two derivatives $x^{(k)}(z)$ and $x^{(k)}(az+bx(z))$ are equal at the point $z = \xi$ since $a\xi + bx(\xi) = \xi$. In other words,

$$x^{(k)}(\xi) = \lambda_{k0}.$$

Furthermore, in view of (6.86), we see that $x^{(k+1)}(z) = (x(az+bx(z)))^{(k)}$ which implies

$$\lambda_{k+1,0} = \lambda_{0,k}.$$

Finally, since

$$\left(x^{(i)}(az+bx(z))\right)^{(j+1)} = \left(x^{(i+1)}(az+bx(z)) \cdot (a+bx'(z))\right)^{(j)}$$
$$= \sum_{k=0}^{j}\binom{j}{k}(a+bx'(z))^{(k)}\left(x^{(i+1)}(az+bx(z))\right)^{(j-k)},$$

we see also that

$$\lambda_{i,j+1} = \sum_{k=0}^{j}\binom{j}{k}\lambda_{i+1,j-k} \cdot (a+bx'(z))^{(k)}\Big|_{z=\xi}$$
$$= \beta\lambda_{i+1,j} + b\sum_{k=1}^{j}\binom{j}{k}\lambda_{i+1,j-k}\lambda_{0,k}$$

for $i,j \in \mathbf{N}$, where we have used the fact that $\lambda_{k+1,0} = \lambda_{0,k}$ in obtaining the last equality. Clearly, if we have obtained the derivatives $x^{(0)}(\xi) = \lambda_{00}, ..., x^{(m)}(\xi) = \lambda_{m0} = \lambda_{0,m-1}$, then by means of the above partial difference equation, we can successively calculate

$$\lambda_{m-1,1}, \lambda_{m-2,1}, \lambda_{m-2,2}, ..., \lambda_{11}, \lambda_{12}, ..., \lambda_{1,m-1}, \lambda_{0m}$$

in a unique manner. In particular, $\lambda_{0m} = \lambda_{m+1,0}$ is the desired derivative $x^{(m+1)}(\xi)$.

This shows that

$$x(z) = \frac{\beta-a}{b} + \frac{1}{b}(\beta-a)\left(z-\frac{\beta-a}{1-a}\right) + \frac{\beta(\beta-a)}{2!b}\left(z-\frac{\beta-a}{1-a}\right)^2$$
$$+ \frac{\beta(\beta-a)(\beta^2+\beta-a)}{3!b}\left(z-\frac{\beta-a}{1-a}\right)^3 + \sum_{i=4}^{\infty}\frac{\lambda_{i,0}}{i!}\left(z-\frac{\beta-a}{1-a}\right)^i.$$

6.2.4 Equation IV

Consider the equation [140]
$$x'(z) = \frac{1}{x(x(z))}. \tag{6.95}$$

If $x^{-1}(z)$ exists, and is substituted into both sides of (6.95), then by
$$x'(z) = \frac{1}{(x^{-1})'(x(z))}$$
we have
$$(x^{-1})'(x(z)) = x(x(z)). \tag{6.96}$$

Furthermore, from
$$\frac{d}{dz}(x^{-1})'(x(z)) = (x^{-1})''(x(z))x'(z) = x'(x(z))x'(z),$$
we see that
$$(x^{-1})''(x(z)) = x'(x(z)).$$

By induction, we see that
$$(x^{-1})^{(r)} \circ x = x^{(r-1)} \circ x \tag{6.97}$$
for $r \geq 2$.

Consider an analytic solution of (6.95) which has a fixed point $\zeta \neq 0$. Such a solution will be denoted by $x_\zeta(z)$ so that
$$x_\zeta(\zeta) = \zeta. \tag{6.98}$$

If such a solution exists, then by (6.95),
$$x'_\zeta(\zeta) = \frac{1}{\zeta}. \tag{6.99}$$

In view of (6.98) and (6.99), we may then let
$$x_\zeta(z) = \sum_{n=0}^{\infty} P_n(\zeta)(z-\zeta)^n, \tag{6.100}$$
where $P_0(\zeta) = \zeta$ and $P_1(\zeta) = 1/\zeta$.

To determine the remaining $P_n(\zeta)$, recall the n-th derivative of the composite function $h \circ g$ is, by Theorem 2.16, given by
$$(h \circ g)^{(n)} = \sum_{r=1}^{n} h^{(r)} \circ g \cdot \frac{n!}{r!} \sum_{n,r} \frac{g^{(p_1)} \cdots g^{(p_r)}}{p_1! \cdots p_r!} \tag{6.101}$$
where, throughout this section, $\sum_{n,r}$ is taken over $P_1 + \cdots + P_r = n$ and $P_1, ..., P_r \in \mathbf{Z}^+$. In (6.101) let $h = x_\zeta^{-1}$ and $g = x_\zeta$. Since $\left(x_\zeta^{-1} \circ x_\zeta\right)^{(n)} = 0$ for $n \geq 2$ and since (6.97) holds, if we evaluate both sides of the resulting equation at ζ, one obtains
$$0 = \sum_{r=1}^{n} \frac{1}{r} P_{r-1}(\zeta) \cdot \sum_{n,r} P_{p_1}(\zeta) \cdots P_{p_r}(\zeta). \tag{6.102}$$

In (6.102), $P_n(\zeta)$ occurs only in the second summation when $r = 1$. Solving for $P_n(\zeta)$ yields the recursion formula

$$P_n(\zeta) = -\frac{1}{\zeta}\sum_{r=2}^{n}\frac{1}{r}P_{r-1}(\zeta) \cdot \sum_{n,r} P_{p_1}(\zeta) \cdots P_{p_r}(\zeta). \tag{6.103}$$

We now note that for real $\zeta > 0$, $P_s(\zeta) = (-1)^{s-1}|P_s(\zeta)|$ for $s \geq 1$. Indeed, since $P_1(\zeta) = 1/\zeta$, our assertion is true for $s = 1$. Assume our assertion is true for $s = 1, 2, ..., n-1$. Then in view of (6.103), since $(P_1 - 1) + \cdots + (P_r - 1) = n - r$, one obtains

$$P_n(\zeta) = \frac{(-1)^{n-1}}{\zeta}\sum_{r=2}^{n}\frac{1}{r}|P_{r-1}(\zeta)| \cdot \sum_{n,r}|P_{p_1}(\zeta)|\cdots|P_{p_r}(\zeta)|. \tag{6.104}$$

Since all terms on the right are positive except for the factor $(-1)^{n-1}$, our conclusion follows.

Next, note that if $\zeta > 0$ and $x_k(z)$ in (6.100) converges for $x \in (\zeta - R_\zeta, \zeta + R_\zeta)$, then for any $\zeta' \geq \zeta$, it converges for $x \in (\zeta' - R_{\zeta'}, \zeta' + R_{\zeta'})$ where $R_{\zeta'} \geq R_\zeta$. Indeed, since $P_1(\zeta) = 1/\zeta$, if $\zeta' \geq \zeta$, then $|P_1(\zeta')| \leq |P_1(\zeta)|$. Assume $|P_s(\zeta')| \leq |P_s(\zeta)|$ for $s = 1, 2, ..., n-1$. Then in view of (6.104), $x_{\zeta'}(z) \ll x_\zeta(z)$ for any real $\zeta' \geq \zeta$.

Let $w_1 = (1 + \sqrt{5})/2, w_2 = (1 - \sqrt{5})/2$. It may be verified that

$$x_{w_1}(z) = \left(\frac{1+\sqrt{5}}{2}\right)^{(3-\sqrt{5})/2} \cdot z^{(-1+\sqrt{5})/2}$$

and

$$x_{w_2}(z) = \left(\frac{1-\sqrt{5}}{2}\right)^{(3+\sqrt{5})/2} \cdot z^{(-1-\sqrt{5})/2}$$

are analytic solutions of (6.95) and (6.96) such that $x_{w_1}(w_1) = w_1$ and $x_{w_2}(w_2) = w_2$. The binomial expansion of x_{w_1} about w_1 will converge for $|z - w_1| < w_1$. If the coefficients in this expansion are denoted by A_n, then by Theorem 3.8, it follows that $P_n(w_1) = A_n$. Hence for any $\zeta \geq w_1$, $x_\zeta(z)$ defined by (6.100) converges for $z \in (\zeta - R_\zeta, \zeta + R_\zeta)$, where $R_\zeta \geq R_{w_1} = w_1$.

Theorem 6.20. Let $w_1 = (1 + \sqrt{5})/2$ and let $\zeta \in \mathbf{C}$ such that $|\zeta| \geq w_1$. Then the function $x_\zeta(z)$ defined by (6.100) converges for $z \in B(\zeta; w_1)$.

Indeed, note that $(-1)^{n-1}P_n(\zeta) = \zeta^s \sum_r a_{n,r}\zeta^r$ where $a_{n,r} > 0$ for $n \geq 1$. This observation follows from $P_1(\zeta) = \zeta^{-1} \cdot 1$ and induction using (6.103). Thus if ζ is complex, then $|P_n(\zeta)| \leq |P_n(|\zeta|)|$.

We remark that if $x_\zeta(z)$ defined by (6.100) converges about $z = \zeta$, then x_ζ^{-1} exists and is analytic about ζ. Let $x_\zeta^{-1}(z) = \sum_{n=0}^{\infty} Q_n(\zeta)(z - \zeta)^n$ where by (6.100), $Q_0(\zeta) = Q_1(\zeta) = \zeta$. We may show that $Q_s(\zeta) = s^{-1}P_{s-1}(\zeta)$. For $s = 1$, this follows from comparison of (6.100) and $Q_1(\zeta)$. To prove the general case, assume our assertion is true for $s = 1, ..., n-1$. Apply (6.102) to $x_n^{-1} \cdot x_\zeta$ and evaluate at ζ, then

$$0 = \sum_{r=1}^{n} Q_r(\zeta) \sum_{n,r} P_{p_1}(\zeta) \cdots P_{p_r}(\zeta).$$

Isolating $Q_n(\zeta)$ yields, since $P_1(\zeta) = 1/\zeta$,

$$Q_n(\zeta) = -\zeta^n \sum_{r=1}^{n-1} Q_r(\zeta) \sum_{n,r} P_{p_1}(\zeta) \cdots P_{p_r}(\zeta).$$

Substituting $Q_r(\zeta) = P_{r-1}(\zeta)/r$, one obtains in view of (6.103), that

$$Q_n(\zeta) = -\zeta^n \left\{ -\zeta P_n(\zeta) - \frac{1}{n} P_{n-1}(\zeta) \frac{1}{\zeta^n} + \zeta P_n(\zeta) \right\} = \frac{1}{n} P_{n-1}(\zeta).$$

Example 6.4. For example, the equation

$$g'(z) = \frac{a}{g(g(z)) - b}$$

has as solutions

$$g_{a\zeta + b} = ax_\zeta \left(\frac{z-b}{a} \right) + b.$$

where $g_{a\zeta+b}(a\zeta + b) = a\zeta + b$.

6.2.5 First Order Neutral Equation

We will be concerned with analytic solutions of a iterative functional differential equation related to a state dependent functional differential equation of the form

$$\alpha z + \beta x'(z) = x(az + bx'(z)), \qquad (6.105)$$

where α, β, a, b are complex numbers.

In case $\alpha = a = 0, \beta = 1$ and $b = 1$, we obtain the functional differential equation

$$x'(z) = x(x'(z)). \qquad (6.106)$$

In case $b = 0$, $a \neq 0$ and $\beta \neq 0$, equation (6.105) changes into the functional differential equation

$$\alpha z + \beta x'(z) = x(az), \qquad (6.107)$$

and in case $b = 0$, $a \neq 0$ and $\beta = 0$, into the functional equation

$$\alpha z = x(az). \qquad (6.108)$$

A distinctive feature of the equation (6.105) when $b \neq 0$ is that the argument of the unknown function is dependent on the state derivative, and this is the case we will emphasize.

It is easy to find some of the analytic solutions of (6.105) in various special cases. For instance, equation (6.106) has the solution $x(z) = pz - p^2 + p$ for any constant

p, while (6.108) has the solution $x(z) = \alpha z/a$. In order to find an analytic solution $x = x(z)$ of (6.107), we formally assume that

$$x(z) = \hat{c}(z) = \sum_{n=0}^{\infty} c_n z^n.$$

Then in view of (6.107), we will obtain

$$\alpha \hbar + \beta Dc = \underline{a} \cdot c.$$

Hence

$$\beta c_1 = c_0, \alpha + 2\beta c_2 = a c_1,$$

and

$$\beta(n+1)c_{n+1} = a^n c_n, \; n \geq 2.$$

This leads to

$$x(z) = c_0 + \frac{c_0}{\beta}z + \frac{ac_0 - \alpha\beta}{2\beta^2}z^2 + (ac_0 - \alpha\beta)\sum_{n=3}^{\infty}\frac{1}{n!\beta^n}a^{(n-2)(n+1)/2}z^n.$$

As can be verified easily, when $0 < |a| \leq 1$, it is an entire solution of (6.107).

We now assume that $b \neq 0$. In order to construct analytic solutions of (6.105) in a systematic manner, we first let

$$y(z) = az + bx'(z). \tag{6.109}$$

Then for any number z_0, we have

$$x(z) = x(z_0) + \frac{1}{b}\int_{z_0}^{z}(y(s) - as)ds, \tag{6.110}$$

and

$$x(y(z)) = x(z_0) + \frac{1}{b}\int_{z_0}^{y(z)}(y(s) - as)ds.$$

Therefore, in view of (6.105), we have

$$x(az + bx'(z)) = x(y(z)) = x(z_0) + \frac{1}{b}\int_{z_0}^{y(z)}(y(s) - as)ds$$

$$= \alpha z + \frac{\beta}{b}(y(z) - az),$$

or

$$bx(z_0) + \int_{z_0}^{y(z)}(y(s) - as)ds = \beta y(z) + (b\alpha - a\beta)z. \tag{6.111}$$

In case z_0 is a fixed point of $y(z)$, i.e., $y(z_0) = z_0$, we see that

$$bx(z_0) + \int_{z_0}^{y(z_0)}(y(s) - as)ds = \beta y(z_0) + (b\alpha - a\beta)z_0,$$

or
$$x(z_0) = \frac{1}{b}(b\alpha + (1-a)\beta) z_0. \qquad (6.112)$$

Furthermore, differentiating both sides of the equation (6.111) with respect to z, we obtain an iterative functional differential equation

$$\{y(y(z)) - ay(z) - \beta\} y'(z) = b\alpha - a\beta. \qquad (6.113)$$

There are two cases to consider: (i) $b\alpha - a\beta = 0$; and (ii) $b\alpha - a\beta \neq 0$. If the first case holds, then we try to find analytic solutions of the equations

$$y'(z) = 0 \qquad (6.114)$$

or

$$y(y(z)) - ay(z) - \beta = 0. \qquad (6.115)$$

If the latter case holds, we try to find analytic solutions of the simultaneous equations

$$y'(z) = b\alpha - a\beta \qquad (6.116)$$

and

$$y(y(z)) - ay(z) - \beta - 1 = 0, \qquad (6.117)$$

or, to find analytic solutions of the single equation (6.113). Once analytic solutions $y(z)$ and their fixed points are found, then analytic solutions of our original equation (6.105) are easily calculated from (6.110) and (6.112).

The solutions of (6.114) are of the form $y(z) = c$. Since $z = c$ is the fixed point of $y(z)$, thus from (6.110) and (6.112), we see that when $b\alpha - a\beta = 0$ and $b \neq 0$,

$$x(z) = \frac{\beta}{b} c + \frac{1}{b} \left\{ \frac{ac^2}{2} - c^2 + cz - \frac{az^2}{2} \right\}$$

is an entire solution of (6.105).

We are now left with the simultaneous equations (6.116) and (6.117), as well as equations (6.113) and (6.115). Sufficient conditions and methods for constructing some of their analytic solutions will be given below under the assumption that $b \neq 0$.

Analytic solutions of the simultaneous equations (6.116) and (6.117) are easily found. We have the following result.

Theorem 6.21. *Suppose $b\alpha - a\beta \neq 0$. Then the simultaneous equations (6.116) and (6.117) has a solution if, and only if, $b\alpha - a\beta = a$. In case $b\alpha - a\beta = a$, the function $y(z) = az + \beta + 1$ is a solution.*

Proof. Suppose $y(z)$ is a solution of (6.116) and (6.117), then in view of (6.116), $y(z) = (b\alpha - a\beta)z + C$. Substituting it into (6.117), we see that

$$\{(b\alpha - a\beta)^2 - a(b\alpha - a\beta)\} z + \{b\alpha - a\beta + 1 - a\} C - \beta - 1 = 0,$$

or, equivalently, that
$$(b\alpha - a\beta)^2 = a(b\alpha - a\beta),$$
$$(b\alpha - a\beta + 1 - a)C = \beta + 1.$$

The above simultaneous equation has a solution if, and only if, $b\alpha - a\beta = a$ and $C = \beta + 1$. The proof is complete.

We remark that the unique fixed point of $y(z) = az + \beta + 1$ is $(\beta+1)/(1-a)$. Therefore, in addition to the conditions $b\alpha - a\beta = a \neq 0$ and $b \neq 0$, the additional condition $a \neq 1$ is needed for constructing the analytic solution

$$x(z) = x\left(\frac{\beta+1}{1-a}\right) + \frac{\beta+1}{b}\left(z - \frac{\beta+1}{1-a}\right) = \frac{\beta+1}{b}\left(\frac{a+\beta}{1-a} + z - \frac{\beta+1}{1-a}\right)$$

of (6.105) from (6.110) and (6.112).

It is easy to see that if $y(z)$ is an analytic solution of (6.115) with a fixed point z_0, then $z_0 = \beta/(1-a)$, provided $a \neq 1$. It is also easy to see that

$$y(z) = \frac{\beta}{1-a}, a \neq 1, \tag{6.118}$$

and
$$y(z) = az + \beta \tag{6.119}$$

are solutions of (6.115). Indeed, these are the only analytic solutions defined in a neighborhood of the fixed point $z_0 = \beta/(1-a)$ when $a \neq 0$.

Theorem 6.22. *Suppose $a \neq 0, 1$. Then the only analytic solutions of (6.115) defined in a neighborhood of the point $z_0 = \beta/(1-a)$ are those defined by (6.118) or (6.119).*

Proof. Let $y(z)$ be an analytic solution of (6.115) such that $y(z_0) = z_0$. In view of (6.115), we see that

$$y'(y(z))y'(z) - ay'(z) = 0,$$

so that
$$(y'(z_0))^2 - ay'(z_0) = 0.$$

Thus either $y'(z_0) = 0$ or $y'(z_0) = a$. Differentiating (6.115) twice, we arrive at

$$y''(y(z))(y'(z))^2 + y'(y(z))y''(z) - ay''(z) = 0,$$

so that
$$y''(z_0)\left[(y'(z_0))^2 + y'(z_0) - a\right] = 0.$$

If $y'(z_0) = 0$, then $y''(z_0) = 0$; while if $y'(z_0) = a \neq 0$, then $y''(z_0) = 0$ also. We assert that $y^{(n)}(z_0) = 0$ for all $n \geq 3$. To see this, let

$$\lambda_{ij}(z) = \left(y^{(i)}(y(z))\right)^{(j)}.$$

Since
$$(y(y(z)))^{(n+1)} = (y'(y(z))\, y'(z))^{(n)} = \sum_{k=0}^{n} C_k^{(n)} \left(y'(y(z))\right)^{(k)} (y'(z))^{(n-k)},$$
thus
$$\lambda_{0,n+1}(z_0) = y'(z_0) y^{(n+1)}(z_0) + 0 + \cdots + 0 + \lambda_{1n}(z_0) y'(z_0).$$
But since
$$(y'(y(z)))^{(n)} = (y''(y(z))\, y'(z))^{(n-1)} = \sum_{k=0}^{n-1} C_k^{(n-1)} \left(y''(y(z))\right)^{(k)} (y'(z))^{(n-1-k)},$$
we see that
$$\lambda_{1n}(z_0) = 0 + \cdots + 0 + \lambda_{2,n-1}(z_0) y'(z_0) = \lambda_{2,n-1}(z_0) y'(z_0).$$
By induction, it is easy to see that
$$\lambda_{1n}(z_0) = \lambda_{2,n-1}(z_0) y'(z_0) = \lambda_{3,n-2}(z_0)(y'(z_0))^2 = \cdots = y^{(n+1)}(z_0)(y'(z_0))^n.$$
Thus we have
$$y^{(n+1)}(z_0)\left[(y'(z_0))^{n+1} + y'(z_0) - a\right] = 0,$$
which shows that $y^{(n+1)}(z_0) = 0$ for $n \geq 2$. The proof is complete.

We may now make use of the solutions just found to construct analytic solutions of (6.105) by means of (6.110) and (6.112). Doing so, under the assumptions that $a \neq 0, a \neq 1, b\alpha - a\beta = 0$ and $b \neq 0$, we see that the solution (6.118) leads to the entire solution
$$x(z) = \frac{\beta^2}{b(1-a)} + \frac{1}{b}\left\{\frac{a}{2}\left(\frac{\beta}{1-a}\right)^2 - \left(\frac{\beta}{1-a}\right)^2 + \frac{\beta z}{1-a} - \frac{az^2}{2}\right\},$$
while the solution (6.119) leads to the entire solution
$$x(z) = \frac{\beta^2}{b(1-a)} + \frac{\beta}{b}\left\{z - \frac{\beta}{1-a}\right\}$$
of (6.105).

To find analytic solutions of (6.113), we first seek an analytic solution $g(z)$ of the auxiliary equation
$$\mu g'(\mu z)\left\{g\left(\mu^2 z\right) - a g(\mu z) - \beta\right\} = g'(z)(b\alpha - a\beta) \tag{6.120}$$
satisfying the condition
$$g(0) = s,$$
where s is to be specified and μ either satisfies $0 < |\mu| < 1$ or is a Siegel number. Then we show that (6.113) has an analytic solution of the form
$$y(z) = g\left(\mu g^{-1}(z)\right)$$

in a neighborhood of the number s.

Theorem 6.23. *Suppose that $0 < |\mu| < 1$ and that $b\alpha - a\beta \neq 0$. Suppose further that when $a = 1$, we have $\beta \neq 0$ and $\mu = (a\beta - b\alpha)/\beta$. Then for any nontrivial complex number η, equation (6.120) has an analytic solution of the form*

$$g(z) = s + \eta z + \sum_{n=2}^{\infty} c_n z^n, \tag{6.121}$$

where s is arbitrary when $a = 1$, and

$$s = \frac{\beta\mu + b\alpha - a\beta}{(1-a)\mu}$$

otherwise.

Proof. We seek a solution of (6.120) in a power series of the form

$$g(z) = \widehat{c}(z) = \sum_{n=0}^{\infty} c_n z^n. \tag{6.122}$$

By letting $c_0 = s$ and then substituting the subsequent power series into (6.120), we see that

$$\mu\underline{\mu} \cdot c * \left[\mu^2 \cdot c - a\underline{\mu} \cdot c - \overline{\beta}\right] = (b\alpha - a\beta)Dc.$$

Hence

$$[\beta\mu + b\alpha - a\beta - (1-a)\mu s]\, c_1 = 0,$$

and

$$(a\beta - b\alpha)(\mu^n - 1)(n+1)c_{n+1} = \sum_{k=0}^{n-1}(k+1)\left(\mu^{2n-k+1} - a\mu^{n+1}\right) c_{k+1}c_{n-k}, n \in \mathbf{Z}^+. \tag{6.123}$$

In view of the definition of s, we see that $\beta\mu + b\alpha - a\beta - (1-a)\mu s = 0$ so that we can choose c_1 to be η. Once c_0 and c_1 are determined, the other terms of the sequence $\{c_n\}$ can be determined successively from (6.123) in a unique manner.

We need to show that the subsequent power series (6.122) converges in a neighborhood of the origin. First of all, note that

$$\left|\frac{(k+1)\left(\mu^{2n-k+1} - a\mu^{n+1}\right)}{(a\beta - b\alpha)(n+1)(\mu^n - 1)}\right| \leq M, \; n \geq 2,$$

for some positive number M. Next recall that the equation,

$$G^2(z) = \frac{1}{M}G(z) - \frac{1}{M}|\eta|\, z,$$

as a special case of (4.10), has a solution

$$G(z) = \frac{1}{2M}\left\{1 - \sqrt{1 - 4M|\eta|\, z}\right\} = \sum_{n=0}^{\infty} B_n z^n$$

on $B(0;1/|4M\eta|)$ where the sequence $B = \{B_n\}_{n\in \mathbf{N}}$ satisfies $B_0 = 0$, $B_1 = |\eta| = |c_1|$ and

$$B_{n+1} = M \sum_{k=0}^{n-1} B_{k+1} B_{n-k}, n \in \mathbf{Z}^+.$$

Then in view of (6.123),

$$|c_n| \leq B_n, n \in \mathbf{Z}^+,$$

which implies that the power series (6.122) is also convergent for $z < 1/(4M|\eta|)$. The proof is complete.

Theorem 6.24. *Suppose that μ is a Siegel number and that $b\alpha - a\beta \neq 0$. Suppose further that when $a = 1$, we have $\beta \neq 0$ and $\mu = (a\beta - b\alpha)/\beta$. Then equation (6.120) has an analytic solution of the form*

$$g(z) = s + z + \sum_{n=2}^{\infty} c_n z^n, \tag{6.124}$$

where s is the same number defined in the previous Theorem 6.23.

Proof. As in the previous proof, we seek a power series solution of the form (6.122). Then defining $c_0 = s$ and $c_1 = 1$, (6.123) holds again so that

$$|c_{n+1}| \leq \frac{1+|a|}{|a\beta - b\alpha|} |\mu^n - 1|^{-1} \sum_{k=0}^{n-1} |c_{k+1}||c_{n-k}|, \ n \in \mathbf{Z}^+.$$

Recall that the equation

$$\frac{1+|a|}{|a\beta - b\alpha|} G^2(z) + z = G(z),$$

as a special case of (4.10), has a solution

$$G(z) = \frac{|a\beta - b\alpha|}{2(1+|a|)} \left\{ 1 - \sqrt{1 - \frac{4(1+|a|)}{|a\beta - b\alpha|} z} \right\}$$

$$= \sum_{n=0}^{\infty} v_n z^n$$

which is analytic on $B(0; |a\beta - b\alpha|/(4 + 4|a|))$ and the sequence $\{v_n\}_{n\in\mathbf{N}}$ satisfies $v_0 = 0$, $v_1 = 1$ and

$$v_{n+1} = \frac{1+|a|}{|a\beta - b\alpha|} \sum_{k=0}^{n-1} v_{k+1} v_{n-k}, \ n \in \mathbf{Z}^+.$$

In view of the Cauchy Estimation (Theorem 3.26), there is some $r > 0$ such that $v_n \leq r^n$ for $n \in \mathbf{Z}^+$. Thus by Theorem 3.32, we may then easily see that there is $\delta > 0$ such that

$$|c_n| \leq r^n \left(2^{5\delta+1}\right)^{n-1} n^{-2\delta}, \ n \in \mathbf{Z}^+.$$

This shows that the power series (6.124) converges on a neighborhood of the origin. The proof is complete.

Theorem 6.25. *Suppose that $0 < |\mu| < 1$ or μ is a Siegel number, and that $b\alpha - a\beta \neq 0$. Suppose further that when $a = 1$, we have $\beta \neq 0$ and $\mu = (a\beta - b\alpha)/\beta$. Then (6.113) has an analytic solution of the form $y(z) = g(\mu g^{-1}(z))$ in a neighborhood of the number s, where s is defined in Theorem 6.23, and g is an analytic solution of the equation (6.120).*

Proof. In view of the previous two results, the equation (6.120) has an analytic solution $g(z)$ in the neighborhood of the origin and $g(0) = s$ as well as $g'(0) \neq 0$. Thus the inverse function $g^{-1}(z)$ is analytic in a neighborhood of the point s, and hence the composite function $y(z) = g(\mu g^{-1}(z))$ is also analytic in a neighborhood of the point s. Finally, note that

$$\{y(y(z)) - ay(z) - \beta\} y'(z)$$
$$= \{g(\mu^2 g^{-1}(z)) - ag(\mu g^{-1}(z)) - \beta\} \frac{\mu g'(\mu g^{-1}(z))}{g'(g^{-1}(z))} = b\alpha - a\beta.$$

This shows that the composite function $y(z)$ is a solution of (6.113) as desired. The proof is complete.

We remark that since $g(0) = s$, the point s is thus a fixed point of $y(z)$.

In the above, we have shown that under the conditions that
(i) if $a = 1$, then $\beta \neq 0$ and $\beta\mu = a\beta - b\alpha$,
(ii) if $a = 1$, then s is arbitrary, and
(iii) if $a \neq 1$, then $s = (\beta\mu + b\alpha - a\beta)/(1-a)\mu$, where $0 < |\mu| < 1$ or μ is a Siegel number,

then equation (6.113) has an analytic solution $y(z) = g(\mu g^{-1}(z))$ in a neighborhood of the number s, where g is an analytic solution of (6.120). Since the function $g(z)$ in (6.122) can be determined by (6.123), it is possible to calculate, at least in theory, the explicit form of $y(z)$ and then under the additional condition that

(iv) $b \neq 0$ and $b\alpha - a\beta \neq 0$,

an explicit analytic solution of (6.105) in a neighborhood of the fixed point s of $y(z)$ by means of (6.110) and (6.112). However, knowing that an analytic solution of (6.105) exists, we can take an alternate route as follows. Assume that $x(z)$ is of the form

$$x(z) = x(s) + x'(s)(z-s) + \frac{x''(s)}{2}(z-s) + \cdots$$
$$= \frac{(b\alpha + (1-a)\beta)s}{b} + x'(s)(z-s) + \frac{x''(s)}{2}(z-s) + \cdots,$$

we need to determine the derivatives $x^{(n)}(s)$ for $n \in \mathbf{Z}^+$. First of all, in view of (6.109), we have

$$x'(s) = \frac{1}{b}(y(s) - as) = \frac{(1-a)s}{b}.$$

Next by differentiating (6.105), we see that
$$\alpha + \beta x''(z) = x'(az + bx'(z)) \cdot (a + bx''(z)),$$
so that
$$\alpha + \beta x''(s) = x'(s)(a + bx''(s)),$$
and
$$x''(s) = \frac{ax'(s) - \alpha}{\beta - bx'(s)} = \frac{a(1-a)s - b\alpha}{b\beta - b(1-a)s} = \frac{\mu - a}{b},$$
where the denominator $\beta - bx'(s)$ cannot be zero in view of our assumptions (i)-(iv).

Similarly, if we differentiate (6.105) twice, we arrive at
$$\beta x'''(z) = bx'(az + bx'(z))x'''(z) + x''(az + bx'(z)) \cdot (a + bx''(z))^2,$$
so that
$$x'''(s) = \frac{x''(s)(a + bx''(s))^2}{\beta - bx'(s)} = \frac{\mu^3(\mu - a)}{b(a\beta - b\alpha)}.$$

In general, we can show that $x^{(n+1)}(s)$, where $n \geq 3$, depends only on the lower derivatives at $z = s$. To see this, note that
$$(x(az + bx'(z))^{(n)} = (x'(az + bx'(z)) \cdot (a + bx''(z)))^{(n-1)}$$
$$= bx'(az + bx'(z))x^{(n+1)}(z)$$
$$+ \sum_{k=1}^{n-1} C_k^{(n-1)}(x'(az + bx'(z)))^{(k)}(a + bx''(z))^{(n-1-k)}.$$

Thus differentiating (6.105) n times at $z = s$, we will end up with
$$\beta x^{(n+1)}(s) = bx'(s)x^{(n+1)}(s) + F(x(s), x'(s), ..., x^{(n)}(s)),$$
where $F(x(s), ..., x^{(n)}(s))$ stands for terms involving the lower derivatives $x(s), ..., x^{(n)}(s)$. This shows that
$$x^{(n+1)}(s) = \frac{F(x(s), ..., x^{(n)}(s))}{\beta - bx'(s)}$$
for $n \geq 2$. By means of this formula, it is then easy to write out the explicit form of our solution $x(z)$:
$$x(z) = \frac{(b\alpha + (1-a)\beta)s}{b} + \frac{(1-a)s}{b}(z-s) + \frac{\mu - a}{2!b}(z-s)^2$$
$$+ \frac{\mu^3(\mu - a)}{3!b(a\beta - b\alpha)}(z-s)^3 + \frac{\mu^5(\mu - a)(\mu^2 + 3\mu - 3a)}{4!b(a\beta - b\alpha)^2}(z-a)^4 + \cdots$$

6.3 Equations with Second Order Derivatives

Iterative functional equations involving the second and higher derivatives of the unknown function are not studied as much as the ones involving the first order derivatives. Indeed, an earlier study of such an equation by Petuhov [151] which appeared back in 1965. In this section, we will be concerned with four such equations that allow analytic solutions.

6.3.1 Equation I

We first consider the following equation [119]

$$x''(z) = \sum_{j=0}^{m} p_j x^{[j]}(z), \qquad (6.125)$$

where m is a positive integer greater than or equal to 2 and $p_0, p_1, ..., p_m$ are complex numbers such that $\sum_{i=0}^{m} |p_i| \neq 0$. We will look for analytic solutions of (6.125) which satisfy the condition

$$x(\alpha) = \alpha,\ x'(\alpha) = \alpha, \qquad (6.126)$$

or the condition

$$x(0) = 0,\ x'(0) = \alpha. \qquad (6.127)$$

where $0 < |\alpha| < 1$ or α is a Siegel number.

In order to seek analytic solutions of (6.125), we first consider a related equation of the form

$$\alpha^2 y''(\alpha z) y'(z) - \alpha y'(\alpha z) y''(z) = (y'(z))^3 \sum_{j=0}^{m} p_j y(\alpha^j z), \qquad (6.128)$$

under the condition

$$y(0) = \alpha,\ y'(0) = \eta, \qquad (6.129)$$

or the condition

$$y(0) = 0,\ y'(0) = \eta. \qquad (6.130)$$

Theorem 6.26. *Suppose $0 < |\alpha| < 1$. Then for any $\eta \neq 0$, equation (6.128) has a solution $y(z)$ of the form*

$$y(z) = \sum_{n=0}^{\infty} b_n z^n,\ b_0 = \alpha,\ b_1 = \eta, \qquad (6.131)$$

which is analytic near 0 (and satisfies (6.129)), where $\{b_n\}_{n=2}^{\infty}$ is defined by the recurrence relation

$$(n+2)(\alpha^{n+2} - \alpha) b_{n+2}$$
$$= \sum_{k=0}^{n} \sum_{i=0}^{n-k} \frac{(i+1)(k+1)}{n-k+1} \left(\sum_{j=0}^{m} p_j \alpha^{j(n-k-i)} \right) b_{k+1} b_{i+1} b_{n-k-i} \qquad (6.132)$$

for $n \in \mathbf{N}$.

Proof. Note that equation (6.128) may be written in the form

$$\frac{\alpha y''(\alpha z) y'(z) - y'(\alpha z) y''(z)}{(y'(z))^2} = \frac{1}{\alpha} y'(z) \sum_{j=0}^{m} p_j y(\alpha^j z),$$

or
$$\left(\frac{y'(\alpha z)}{y'(z)}\right)' = \frac{1}{\alpha} y'(z) \sum_{j=0}^{m} p_j y\left(\alpha^j z\right).$$

Since we have assumed that $y'(0) = \eta \neq 0$, by integration, we obtain
$$y'(\alpha z) = y'(z) + \frac{1}{\alpha} y'(z) \int_0^z y'(s) \sum_{j=0}^{m} p_j y\left(\alpha^j s\right) ds. \qquad (6.133)$$

Let
$$y(z) = \widehat{b}(z) = \sum_{n=0}^{\infty} b_n z^n$$
be a formal solution of (6.133). Then in view of (6.133), we see that
$$\underline{\alpha} \cdot Db = Db + \frac{1}{\alpha} (Db) * \left[\sum_{j=0}^{m} p_j \int (Db) * \left(\underline{\alpha}^j \cdot b\right) \right].$$

Since
$$(Db) * \left(\underline{\alpha}^j \cdot b\right) = \{(n+1) b_{n+1}\}_{n \in \mathbf{N}} * \{\alpha^{jn} b_n\}_{n \in \mathbf{N}}$$
$$= \left\{ \sum_{i=0}^{n} (i+1) b_{i+1} \alpha^{j(n-i)} b_{n-i} \right\}_{k \in \mathbf{N}},$$
we see that
$$Db * \int (Db) * \left(\underline{\alpha}^j \cdot b\right)$$
$$= \hbar * \{(n+1) b_n\}_{n \in \mathbf{N}} * \left\{ p_j \frac{1}{n+1} \sum_{i=0}^{n} (i+1) b_{i+1} \alpha^{j(n-i)} b_{n-i} \right\}_{n \in \mathbf{N}}$$
$$= \hbar * \left\{ \sum_{t=0}^{n} (t+1) b_t p_j \frac{1}{n-t+1} \sum_{i=0}^{n-t} (i+1) b_{i+1} \alpha^{j(n-t-i)} b_{n-t-i} \right\}_{n \in \mathbf{N}}.$$

Hence b_1 is arbitrary and
$$(n+2) \left(\alpha^{n+2} - \alpha\right) b_{n+2}$$
$$= \sum_{k=0}^{n} \sum_{i=0}^{n-k} \frac{(i+1)(k+1)}{n-k+1} \left(\sum_{j=0}^{m} p_j \alpha^{j(n-k-i)} \right) b_{k+1} b_{i+1} b_{n-k-i} \qquad (6.134)$$
for $n \in \mathbf{N}$. If we set $b_0 = \alpha$ and $b_1 = y'(0) = \eta \neq 0$, then by (6.134), we may determine $\{b_n\}_{n=0}^{\infty}$ uniquely in a recursive manner.

We need to show that $y(z)$ defined by (6.131) has a positive radius of convergence. To this end, note that
$$\left| \frac{(i+1)(k+1)}{(n+2)(n-k+1)(\alpha^{n+2} - \alpha)} \left(\sum_{j=0}^{m} p_j \alpha^{j(n-k-i)} \right) \right| < \frac{1}{|\alpha| - |\alpha|^{n+2}} \left(\sum_{j=0}^{m} |p_j| \right),$$

for $0 \le k \le n$ and $0 \le i \le n-k$. Furthermore, since

$$\lim_{n\to\infty} \frac{1}{|\alpha| - |\alpha|^{n+2}} = \frac{1}{|\alpha|},$$

there is $M > 0$ such that

$$\left| \frac{(i+1)(k+1)}{(n+2)(n-k+1)(\alpha^{n+2} - \alpha)} \left(\sum_{j=0}^{m} p_j \alpha^{j(n-k-i)} \right) \right| \le M \qquad (6.135)$$

for $n \ge 0$.

Note that Example 4.15 asserts that the equation

$$MH^3(z) - 2M|\alpha|H^2(z) + \left(M|\alpha|^2 - 1\right)H(z) + |\eta|z + |\alpha| = 0$$

has a solution

$$H(z) = \sum_{n=0}^{\infty} h_n z^n$$

which is analytic on a neighborhood of the origin and the sequence $h = \{h_n\}_{n \in \mathbf{N}}$ is determined by $h_0 = |\alpha|$, $h_1 = |\eta|$ and

$$h_{n+2} = M \sum_{k=0}^{n} \left(\sum_{i=0}^{n-k} h_{k+1} h_{i+1} h_{n-k-i} \right), \quad n \in \mathbf{N}. \qquad (6.136)$$

In view of (6.132), (6.135) and (6.136), we may show by induction that

$$|b_n| \le h_n, \ n \ge 0. \qquad (6.137)$$

Therefore, $y(z) \ll H(z)$ so that $y(z)$ is analytic on a neighborhood of the origin. The proof is complete.

Theorem 6.27. *Suppose $0 < |\alpha| < 1$. Then for any $\eta \ne 0$, equation (6.128) has a solution $y(z)$ of the form*

$$y(z) = \sum_{n=0}^{\infty} b_n z^n, \ b_0 = 0, b_1 = \eta, \qquad (6.138)$$

which is analytic near 0 (and satisfies (6.130)) and $\{b_n\}_{n=2}^{\infty}$ is determined by

$$(n+2)\left(\alpha^{n+2} - \alpha\right) b_{n+2}$$
$$= \sum_{k=1}^{n} \sum_{i=1}^{n-k+1} \frac{k \cdot i}{n-k+2} \left(\left(\sum_{j=0}^{m} p_j \alpha^{j(n-k-i+2)} \right) b_k b_i b_{n-k-i+2} \right) \qquad (6.139)$$

for $n \in \mathbf{N}$ and $b_{2m} = 0$ for $m \in \mathbf{Z}^+$.

Proof. Let
$$y(z) = \sum_{n=0}^{\infty} b_n z^n$$
be a formal solution of (6.133). As in the previous proof, we may see that b_1 is arbitrary, $2b_2\alpha = 2b_2$ and
$$(n+2)\left(\alpha^{n+2} - \alpha\right) b_{n+2} = \sum_{k=1}^{n} \sum_{i=1}^{n-k+1} \frac{k \cdot i}{n-k+2} \left(\sum_{j=0}^{m} P_j \alpha^{j(n-k-i+2)} \right) b_k b_i b_{n-k-i+2}$$
(6.140)
for $n \geq 1$. If we set $b_0 = 0$ and $b_1 = \eta \neq 0$, then $2b_2\alpha = 2b_2$ implies $b_2 = 0$. By induction, we may then infer from (6.140) that $b_{2m} = 0$ for $m \geq 1$.

We now show that the formal solution $y(z)$ has a positive radius of convergence. To this end, note that Example 4.15 asserts that the equation
$$MH^3(z) - H(z) + |\eta| z = 0$$
has a solution
$$H(z) = \sum_{n=0}^{\infty} h_n z^n$$
which is analytic on a neighborhood of the origin and the sequence $\{h_n\}_{n \in \mathbf{N}}$ is determined by $h_0 = 0$, $h_1 = |\eta|$, $h_2 = 0$ and
$$h_{n+2} = M \sum_{k=1}^{n} \sum_{i=1}^{n-k+1} h_k h_i h_{n-k-i+2}, \ n \in \mathbf{Z}^+.$$
for $n \geq 1$. By induction, it is easy to see that
$$|b_n| \leq h_n, \ n \geq 1.$$
Thus $y(z)$ also has a positive radius of convergence. The proof is complete.

Theorem 6.28. *Suppose α is a Siegel number. Suppose further that $0 < \sum_{j=0}^{m} |p_j| \leq 1$. Then for $0 < |\eta| \leq 1$, equation (6.128) has a solution $y(z)$ of the form (6.131) which is analytic near 0 (and satisfies (6.129)), where $\{b_n\}_{n=2}^{\infty}$ is determined by (6.132).*

Proof. As in the previous proof, equation (6.128) has a formal solution $y(z)$ of the form (6.131). We need to show that $y(z)$ has a positive radius of convergence. To the end, let us consider
$$\psi(z) = \sum_{n=0}^{\infty} u_n z^n, \qquad (6.141)$$
where $u_0 = 1$, $u_1 = 1$,
$$u_{n+2} = s_{n+1} \sum_{k=0}^{n} \sum_{i=0}^{n-k} u_{k+1} u_{i+1} u_{n-k-i},$$

for $n \geq 0$. Since $0 < \sum_{j=0}^{m} |P_j| \leq 1$, we may show that
$$|b_n| \leq u_n, \quad n \geq 0. \tag{6.142}$$
Indeed, $|b_0| = |\alpha| = 1 = u_0$ and $|b_1| = |\eta| \leq 1 = u_1$. Assume by induction that $|b_j| \leq u_j$ for $j = 2, 3, \ldots, n+1$. Then from (6.132) and the fact that
$$\left| \frac{(i+1)(k+1)}{(n+2)(n-k+1)} \sum_{j=0}^{m} p_j \alpha^{j(n-k-i)} \right| \leq \sum_{j=0}^{m} |p_j| \leq 1,$$
for $0 \leq k \leq n$ and $0 \leq i \leq n-k$, we see that
$$|b_{n+2}| \leq s_{n+1} \sum_{k=0}^{n} \sum_{i=0}^{n-k} u_{k+1} u_{i+1} u_{n-k-i},$$
as desired. In other words, we have shown that $y(z) \ll \psi(z)$. Therefore we only need to show that $\psi(z)$ has a positive radius of convergence. To this end, note that Example 4.15 asserts that the equation
$$\varphi^3(z) - 2\varphi^2(z) + z + 1 = 0 \tag{6.143}$$
has a solution
$$\varphi(z) = \sum_{n=0}^{\infty} v_n z^n \tag{6.144}$$
which is analytic on a neighborhood of the origin and the sequence $\{v_n\}_{n \in \mathbf{N}}$ satisfies $v_0 = 1$, $v_1 = 1$ and
$$v_{n+2} = \sum_{k=0}^{n} \sum_{j=0}^{n-k} v_{k+1} v_{j+1} v_{n-k-j}, \quad n \in \mathbf{N}. \tag{6.145}$$
In view of the Cauchy Estimation (Theorem 3.26), there is $r > 0$ such that $v_n \leq r^n$ for $n \in \mathbf{Z}^+$. As in the proof of Theorem 3.32, we may apply Seigel's Lemma (Theorem 3.31) to conclude that there is some $\delta > 0$ such that
$$u_n \leq r^n \left(2^{5\delta+1}\right)^{n-1} n^{-2\delta}, \quad n \geq 2.$$
The proof is complete.

Theorem 6.29. *Suppose α is a Siegel number. Suppose further that $0 < |\eta| \leq 1$. Then (6.128) has a solution $y(z)$ of the form (6.138) which is analytic near 0 (and satisfies (6.130)) where $\{b_n\}_{n=2}^{\infty}$ is determined by (6.139) for $n \in \mathbf{N}$ and $b_{2m} = 0$ for $m \in \mathbf{Z}^+$.*

Proof. As in the previous proof, equation (6.128) has a formal solution of the form (6.138). We will show that the formal solution has a positive radius of convergence. Consider
$$\psi(z) = \sum_{n=1}^{\infty} u_n z^n$$

where $u_1 = 1$, $u_2 = 0$ and

$$u_{n+2} = s_{n+1} \sum_{k=0}^{n} \sum_{i=0}^{n-k} u_{k+1} u_{i+1} u_{n-k-i}, \ n \in \mathbf{Z}^+.$$

It is easy to see that $u_{2m} = 0$ for $m \in \mathbf{Z}^+$, and that

$$|b_n| \leq u_n, \ n \geq 1. \tag{6.146}$$

In other words, $y(z) \ll \psi(z)$. It suffices to show that $\psi(z)$ converges on a neighborhood of the origin. To this end, note that Example 4.15 asserts that the equation

$$\varphi^3(z) - \varphi(z) + z = 0 \tag{6.147}$$

has a solution

$$\varphi(z) = \sum_{n=1}^{\infty} v_n z^n$$

which is analytic on a neighborhood of the origin and the sequence $\{v_n\}_{n \in \mathbf{N}}$ satisfies $v_0 = 0$, $v_1 = 1$, $v_2 = 0$ and

$$v_{n+2} = \sum_{k=0}^{n} \sum_{i=0}^{n-k} v_{k+1} v_{i+1} v_{n-k-i}, \ n \in \mathbf{Z}^+.$$

In view of the Cauchy Estimation (Theorem 3.26), there is some $r > 0$ such that $v_n \leq r^n$ for $n \in \mathbf{Z}^+$. As in the proof of Theorem 3.32, we may apply Seigel's Lemma (Theorem 3.31) to conclude that there is some $\delta > 0$ such that

$$u_n \leq r^n \left(2^{5\delta+1}\right)^{n-1} n^{-2\delta}, \ n \geq 2.$$

The proof is complete.

Theorem 6.30. *Suppose $0 < |\alpha| < 1$. Then equation (6.125) has a solution $x(z)$ which is analytic near 0 and satisfies (6.126).*

Proof. By Theorem 6.26, for any $\eta \neq 0$, equation (6.128) has a solution $y(z)$ which is analytic near 0 and satisfies (6.129). This solution is of the form (6.131) where $\{b_n\}_{n=2}^{\infty}$ is defined by the recurrence relation (6.132). Since $y'(0) = \eta \neq 0$, thus by the Analytic Inverse Function Theorem 4.2, the inverse function $y^{-1}(z)$ is analytic in a neighborhood of the origin. Let

$$x(z) = y\left(\alpha y^{-1}(z)\right). \tag{6.148}$$

Then

$$x'(z) = \frac{\alpha y'\left(\alpha y^{-1}(z)\right)}{y'\left(y^{-1}(z)\right)},$$

and

$$x^{[j]}(z) = y\left(\alpha^j y^{-1}(z)\right), \ j = 1, 2, ..., m.$$

Thus,

$$x''(z) = \frac{1}{[y'(y^{-1}(z))]^3}\left\{\alpha^2 y''(\alpha y^{-1}(z))\cdot y'(y^{-1}(z)) - \alpha y'(\alpha y^{-1}(z))y''(y^{-1}(z))\right\}$$

$$= \sum_{j=0}^{m} p_j y(\alpha^j y^{-1}(z))$$

$$= \sum_{j=0}^{m} p_j x^{[j]}(z).$$

Furthermore, note that $y^{-1}(\alpha) = 0$, $y'(0) = \eta \neq 0$ and

$$x(\alpha) = y(\alpha y^{-1}(\alpha)) = y(0) = \alpha,$$

$$x'(\alpha) = \frac{\alpha y'(\alpha y^{-1}(\alpha))}{y'(y^{-1}(\alpha))} = \frac{\alpha\eta}{\eta} = \alpha.$$

These show that $x(z)$ is an analytic solution of (6.125). The proof is complete.

Theorem 6.31. *Suppose $0 < |\alpha| < 1$. Then equation (6.125) has a solution which is of the form*

$$x(z) = y(\alpha y^{-1}(z))$$

and is analytic in a neighborhood of the origin and satisfies (6.127), where $y(z)$ is an analytic solution (6.138) of (6.128) under the additional conditions $\eta \neq 0$ and (6.130).

The proof is similar to that above. Note that $y^{-1}(0) = 0$, $y'(0) = \eta \neq 0$ imply

$$x(0) = y(\alpha y^{-1}(0)) = y(0) = 0,$$

$$x'(0) = \alpha y'(\alpha y^{-1}(0))/y'(y^{-1}(0)) = \frac{\alpha\eta}{\eta} = \alpha.$$

Theorem 6.32. *Suppose α is a Siegel number. Suppose further that $0 < \sum_{j=0}^{m}|p_j| \leq 1$. Then equation (6.125) has a solution of the form*

$$x(z) = y(\alpha y^{-1}(z)),$$

which is analytic in a neighborhood of α and satisfies (6.126), where $y(z)$ is an analytic solution (6.131) of (6.128) under the additional conditions $0 < |\eta| < 1$ and (6.129).

Theorem 6.33. *Suppose α is a Siegel number. Suppose further that $0 < \sum_{j=0}^{m}|p_j| \leq 1$. Then equation (6.125) has a solution of the form*

$$x(z) = y(\alpha y^{-1}(z)),$$

which is analytic in a neighborhood of the origin and satisfies (6.127), where $y(z)$ is an analytic solution (6.138) of (6.128) under the additional conditions $0 < |\eta| < 1$ and (6.130).

Example 6.5. Consider the equation
$$x''(z) = px(x(z)), \quad z \in \mathbf{C}, \tag{6.149}$$
under the condition
$$x(0) = 0, \, x'(0) = \alpha \tag{6.150}$$
where p is a nonzero complex number. If $0 < |\alpha| < 1$, or, α is a Siegel number and $0 < |p| \leq 1$, then Theorem 6.31 or Theorem 6.33 assert that (6.149) has a solution which is analytic on a neighborhood of the origin and satisfies (6.150). We may let
$$x(z) = \widehat{c}(z),$$
where $c_0 = 0$ and $c_1 = \alpha$. Substituting it into (6.149), we see that $c_2 = 0$ and
$$(n+2)(n+1)c_{n+2} = p \sum_{m=2}^{n} c_m c_n^{\langle m \rangle}, \quad n \in \mathbf{Z}^+. \tag{6.151}$$
We remark that $c_{2k} = 0$ for $k \geq 1$, as can be easily checked by induction, so that $x(z)$ is an odd function.

6.3.2 Equation II

We will be concerned with a class of iterative functional differential equation of the form [189]
$$x''(z) = x(az + bx(z)). \tag{6.152}$$
When $a = 0$ and $b = 1$, equation (6.152) reduces to the second-order iterative functional differential equation
$$x''(z) = x(x(z)).$$
When $b = 0$, equation (6.152) changes into
$$x''(z) = x(az). \tag{6.153}$$
In order to find an analytic solution $x = x(z)$ of (6.153), we formally assume that
$$x(z) = \widehat{c}(z) = \sum_{n=0}^{\infty} c_n z^n.$$
Then in view of (6.153), we may obtain
$$D^2 c = \underline{a} \cdot c.$$
Hence
$$2c_2 = c_0, \, 3 \cdot 2 c_3 = a c_1$$
and
$$(n+2)(n+1)c_{n+2} = a^n c_n, \, n = 2, 3, \ldots.$$

This leads to

$$x(z) = c_0 + c_1 z + \frac{c_0}{2!}z^2 + \frac{c_1 a}{3!}z^3 + c_0 \sum_{k=0}^{\infty} \frac{a^{k(k-1)}}{(2k)!} z^{2k} + c_1 \sum_{k=2}^{\infty} \frac{a^{k^2}}{(2k+1)!} z^{2k+1}$$

which, as can be verified by means of the ratio test, is an entire solution of (6.153) when $|a| \leq 1$.

We now assume that $b \neq 0$. To find analytic solutions of (6.152), we first seek an analytic solution $y(z)$ of the auxiliary equation

$$\alpha^2 y''(\alpha z) y'(z) = \alpha y'(\alpha z) y''(z) + (y'(z))^3 [y(\alpha^2 z) - a y(\alpha z)] \qquad (6.154)$$

satisfying the initial value conditions

$$y(0) = \gamma, \ y'(0) = \eta \neq 0, \qquad (6.155)$$

where γ, η are complex numbers, and α is either Siegel number or it satisfies $0 < |\alpha| < 1$.

Then we show that (6.152) has an analytic solution of the form

$$x(z) = \frac{1}{b}\left(y(\alpha y^{-1}(z)) - az \right) \qquad (6.156)$$

in a neighborhood of the number γ.

Theorem 6.34. *Suppose $0 < |\alpha| < 1$. Then for any given complex numbers γ and $\eta \neq 0$, equation (6.154) has a solution of the form*

$$y(z) = \widehat{b}(z) = \sum_{n=0}^{\infty} b_n z^n, \ b_0 = \gamma, b_1 = \eta, \qquad (6.157)$$

which is analytic on a neighborhood of the origin.

Proof. If $y(z)$ given by (6.157) is such a solution, then we may rewrite (6.154) in the form

$$\frac{\alpha y''(\alpha z) y'(z) - \alpha y'(\alpha z) y''(z)}{(y'(z))^2} = y'(z)[y(\alpha^2 z) - a y(\alpha z)],$$

or

$$\alpha \left(\frac{y'(\alpha z)}{y'(z)} \right)' = y'(z)[y(\alpha^2 z) - a y(\alpha z)].$$

Since $y'(0) = \eta \neq 0$, we see further that

$$y'(\alpha z) = y'(z) \left[1 + \frac{1}{\alpha} \int_0^z y'(s) \left(y(\alpha^2 s) - a y(\alpha s) \right) ds \right]. \qquad (6.158)$$

Substituting (6.157) into (6.158), we see that

$$\underline{\alpha} \cdot Db = Db + \frac{1}{\alpha} (Db) * \int (Db) * (\underline{\alpha}^2 \cdot b - a\underline{\alpha} \cdot b).$$

Hence the sequence $\{b_n\}_{n=2}^{\infty}$ is successively determined by the condition

$$(\alpha^{n+2} - \alpha)(n+2)b_{n+2}$$
$$= \sum_{k=0}^{n}\sum_{j=0}^{n-k} \frac{(k+1)(j+1)(\alpha^{2(n-k-j)} - a\alpha^{n-k-j})}{n-k+1} b_{k+1}b_{j+1}b_{n-k-j}, \quad (6.159)$$

for $n \geq 0$ in a unique manner.

It suffices now to show that the power series just determined converges in a neighborhood of the origin. First of all, note that

$$\left|\frac{(k+1)(j+1)(\alpha^{2(n-k-j)} - a\alpha^{n-k-j})}{(n+2)(n-k+1)(\alpha^{n+2} - \alpha)}\right| \leq \frac{1+|a|}{|\alpha| - |\alpha|^{n+2}} \leq M$$

for some positive number M. Thus if we define a sequence $\{B_n\}_{n \in \mathbf{N}}$ by $B_0 = |\gamma|, B_1 = |\eta|$ and

$$B_{n+2} = M \sum_{k=0}^{n}\sum_{j=0}^{n-k} B_{k+1}B_{j+1}B_{n-k-j}, n \in \mathbf{N},$$

then in view of (6.159),

$$|b_n| \leq B_n, n \in \mathbf{N},$$

so that $\{B_n\}_{n \in \mathbf{N}}$ is a majorant of $\{b_n\}_{n \in \mathbf{N}}$. Next, note that Example 4.15 asserts that the equation

$$G^3(z) - 2|\gamma|G^2(z) - \left(\frac{1}{M} - |\gamma|^2\right)G(z) + \frac{1}{M}(|\eta|z + |\gamma|) = 0. \quad (6.160)$$

has a solution

$$G(z) = \sum_{n=0}^{\infty} g_n z^n \quad (6.161)$$

which is analytic on a neighborhood of the origin and $g_0 = |\gamma|$, $g_1 = |\eta|$ and

$$B_{n+2} = M \sum_{k=0}^{n}\sum_{j=0}^{n-k} B_{k+1}B_{j+1}B_{n-k-j}, n \in \mathbf{N}.$$

Therefore, $B_n = g_n$ for $n \in \mathbf{N}$ and hence $\{B_n\}_{n \in \mathbf{N}}$ has a positive radius of convergence. The proof is complete.

We remark that if $\gamma = 0$ in (6.155), then by induction, it is not difficult to see from (6.159) that

$$b_{2k} = 0, \ k \in \mathbf{Z}^+.$$

This shows that the desired solution (6.157) is an odd function.

Theorem 6.35. *Suppose α is a Siegel number. Then given any complex number γ and η that satisfies $0 < |\eta| \leq 1$, equation (6.154) has a solution of the form (6.157) which is analytic on a neighborhood of the origin and satisfies $b_0 = \gamma$ and $b_1 = \eta$.*

Proof. As in the previous proof, we seek a power series solution of the form (6.157) with $b_0 = \gamma$ and $b_1 = \eta$. (6.159) holds again so that

$$|b_{n+2}| \leq \frac{1+|a|}{|a^{n+1}-1|} \sum_{k=0}^{n} \sum_{j=0}^{n-k} |b_{k+1}||b_{j+1}||b_{n-k-j}|, \ n \in \mathbf{N}. \tag{6.162}$$

Note that Example 4.15 asserts that

$$G^3(z) - 2|\gamma|G^2(z) - \left(\frac{1}{1+|a|} - |\gamma|^2\right)G(z) + \frac{1}{1+|a|}(z+|\gamma|) = 0. \tag{6.163}$$

has a solution

$$G(z) = \sum_{n=0}^{\infty} C_n z^n, \tag{6.164}$$

which is analytic on a neighborhood of the origin and $C_0 = |\gamma|$, $C_1 = 1$ and

$$C_{n+2} = (1+|a|) \sum_{k=0}^{n} \sum_{j=0}^{n-k} C_{k+1} C_{j+1} C_{n-k-j}, \ n \in \mathbf{N}. \tag{6.165}$$

In view of the Cauchy Estimation (Theorem 3.26), there exists a positive constant r such that

$$C_n < r^n \tag{6.166}$$

for $n \in \mathbf{Z}^+$. As in the proof of Theorem 3.32, we may apply Siegel's Lemma (Theorem 3.31) to deduce a positive δ such that

$$|b_n| \leq r^n (2^{5\delta+1})^{n-1} n^{-2\delta}, \ n \in \mathbf{Z}^+.$$

This shows that $\{b_n\}_{n \in \mathbf{N}}$ has a positive radius of convergence. The proof is complete.

Theorem 6.36. *Suppose $0 < |\alpha| < 1$ or α is a Siegel number. Then equation (6.152) has an analytic solution of the form (6.156) near γ, where $y(z)$ is an analytic solution of the initial value problem (6.154) and (6.155).*

Proof. In view of the previous two results, we may find a sequence $\{b_n\}_{n=2}^{\infty}$ such that the function $y(z)$ of the form (6.157) is an analytic solution of (6.154) in a neighborhood of the origin. Since $y'(0) = \eta \neq 0$, the function $y^{-1}(z)$ is analytic in a neighborhood of the $y(0) = \gamma$. If we now define $x(z)$ by means of (6.156), then

$$x'(z) = \frac{\alpha}{b} \cdot \frac{y'(\alpha y^{-1}(z))}{y'(y^{-1}(z))} - \frac{a}{b},$$

$$x''(z) = \frac{\alpha}{b} \cdot \frac{\alpha y''(\alpha g^{-1}(z))y'(y^{-1}(z)) - y'(\alpha y^{-1}(z))y''(y^{-1}(z))}{(y'(y^{-1}(z)))^3}$$

$$= \frac{1}{b} \left[y(\alpha^2 y^{-1}(z)) - ay(\alpha y^{-1}(z)) \right],$$

and
$$x(az+bx(z)) = x\left\{az+b\left[\frac{1}{b}(y(\alpha y^{-1}(z))-az)\right]\right\}$$
$$= x(y\alpha y^{-1}(z))) = \frac{1}{b}\left[y(\alpha y^{-1}(y(\alpha y^{-1}(z))))-ay(\alpha y^{-1}(z))\right]$$
$$= \frac{1}{b}\left[y(\alpha^2 y^{-1}(z))-ay(\alpha y^{-1}(z))\right]$$
as required.

We now show how to explicitly construct an analytic solution of (6.152) by means of (6.156) in a neighborhood of γ. Since
$$x(z) = \frac{1}{b}\left(y(\alpha y^{-1}(z))-az\right),$$
thus
$$x(\gamma) = \frac{1}{b}\left(y(\alpha y^{-1}(\gamma))-a\gamma\right) = \frac{(1-a)\gamma}{b}.$$
Furthermore,
$$x'(z) = \frac{\alpha}{b} \cdot \frac{y'(\alpha y^{-1}(z))}{y'(y^{-1}(z))} - \frac{a}{b},$$
$$x''(z) = x(az+bx(z)),$$
thus
$$x'(\gamma) = \frac{\alpha-a}{b}$$
and
$$x''(\gamma) = \frac{(1-a)\gamma}{b}.$$
By calculating the derivatives of both sides of (6.152), we obtain
$$x'''(z) = x'(az+bx(z))(a+bx'(z)),$$
$$x^{(4)}(z) = x''(az+bx(z))(a+bx'(z))^2 + x'(az+bx(z))bx''(z),$$
so that
$$x'''(\gamma) = \frac{\alpha(\alpha-a)}{b}$$
and
$$x^{(4)}(\gamma) = \frac{1}{b}(1-a)(\alpha^2+\alpha-a)\gamma.$$
In general, we can use the Formula of Faà di Bruno to calculate
$$\Gamma_k := x^{(k+2)}(\gamma)$$
for $k \geq 1$. By means of this formula, it is then easy to write out the explicit form of our solution $x(z)$:
$$x(z) = \frac{(1-a)\gamma}{b} + \frac{\alpha-a}{b}(z-\gamma) + \frac{(1-a)\gamma}{2!b}(z-\gamma)^2$$
$$+ \frac{\alpha(\alpha-a)}{3!b}(z-\gamma)^3 + \frac{(1-a)(\alpha^2+\alpha-a)\gamma}{4!b}(z-\gamma)^4$$
$$+ \sum_{k=3}^{\infty} \frac{\Gamma_k}{(k+2)!}(z-\gamma)^{k+2}.$$

6.3.3 Equation III

Consider the iterative functional differential equation [187]
$$x''(z) = x(az + bx'(z)), \qquad (6.167)$$
where a and $b \neq 0$ are complex numbers. To find analytic solutions of (6.167), we formally let
$$y(z) = az + bx'(z). \qquad (6.168)$$
Then
$$y'(z) - a = bx''(z),$$
and for any number z_0, we have
$$x(z) = x(z_0) + \frac{1}{b}\int_{z_0}^{z}(y(s) - as)\,ds. \qquad (6.169)$$
Thus
$$x''(z) = x(y(z)) = x(z_0) + \frac{1}{b}\int_{z_0}^{y(z)}(y(s) - as)\,ds,$$
or
$$\frac{1}{b}(y'(z) - a) = x(z_0) + \frac{1}{b}\int_{z_0}^{y(z)}(y(s) - as)\,ds. \qquad (6.170)$$

If z_0 is a fixed point of $y(z)$, i.e., $y(z_0) = z_0$, then substituting z_0 into the above equality, we see that
$$x(z_0) = \frac{1}{b}(y'(z_0) - a). \qquad (6.171)$$

Furthermore, differentiating both sides of (6.170) with respect to z, we obtain
$$y''(z) = [y(y(z)) - ay(z)]y'(z). \qquad (6.172)$$

Next, we first seek an analytic solution $g(z)$ of the auxiliary equation
$$\alpha g''(\alpha z)g'(z) = g'(\alpha z)g''(z) + (g'(z))^2 g'(\alpha z)[g(\alpha^2 z) - ag(\alpha z)] \qquad (6.173)$$
satisfying the initial value conditions
$$g(0) = \mu, \ g'(0) = \eta \neq 0, \qquad (6.174)$$
where μ, η are complex numbers, and α is either a Siegel number of it satisfies $0 < |\alpha| < 1$. Then we show that (6.172) has an analytic solution of the form
$$g(z)g(\alpha g^{-1}(z)) \qquad (6.175)$$
in a neighborhood of μ.

Theorem 6.37. *Suppose $0 < |\alpha| < 1$. Then given any complex numbers μ and $\eta \neq 0$, equation (6.173) has a solution of the form*
$$g(z) = \widehat{b}(z) = \sum_{n=0}^{\infty} b_n z^n, \ b_0 = \mu, b_1 = \eta, \qquad (6.176)$$
which is analytic near 0.

Proof. Rewrite (6.173) in the form
$$\frac{\alpha g''(\alpha z) g'(z) - g'(\alpha z) g''(z)}{(g'(z))^2} = g'(\alpha z) \left[g(\alpha^2 z) - ag(\alpha z) \right],$$
or
$$\left(\frac{g'(\alpha z)}{g'(z)} \right)' = g'(\alpha z) \left[g(\alpha^2 z) - ag(\alpha z) \right].$$
Therefore, if $g'(0) = \eta \neq 0$, we have
$$g'(\alpha z) = g'(z) \left[1 + \int_0^z g'(\alpha s) \left(g(\alpha^2 s) - ag(\alpha s) \right) ds \right]. \qquad (6.177)$$

By substituting (6.176) into (6.177), we see that
$$\underline{\alpha} \cdot Db = Db + (Db) * \int (\underline{\alpha} \cdot Db) * (\underline{\alpha}^2 \cdot b - a\underline{\alpha} \cdot b).$$

Hence the sequence $\{b_n\}_{n=2}^\infty$ is successively determined by the condition
$$(\alpha^{n+1} - 1)(n+2) b_{n+2}$$
$$= \sum_{k=0}^{n} \sum_{j=0}^{n-k} \frac{(k+1)(j+1)\left(\alpha^{2(n-k)-j} - a\alpha^{n-k} \right)}{n-k+1} b_{k+1} b_{j+1} b_{n-k-j}, \qquad (6.178)$$

for $n \in \mathbf{N}$ in a unique manner. We need to show that the resulting power series (6.176) converges in a neighborhood of the origin. First of all, note that
$$\left| \frac{(k+1)(j+1)\left(\alpha^{2(n-k)-j} - a\alpha^{n-k} \right)}{(n+2)(n-k+1)(\alpha^{n+1}-1)} \right| \leq \frac{1+|\alpha|}{|\alpha^{n+1}-1|} \leq M$$

for some positive number M.

Next note that Example 4.15 asserts the implicit equation
$$G^3(z) - 2|\mu| G^2(z) - \left(\frac{1}{M} - |\mu|^2 \right) G(z) + \frac{1}{M}(|\eta| z + |\mu|) = 0 \qquad (6.179)$$

has a solution of the form
$$G(z) = \sum_{n=0}^\infty B_n z^n, \quad B_0 = |\mu|, B_1 = |\eta|, \qquad (6.180)$$

which is analytic near 0 and $\{B_n\}_{n=2}^\infty$ is determined by
$$B_{n+2} = M \sum_{k=0}^n \sum_{j=0}^{n-k} B_{k+1} B_{j+1} B_{n-k-j}, \quad n \in \mathbf{N}.$$

In view of (6.178),
$$|b_n| \leq B_n, \ n \in \mathbf{N},$$

which shows that the sequence $\{b_n\}$ is majorized by $\{B_n\}$. Thus $\{b_n\}$ has a positive radius of convergence. The proof is complete.

Theorem 6.38. *Suppose α is a Siegel number. Then if $0 < |\eta| \leq 1$, equation (6.173) has a solution of the form (6.176) which is analytic near 0.*

Proof. As in the previous proof, we seek a power series solution of the form (6.176). Set $b_0 = \mu$ and $b_1 = \eta$. Then (6.178) again holds so that

$$|b_{n+2}| \leq \frac{1+|a|}{|\alpha^{n+1}-1|} \sum_{k=0}^{n} \sum_{j=0}^{n-k} |b_{k+1}| \cdot |b_{j+1}| \cdot |b_{n-k-j}|, \ n \in \mathbf{N}. \tag{6.181}$$

Next note that Example 4.15 asserts that the implicit relation

$$\omega^3 - 2|\mu|\omega^2 - \left(\frac{1}{1+|a|} - |\mu|^2\right)\omega + \frac{1}{1+|a|}(z+|\mu|) = 0 \tag{6.182}$$

has a solution of the form

$$\omega(z) = |\mu| + z + \sum_{n=2}^{\infty} C_n z^n, \tag{6.183}$$

which is analytic near 0 and

$$C_{n+2} = (1+|a|) \sum_{k=0}^{n} \sum_{j=0}^{n-k} C_{k+1} C_{j+1} C_{n-k-j}, \ n \in \mathbf{N}. \tag{6.184}$$

Thus there is a positive constant T such that

$$C_n < T^n, \ n \in \mathbf{Z}^+. \tag{6.185}$$

Now by induction, we may prove that

$$|b_n| \leq C_n d_n, \ n \in \mathbf{Z}^+.$$

where the sequence $d = \{d_n\}_{n=0}^{\infty}$ is defined in Siegel's Lemma (Theorem 3.31). In fact,

$$\begin{aligned}
|b_1| &= |\eta| \leq 1 = C_1 d_1, \\
|b_2| &= (1+|a|)|\alpha - 1|^{-1} |b_1| \cdot |b_1| \cdot |b_0| \\
&\leq (1+|a|)|\alpha - 1|^{-1} C_1 d_1 \cdot C_1 d_1 \cdot C_0 \\
&\leq C_2 |\alpha - 1|^{-1} \Upsilon_2(d) \\
&= C_2 d_2.
\end{aligned}$$

Assume that the above inequality holds for $n = 1, ..., m$. Then

$$|b_{m+1}| \leq (1+|a|)|\alpha^m - 1|^{-1} \sum_{k=0}^{m-1} \sum_{j=0}^{m-1-k} |b_{k+1}| \cdot |b_{j+1}| \cdot |b_{m-1-k-j}|$$

$$= (1+|a|)|\alpha^m - 1|^{-1} \left(\sum_{k=0}^{m-1} |b_{k+1}| \cdot |b_{m-k}| \cdot |b_0| \right.$$

$$\left. + \sum_{k=0}^{m-2} \sum_{j=0}^{m-2-k} |b_{k+1}| \cdot |b_{j+1}| \cdot |b_{m-1-k-j}| \right)$$

$$\leq (1+|a|)|\alpha^m - 1|^{-1} \left(\sum_{k=0}^{m-1} C_{k+1} d_{k+1} C_{m-k} d_{m-k} C_0 \right.$$

$$\left. + \sum_{k=0}^{m-2} \sum_{j=0}^{m-2-k} C_{k+1} d_{k+1} C_{j+1} d_{j+1} C_{m-1-k-j} d_{m-1-k-j} \right)$$

$$\leq (1+|a|)|\alpha^m - 1|^{-1} \Upsilon_{m+1}(d)$$

$$\times \left(\sum_{k=0}^{m-1} C_{k+1} C_{m-k} C_0 + \sum_{k=0}^{m-2} \sum_{j=0}^{m-2-k} C_{k+1} C_{j+1} C_{m-1-k-j} \right)$$

$$= C_{m+1} d_{m+1}.$$

as desired. In view of (6.185) and Siegel's Lemma (Theorem 3.31), we finally see that there is $\delta > 0$ such that

$$|b_n| \leq T^n \left(2^{5\delta+1}\right)^{n-1} n^{-2\delta}, \; n \in \mathbf{Z}^+,$$

which shows that the power series (6.176) converges for

$$|z| < \frac{1}{T 2^{5\delta+1}}.$$

The proof is complete.

Theorem 6.39. *Suppose $0 < |\alpha| < 1$ or α is a Siegel number. Then equation (6.172) has an analytic solution of the form (6.175) in a neighborhood of the number μ, where $g(z)$ is an analytic solution of (6.173).*

Proof. In view of Theorems (6.37) and 6.38, we may find a sequence $\{b_n\}_{n=2}^{\infty}$ such that the function $g(z)$ of the form (6.176) is an analytic solution of (6.173) in a neighborhood of the origin. Since $g'(0) = \eta \neq 0$, the function $g^{-1}(z)$ is analytic in a neighborhood of $g(0) = \mu$. If we now define $y(z)$ by means of (6.175), then

$$y'(z) = \alpha g'\left(\alpha g^{-1}(z)\right)\left(g^{-1}(z)\right)' = \frac{\alpha g'\left(\alpha g^{-1}(z)\right)}{g'\left(g^{-1}(z)\right)},$$

$$y''(z) = \frac{\alpha^2 g''\left(\alpha g^{-1}(z)\right) - \alpha g'\left(\alpha g^{-1}(z)\right) g''\left(g^{-1}(z)\right) \cdot \frac{1}{g'(g^{-1}(z))}}{\left(g'\left(g^{-1}(z)\right)\right)^2}$$

$$= \frac{\alpha g'\left(\alpha g^{-1}(z)\right) \left[g\left(\alpha^2 g^{-1}(z)\right) - a g\left(\alpha g^{-1}(z)\right)\right]}{g'\left(g^{-1}(z)\right)},$$

and
$$[y(y(z)) - ay(z)]y'(z) = [g(\alpha^2 g^{-1}(z)) - ag(\alpha g^{-1}(z))]\frac{\alpha g'(\alpha g^{-1}(z))}{g'(g^{-1}(z))}$$
$$= \frac{\alpha g'(\alpha g^{-1}(z))[g(\alpha^2 g^{-1}(z)) - ag(\alpha g^{-1}(z))]}{g'(g^{-1}(z))}$$

as required. The proof is complete.

Knowing that an analytic solution of (6.167) exists, we may assume that $x(z)$ is of the form
$$x(z) = x(\mu) + x'(\mu)(z - \mu) + \frac{x''(\mu)}{2!}(z - u) + \cdots ;$$
we need to determine the derivatives $x^{(n)}(\mu)$, $n \in \mathbf{N}$. First of all, in view of (6.171) and (6.168), we have
$$x(\mu) = \frac{1}{b}(y'(\mu) - a) = \frac{1}{b}\left(\frac{\alpha g'(\alpha g^{-1}(\mu))}{g'(g^{-1}(\mu))} - a\right) = \frac{\alpha - a}{b}$$
and
$$x'(\mu) = \frac{1}{b}(y(\mu) - a\mu) = \frac{(1-a)\mu}{b},$$
respectively. Furthermore,
$$x''(\mu) = x(a\mu + bx'(\mu)) = x\left(a\mu + b\frac{(1-a)\mu}{b}\right) = x(\mu) = \frac{\alpha - a}{b}.$$
Next by calculating the derivatives of both sides of (6.167), we obtain successively
$$x'''(z) = x'(az + bx'(z))(a + bx''(z)),$$
$$x^{(4)}(z) = x''(az + bx'(z))(a + bx''(z))^2 + x'(az + bx'(z))(bx'''(z)),$$
so that
$$x'''(\mu) = x'(a\mu + bx'(\mu))(a + bx''(\mu)),$$
$$x^{(4)}(\mu) = x''(\mu)\alpha^2 + x'(\mu)[\alpha\mu(1-a)]$$
$$= \frac{\alpha}{b}\left[(\alpha - a)\alpha + ((1-a)\mu)^2\right].$$

In general, we can use the Formula of Faa di Bruno to find $(x(az + bx'(z)))^{(m)}$ and then calculate
$$\Gamma_m := x^{(m+2)}(\mu)$$
for $m \in \mathbf{Z}^+$. It is then easy to write out the explicit form of our solution $x(z)$:
$$x(z) = \frac{\alpha - a}{b} + \frac{(1-a)\mu}{b}(z - \mu) + \frac{\alpha - a}{2!b}(z - \mu)^2 + \frac{\alpha\mu(1-a)}{3!b}(z - \mu)^3$$
$$+ \frac{\alpha}{4!b}\left[(\alpha - a)\alpha + ((1-a)\mu)^2\right](z - \mu)^4$$
$$+ \sum_{m=3}^{\infty} \frac{\Gamma_m}{(m+2)!}(z - \mu)^{m+2}.$$

6.3.4 Equation IV

In this section, we consider a class of iterative functional differential equations of the form [188]

$$x''\left(x^{[r]}(z)\right) = c_0 z + c_1 x(z) + \cdots + c_m x^{[m]}(z), \qquad (6.186)$$

where r and m are nonnegative integers, $c_0, c_1, ..., c_m$ are complex constants, $\sum_{i=0}^{m} |c_i| \neq 0$, and $x^{[i]}$ denotes the i-th iterate of x. When $r = 0$, $c_2 \neq 0$ and $c_i = 0$ $(0 \leq i \leq m, i \neq 2)$, Eq. (6.186) reduces to the second-order iterative functional differential equation $x''(z) = c_2 x(x(z))$ which has been discussed before.

To find analytic solution of (6.186), we first seek the analytic solution $y(z)$ of the companion equation

$$\alpha^2 y''\left(\alpha^{r+1} z\right) y'(\alpha^r z) = \alpha y'\left(\alpha^{r+1} z\right) y'(\alpha^r z) + [y'(\alpha^r z)]^3 \left[\sum_{i=0}^{m} c_i y(\alpha^i z)\right], \qquad (6.187)$$

satisfying the initial value conditions

$$y(0) = \mu, \ y'(0) = \eta \neq 0, \qquad (6.188)$$

where μ, η are complex numbers, and α satisfies $|\alpha| > 1$ or $0 < |\alpha| < 1$ or is a Siegel number. Then we show that (6.186) has an analytic solution of the form

$$x(z) = y\left(\alpha y^{-1}(z)\right), \qquad (6.189)$$

in a neighborhood of the number u. Finally, we make use of (6.189) to show how to derive an explicit power series solution. First of all, suppose (6.187) has a solution of the form

$$y(z) = \sum_{n=0}^{\infty} b_n z^n, \qquad (6.190)$$

which is analytic on a neighborhood of the origin and $b_0 = \mu$ and $b_1 = \eta \neq 0$. Then we may rewrite (6.187) as

$$\frac{\alpha^2 y''\left(\alpha^{r+1} z\right) y'(\alpha^r z) - \alpha y'\left(\alpha^{r+1} z\right) y'(\alpha^r z)}{[y'(\alpha^r z)]^2} = y'(\alpha^r z) \left[\sum_{i=0}^{m} c_i y(\alpha^i z)\right]$$

or

$$\left(\frac{y'(\alpha^{r+1} z)}{y'(\alpha^r z)}\right)' = \alpha^{r-1} y'(\alpha^r z) \left[\sum_{i=0}^{m} c_i y(\alpha^i z)\right].$$

Since $y'(0) = \eta \neq 0$, we have

$$y'(\alpha^{r+1} z) = y'(\alpha^r z) \left[1 + \alpha^{r-1} \int_0^z y'(\alpha^r s) \sum_{i=0}^{m} c_i y(\alpha^i s) \, ds\right]. \qquad (6.191)$$

By substituting (6.190) into (6.191), we see that

$$\alpha^{rn+1} \left(\alpha^{n+1} - 1\right) (n+2) b_{n+2}$$
$$= \sum_{k=0}^{n} \sum_{j=0}^{n-k} \frac{(k+1)(j+1) \alpha^{r(k+j)} \sum_{i=0}^{m} c_i \alpha^{i(n-k-j)}}{n-k+1} b_{k+1} b_{j+1} b_{n-k-j} \qquad (6.192)$$

for $n \geq 1$. So for $b_0 = \mu$ and $b_1 = \eta$, the sequence $\{b_n\}_{n=2}^{\infty}$ is successively determined by (6.192) in a unique manner. It suffices now to show that the power series just derived is analytic at 0. To this end, we show that such a power series solution is majorized by a convergent power series.

Theorem 6.40. *Suppose α is a Siegel number. Then for any given complex numbers μ and η that satisfies $0 < |\eta| \leq 1$, Eq. (6.187) has a solution of the form*

$$y(z) = \sum_{n=0}^{\infty} b_n z^n, \quad b_0 = \mu, b_1 = \eta,$$

which is analytic near 0.

Proof. From (6.192), it follows that

$$\left| \frac{(k+1)(j+1)\alpha^{r(k+j)} \sum_{i=0}^{m} c_i \alpha^{i(n-k-j)}}{(n+2)(n-k+1)\alpha^{rn+1}(\alpha^{n+1}-1)} \right| \leq \frac{\sum_{i=0}^{m} |c_i|}{|\alpha^{n+1}-1|}, \quad n \in \mathbf{N}.$$

Thus

$$|b_{n+2}| \leq \frac{\sum_{i=0}^{m} |c_i|}{|\alpha^{n+1}-1|} \sum_{k=0}^{n} \sum_{j=0}^{n-k} |b_{k+1}| |b_{j+1}| |b_{n-k-j}|, \quad n \in \mathbf{N}. \tag{6.193}$$

Note that Example 4.15 asserts that

$$G^3(z) - 2|\mu| G^2(z) + \left(\frac{1}{\sum_{i=0}^{m} |c_i|} - |\mu|^2 \right) G(z) - \frac{1}{\sum_{i=0}^{m} |c_i|} (C_1 z + |\mu|) = 0$$

has a solution

$$G(z) = \sum_{n=0}^{\infty} C_n z^n$$

which is analytic on a neighborhood of the origin and $C_0 = |\mu|$, $C_1 = 1$ and

$$C_{n+2} = \left(\sum_{i=0}^{m} |c_i| \right) \sum_{k=0}^{n} \sum_{j=0}^{n-k} C_{k+1} C_{j+1} C_{n-k-j}, \quad n \in \mathbf{N}, \tag{6.194}$$

In view of the Cauchy's Estimation (Theorem 3.26), there exists a positive constant r such that

$$C_n < r^n, \quad n \in \mathbf{Z}^+. \tag{6.195}$$

As in the proof of Theorem 3.32, we may apply Siegel's Lemma (Theorem 3.31) to deduce a $\delta > 0$ such that

$$|b_n| \leq r^n \left(2^{5\delta+1} \right)^{n-1} n^{-2\delta}, \quad n \in \mathbf{Z}^+,$$

so that $\{b_n\}_{n \in \mathbf{N}}$ has a positive radius of convergence. The proof is complete.

Theorem 6.41. *Suppose $|\alpha| > 1$ and $r \geq m$. Then for any given complex numbers μ and $\eta \neq 0$, Eq. (6.187) has a solution of the form (6.190) which is analytic on a neighborhood of the origin and $b_0 = \mu$ as well as $b_1 = \eta$.*

Proof. First of all, for $r \geq m$ and $0 \leq k+j \leq n$, we have

$$\left| \frac{(k+1)(j+1)\alpha^{r(k+i)} \sum_{i=0}^{m} c_i \alpha^{i(n-k-j)}}{(n+2)(n-k+1)\alpha^{rn+1}(\alpha^{n+1}-1)} \right| = \left| \frac{(k+1)(j+1) \sum_{i=0}^{m} c_i \alpha^{i(n-k-j)}}{(n+2)(n-k+1)\alpha(\alpha^{n+1}-1)} \right|$$
$$\leq \frac{\sum_{i=0}^{m}|c_i|}{|\alpha^{n+1}-1|} \leq M$$

for $n \in \mathbf{N}$, where M is some positive number. Note that Example 4.15 asserts that

$$G^3(z) - 2|\mu| G^2(z) + \left(\frac{1}{M} - |\mu|^2\right) G(z) - \frac{1}{M}(|\eta|z + |\mu|) = 0 \qquad (6.196)$$

has a solution

$$G(z) = \sum_{n=0}^{\infty} D_n z^n$$

which is analytic on a neighborhood of the origin and $D_0 = |u|$, $D_1 = |\eta|$, as well as

$$D_{n+2} = M \sum_{k=0}^{n} \sum_{j=0}^{n-k} D_{k+1} D_{j+1} D_{n-k-j}, \ n \in \mathbf{N}. \qquad (6.197)$$

Clearly, in view of (6.192)

$$|b_n| \leq D_n, \ n \in \mathbf{N}.$$

Therefor the sequence $\{b_n\}_{n \in \mathbf{N}}$ has a positive radius of convergence. The proof is complete.

Theorem 6.42. *Suppose $0 < |\alpha| < 1$. Suppose further that $0 < r \leq m$ and $c_0 = 0, ..., c_{r-1} = 0$, or, $r = 0$. Then for any given complex numbers μ and $\eta \neq 0$, (6.187) has a solution of the form (6.190) which is analytic on a neighborhood of the origin and satisfies $b_0 = \mu$ and $b_1 = \eta$.*

The proof is similar to that of the previous Theorem and hence skipped.

Consider the following three hypotheses:
(i) α is a Siegel number;
(ii) $|\alpha| > 1$ and $r \geq m$;
(iii) $0 < |\alpha| < 1$, $0 < r \leq m$ and $c_0 = 0, ..., c_{r-1} = 0$;
(iv) $0 < |\alpha| < 1$ and $r = 0$.

Theorem 6.43. *Suppose one of the above conditions (i)-(iv) is fulfilled. Then for any given complex number μ, Eq. (6.186) has a solution $x(z)$ which is analytic on a neighborhood of μ and satisfies the conditions $x(\mu) = \mu$ and $x'(\mu) = \alpha$. This solution has the form $x(z) = y(\alpha y^{-1}(z))$, where $y(z)$ is an analytic solution of the initial value problem (6.187), (6.188).*

Proof. In view of Theorems 6.40, 6.41 and 6.42, we may find a sequence $\{b_n\}_{n=2}^{\infty}$ such that the function $y(z)$ of the form (6.190) is an analytic solution of (6.187) in a neighborhood of the origin. Since $y'(0) = \eta \neq 0$, the function $y^{-1}(z)$ is analytic in a neighborhood of the $y(0) = \mu$. If we now define $x(z)$ by means of (6.189), then

$$x''\left(x^{[r]}(z)\right) = \frac{\alpha^2 y''\left(\alpha^{r+1}y^{-1}(z)\right)y'\left(\alpha^r y^{-1}(z)\right) - \alpha y'\left(\alpha^{r+1}y^{-1}(z)\right)y''\left(\alpha^r y^{-1}(z)\right)}{[y'(\alpha^r y^{-1}(z))]^3}$$

$$= \sum_{i=0}^{m} c_i y\left(\alpha^i y^{-1}(z)\right) = \sum_{i=0}^{m} c_i x^{[i]}(z),$$

as required. The proof is complete.

We may now construct an analytic solution of (6.186) by means of (6.189) in a neighborhood of μ. Since

$$x(\mu) = y\left(\alpha y^{-1}(\mu)\right) = y(\alpha \cdot 0) = \mu,$$

we may assume that $x(z)$ is of the form

$$x(z) = \mu + x'(\mu)(z - \mu) + \frac{x''(\mu)}{2!}(z - \mu)^2 + \frac{x'''(\mu)}{3!}(z - \mu)^3 + \cdots.$$

We need to determine the derivatives $x^{(n)}(\mu)$ for $n \in \mathbf{Z}^+$. First of all, in view of (6.189), we have

$$x'(z) = \frac{\alpha y'\left(\alpha y^{-1}(z)\right)}{y'\left(y^{-1}(z)\right)}.$$

Thus

$$x'(\mu) = \frac{\alpha y'\left(\alpha y^{-1}(\mu)\right)}{y'\left(y^{-1}(\mu)\right)} = \frac{\alpha y'(0)}{y'(0)} = \alpha,$$

and

$$x''\left(x^{[r]}(\mu)\right) = c_0\mu + c_1 x(\mu) + \cdots + c_m x^{[m]}(\mu) = \left(\sum_{i=0}^{m} c_i\right)\mu..$$

Next, by calculating the derivatives of both side of (6.186), we obtain

$$x'''\left(x^{[r]}(z)\right) x'\left(x^{[r-1]}(z)\right) \cdots x'(x(z)) x'(z) = \sum_{i=0}^{m} c_i x'\left(x^{[i-1]}(z)\right) \cdots x'(x(z)) x'(z)$$

so that

$$x'''(\mu) [x'(\mu)]^r = \sum_{i=0}^{m} c_i [x'(\mu)]^i$$

and

$$x'''(\mu) = \frac{\sum_{i=0}^{m} c_i [x'(\mu)]^i}{[x'(\mu)]^r} = \frac{\sum_{i=0}^{m} c_i \alpha^i}{\alpha^r} = \sum_{i=0}^{m} c_i \alpha^{i-r}.$$

In general, we can show by induction that

$$\left(x''\left(x^{[r]}(z)\right)\right)^{(n-2)} = \left(\left(x^{[r]}(z)\right)'\right)^{n-2} x^{(n)}\left(x^{[r]}(z)\right)$$

$$+ \sum_{k=1}^{n-3} P_{k,n-2}\left(\left(x^{[r]}(z)\right)', ..., \left(x^{[r]}(z)\right)^{(n-2)}\right) x^{(k+2)}\left(x^{[r]}(z)\right)$$

for $n \geq 3$, and

$$\left(x^{[j]}(z)\right)^{(l)} = Q_{jl}\left(x_{10}(z), ..., x_{1,j-1}(z); ...; x_{10}(z), ..., x_{l,j-1}(z)\right),$$

respectively, where $x_{ij}(z) = x^{(i)}\left(x^{[j]}(z)\right)$ and P_{jk} and Q_{jl} are some polynomials with nonnegative coefficients.

Moreover, if we write

$$\beta_{jl} = Q_{lj}\left(x'(\mu), ..., x'(\mu); ...; x^{(j)}(\mu), ..., x^{(j)}(\mu)\right),$$

then differentiating (6.186) $n-2$ times at $z = u$, we will end up with

$$(x'(u))^{r(n-2)} x^{(n)}(\mu) + \sum_{k=1}^{n-3} P_{k,n-2}(\beta_{r1}, ..., \beta_{r,n-2}) x^{(k+2)}(\mu) = \sum_{i=0}^{m} c_i \beta_{i,n-2}, \ n \geq 3.$$

This shows that

$$x^{(n)}(\mu) = \frac{1}{\alpha^{r(n-2)}} \left[\sum_{i=0}^{m} c_i \beta_{i,n-2} - \sum_{k=1}^{n-3} P_{k,n-2}(\beta_{r1}, ..., \beta_{r,n-2}) x^{(k+2)}(\mu)\right]$$

for $n \geq 3$. By means of this formula, we may then write

$$x(z) = \mu + \alpha(z - \mu) + \frac{\mu \sum_{i=0}^{m} c_i}{2!}(z - \mu)^2 + \frac{\sum_{i=0}^{m} c_i \alpha^{i-t}}{3!}(z - \mu)^3 + \cdots.$$

6.4 Equations with Higher Order Derivatives

In some iterative functional differential equations, solutions in the form of elementary functions may exist. In this section, we show how power function solutions of the form

$$x(z) = \lambda z^{\mu} \qquad (6.198)$$

can be computed. We first illustrate this by considering an equation of the form

$$x^{(n)}(z) = az^{j}\left(x^{[m]}(z)\right)^{k}, \qquad (6.199)$$

where k, m, n are positive integers, j is a nonnegative integer, a is a complex number, $x^{(n)}(z)$ is the n-th derivative of $x(z)$ and $x^{[m]}(z)$ is the m-th iterate of $x(z)$. We assume $m \geq 2$ and $a \neq 0$ to avoid trivial cases.

In the following discussions, recall the notation (in the first Chapter)

$$\lfloor \mu \rfloor_n = \mu(\mu - 1) \cdots (\mu - n + 1).$$

Substituting (6.198) into (6.199), we obtain

$$\lambda \lfloor \mu \rfloor_n z^{\mu-n} = a\lambda^{k(1+\mu+\cdots+\mu^{m-1})} z^{k\mu^m+j}.$$

This prompts us to consider the equations

$$\lambda \lfloor \mu \rfloor_n = a\lambda^{k(1+\mu+\cdots+\mu^{m-1})}, \tag{6.200}$$

and

$$k\mu^m + j = \mu - n. \tag{6.201}$$

First of all, we assert that the polynomial

$$f(z) = kz^m - z + n + j$$

does not have any real roots if m is even, and has exactly one real root if m is odd. Indeed, for even m and real z, by solving

$$f'(z) = kmz^{m-1} - 1 = 0,$$

we see that the minimum of f occurs at the root $\rho = 1/\sqrt[m-1]{km} \in (0,1)$. Hence

$$f(z) \geq f(\rho) = \rho\left(\frac{1}{m} - 1\right) + n + j > \left(\frac{1}{m} - 1\right) + n + j > 0$$

for all real z. If m is odd, then f' has two zeros $\pm \rho$. Since

$$\min_{z \geq -\rho} f(z) = \min\{f(\rho), f(-\rho)\} = f(\rho) > 0,$$

f does not have any real roots greater than or equal to $-\rho$. Furthermore, since $f(-\rho)$ and $f(-\infty)$ have opposite signs, f has at least one real root in $(-\infty, -\rho)$. Finally, since $f'(z) > 0$ for all $z < -\rho$, f is increasing in $(-\infty, -\rho)$. So f has exactly one real root which is negative. As a consequence, the roots of f in either case cannot be $0, 1, \ldots,$ nor $n-1$.

Next, we assert that $f(z)$ has simple roots only. Suppose not, let r be a double root of f, then it is a root of f' and

$$f(z) - \frac{z}{m}f'(z) = \frac{1-m}{m}z + n + j. \tag{6.202}$$

Hence (6.202) implies that $r = m(n+j)/(m-1)$ is real and positive, which is impossible by our previous assertion.

Let $\mu_1, \ldots, \mu_m$ be the roots of (6.201). In view of the above results, $\mu_1, \ldots, \mu_m$ are pairwise distinct and each one of them is different from $0, 1, \ldots,$ or $n-1$. Furthermore, in view of (6.200) and (6.201), we have

$$\lambda \lfloor \mu_i \rfloor_n = a\lambda^{k(1-\mu_i^m)/(1-\mu_i)} = a\lambda^{(k+n+j-\mu_i)/(1-\mu_i)}$$

for $i = 1, \ldots, m$, from which we obtain

$$\lambda_i = \left[\frac{\lfloor \mu_i \rfloor_n}{a}\right]^{(1-\mu_i)/(k+n+j-1)}, \quad i = 1, \ldots, m. \tag{6.203}$$

In other words, we have found m distinct solutions of the form:

$$x_i(z) = \lambda_i z^{\mu_i}, \quad i = 1, 2, ..., m, \tag{6.204}$$

where $\mu_1, ..., \mu_m$ are roots of (6.201) and $\lambda_1, ..., \lambda_m$ are defined by (6.203).

To summarize, there exist m distinct (single valued and analytic) power functions of the form (6.204) which are solutions of (6.199) defined on $\mathbf{C}\setminus(-\infty, 0]$.

We remark that each solution $x_i(z) = \lambda_i z^{\mu_i}$ has a nontrivial fixed point α_i. Indeed, from $\lambda_i \alpha_i^{\mu_i} = \alpha_i$, we find

$$\alpha_i = \lambda_i^{1/(1-\mu_i)} = (\lfloor \mu_i \rfloor_n)^{1/(k+n+j-1)} \neq 0. \tag{6.205}$$

In terms of the fixed point α_i, we may therefore write $x_i(z)$ in the form

$$x_i(z) = \alpha_i^{1-\mu_i} z^{\mu_i}. \tag{6.206}$$

As a corollary, let $\mu_1, ..., \mu_m$ be the roots of (6.201), and $\alpha_1, ..., \alpha_m$ given by (6.205). Then in a neighborhood of each point α_i, $i = 1, ..., m$, equation (6.199) has an analytic solution of the form

$$x_i(z) = \alpha_i + \mu_i(z - \alpha_i) + \frac{\mu_i(\mu_i - 1)}{2!\alpha_i}(z - \alpha_i)^2 + \cdots$$

$$+ \frac{\lfloor \mu_i \rfloor_n}{n!\alpha_i^{n-1}}(z - \alpha_i)^n + \cdots .$$

Indeed, in view of (6.206),

$$x_i(z) = \alpha_i^{1-\mu_i} z^{\mu_i} = \alpha_i \left(1 + \frac{z - \alpha_i}{\alpha_i}\right)^{\mu_i}$$

$$= \alpha_i \left[1 + \frac{\mu_i}{1!}\left(\frac{z - \alpha_i}{\alpha_i}\right) + \frac{\mu_i(\mu_i - 1)}{2!}\left(\frac{z - \alpha_i}{\alpha_i}\right)^2 + \cdots \right]$$

as required.

Example 6.6. Consider the equation

$$x'(z) = x(x(z)).$$

Then (6.201) is reduced to

$$\mu^2 - \mu + 1 = 0,$$

which has roots $\mu_\pm = (1 \pm \sqrt{3}i)/2$. And from (6.200), we find $\lambda_- = \mu_-^{1/\mu_-} \approx 2.145 - 1.238i$, $\lambda_+ = \mu_+^{1/\mu_+} \approx 2.145 + 1.238i$. Since $|\mu_\pm| = 1$ and $\mu_\pm^6 = 1$, $\alpha_\pm = \mu_\pm$ are roots of unity.

6.4.1 Equation I

In this section, we prove the existence of power solutions for the more general equation

$$\left(x^{(n_1)}(p_1 z)\right)^{N_1} \cdots \left(x^{(n_a)}(p_a z)\right)^{N_a} = Az^j \left(x^{[m_1]}(q_1 z)\right)^{M_1} \cdots \left(x^{[m_b]}(q_b z)\right)^{M_b} \tag{6.207}$$

where $a, b, N_1, \ldots, N_a, M_1, \ldots, M_b$ and $n_1, \ldots, n_a, m_1, \ldots, m_b$ are positive integers such that $n_1 > n_2 > \cdots > n_a$ and $m_1 > m_2 > \cdots > m_b$. The number j is a nonnegative integer and $A, p_1, \ldots, p_a, q_1, \ldots, q_b$ are nonzero complex numbers. Note that by taking $a = b = 1$, $N_1 = 1$, $M_1 = k$ and $p_1 = q_1 = 1$ in (6.207), we obtain (6.199).

Theorem 6.44. *Put $s(N, a) = N_1 + \cdots + N_a$, $s(M, b) = M_1 + \cdots + M_b$ and $s(Nn, a) = N_1 n_1 + \cdots + N_a n_a$. Let $\mu_1, \ldots, \mu_m$, where $1 \leq m \leq m_1$, be distinct roots of the polynomial*

$$f(z) = M_1 z^{m_1} + \cdots + M_b z^{m_b} - s(N, a) z + s(Nn, a) + j. \tag{6.208}$$

If $s(N, a) \leq s(M, b)$, then (6.207) has m distinct, single-valued, nonzero, analytic power solutions of the form

$$x_i(z) = \lambda_i z^{\mu_i}, \quad i = 1, 2, \ldots, m; z \in \mathbf{C} \setminus (-\infty, 0],$$

where

$$\lambda_i = \left[\frac{\prod_{l=1}^{a} p_l^{N_l \mu_i}}{A \prod_{l=1}^{b} q_l^{M_l \mu_i^{m_l}}} (\lfloor \mu_i \rfloor_{n_a})^{s(N, a)} \right.$$

$$\left. \times (\lfloor \mu_i - n_a \rfloor_{n_{a-1} - n_a})^{s(N, a-1)} \cdots (\lfloor \mu_i - n_2 \rfloor_{n_1 - n_2})^{s(N, 1)} \right]^{B_i} \tag{6.209}$$

and

$$B_i = \frac{1 - \mu_i}{s(M, b) + s(Nn, a) - s(N, a) + j}.$$

Proof. Substituting $x(z) = \lambda z^\mu$ into (6.207), we obtain

$$Q_\mu A \lambda^c z^r = P_\mu \lambda^{s(N,a)} (\lfloor \mu \rfloor_{n_a})^{s(N,a)}$$

$$\times (\lfloor \mu - n_a \rfloor_{n_{a-1} - n_a})^{s(N, a-1)} \cdots (\lfloor \mu - n_2 \rfloor_{n_1 - n_2})^{s(N, 1)} z^{s(N,a)\mu - s(Nn,a)}$$

where

$$c = \sum_{l=1}^{b} M_l (1 + \mu + \cdots + \mu^{m_l - 1}),$$

$$r = \sum_{l=1}^{b} M_l \mu^{m_l} + j,$$

$$P_\mu = \prod_{l=1}^{a} p_l^{N_l \mu},$$

and

$$Q_\mu = \prod_{l=1}^{b} q_l^{M_l \mu^{m_l}}.$$

This leads to two requirements

$$Q_\mu A \lambda^c = P_\mu \lambda^{s(N,a)} \left(\lfloor \mu \rfloor_{n_a}\right)^{s(N,a)}$$
$$\times (\lfloor \mu - n_a \rfloor_{n_{a-1}-n_a})^{s(N,a-1)} \cdots (\lfloor \mu - n_2 \rfloor_{n_1-n_2})^{s(N,1)} \quad (6.210)$$

and

$$s(N,a)\mu - s(Nn,a) = \sum_{l=1}^{b} M_l \mu^{m_l} + j, \quad (6.211)$$

or

$$f(\mu) = 0.$$

Note that the polynomial $f(z)$ does not have any nonnegative real roots if $s(N,a) \leq s(M,b)$. Indeed, $f(0) = s(Nn,a) + j > 0$. For real $z \geq 1$, from $s(N,a) \leq s(M,b)$, we get $s(N,a) z \leq s(M,b) z \leq M_1 z^{m_1} + \cdots + M_b z^{m_b}$ and so $f(z) \geq s(Nn,a) + j > 0$. For real $z \in (0,1)$, we have $f(z) > 0 - s(N,a) + s(Nn,a) + j \geq 0$. Thus none of $\mu_1, \ldots, \mu_m$ is a nonnegative real number. Substitute $\mu = \mu_i$ into (6.210), we may then solve for $\lambda = \lambda_i \neq 0$ and conclude that $\lambda_i z^{\mu_i}$ is a desired solution. The proof is complete.

We remark that if the condition $s(N,a) \leq s(M,b)$ fails to hold, the theorem is not true as can be seen from the following example.

Example 6.7. Consider the equation

$$\left(x^{(3)}(z)\right)\left(x^{(1)}(z)\right)^3 = x^{[1]}(z).$$

Here $s(N,2) = 4 > s(M,1) = 1$, $f(z) = z - 4z + 6$ has a unique root $\mu = 2$ with $\lambda = 0$, yielding only the trivial power function solution.

In certain cases, the number of solutions can be strengthened to m_1 as follows: In addition to the hypotheses in Theorem 1, suppose $m_1, \ldots, m_b$ are all even, or, m_1 is odd but $m_2, \ldots, m_b$ are even. Then there exist m_1 distinct, single-valued, nonzero, analytic power function solutions.

Indeed, in the proof above, we already have $f(z) > 0$ for each $z \geq 0$. If $m_1, \ldots, m_b$ are even, then Descartes' rule of sign (see e.g. page 171 in Barbeau [14]), tells us that $f(z)$ has no negative real root, while if m_1 is odd but $m_2, \ldots, m_b$ are even, then $f(z)$ has at most one negative real root. In either case, $f(z)$ cannot have repeated roots, other roots being complex conjugates. Hence, all m_1 roots of $f(z)$ are distinct.

Observe that each solution $x_i(z) = \lambda_i z^{\mu_i}$ has a nontrivial fixed point α_i of the form
$$\alpha_i = \lambda_i^{\frac{1}{1-\mu_i}} = \lambda_i^{\frac{1}{s(M,b)+s(Nn,a)-s(N,a)+j}} \neq 0,$$
thus we may write each solution $x_i(z)$ as
$$x_i(z) = \alpha_i^{1-\mu_i} z^{\mu_i}.$$

Expanding such solution about its fixed point, we immediately get the following consequence: Let $\mu_1, \ldots, \mu_m$ be the distinct roots of (6.211), and
$$\alpha_i = \lambda_i^{\frac{1}{1-\mu_i}}, \ i = 1, 2, \ldots, m,$$
where λ_i is defined by (6.209). Then in a neighborhood of each point α_i, the iterative functional differential equation (6.207) has an analytic solution of the form
$$x_i(z) = \alpha_i + \frac{\lfloor \mu_i \rfloor_1}{1!}(z-\alpha_i) + \frac{\lfloor \mu_i \rfloor_2}{2!\alpha_i}(z-\alpha_i)^2 + \cdots + \frac{\lfloor \mu_i \rfloor_n}{n!\alpha_i^{n-1}}(z-\alpha_i)^n + \cdots.$$

6.4.2 Equation II

The following equation
$$x'(z) = \frac{1}{x^{[2]}(z)} \tag{6.212}$$
has been discussed earlier. Nontrivial power function solutions can also be found for (6.212). Indeed, let us seek solutions of the form $x(z) = \lambda z^\mu$ where $\lambda \neq 0$. Setting it into (6.212), we obtain
$$\lambda \mu z^{\mu-1} = \lambda^{-(1+\mu)} z^{-\mu^2}.$$

We are led to the equations
$$\mu \lambda^{2+\mu} = 1, \tag{6.213}$$
and
$$\mu^2 + \mu - 1 = 0. \tag{6.214}$$

Since the two distinct roots μ_+ and μ_- are
$$\mu_\pm = \frac{-1 \pm \sqrt{5}}{2},$$
we can then solve from (6.213) to find
$$\lambda_\pm = \mu_\pm^{(\mu_\pm - 1)/(2 - \mu_\pm - \mu_\pm^m)} = \mu_\pm^{\mu_\pm - 1}, \tag{6.215}$$
and two corresponding distinct solutions of (6.212) of the form
$$x_\pm(z) = \lambda_\pm z^{\mu_\pm}. \tag{6.216}$$

On the other hand, we may try to find power solutions of the form $x(z) = \lambda z^\mu$, where $\lambda \neq 0$, for the iterative functional differential equation

$$x'(z) = Ax(z), \quad A \neq 0. \tag{6.217}$$

Substituting $x(z) = \lambda z^\mu$ into (6.217), we see that $\lambda \mu z^{\mu-1} = A\lambda z^\mu$, so that

$$\mu = \mu - 1,$$
$$\lambda \mu = A\lambda.$$

But then μ cannot exist as expected.

The above approach prompts us to consider more general iterative functional differential equations. Let

$$N := (N_1, N_2, ..., N_a), \quad M := (M_1, M_2, ..., M_b), \quad T := (T_1, T_2, ..., T_c)$$

be vectors with nonnegative integer components. Let

$$n := (n_1, n_2, ..., n_a), \quad m := (m_1, m_2, ..., m_b), \quad t := (t_1, t_2, ..., t_c)$$

be vectors with nonnegative integer components such that

$$n_1 > n_2 > \cdots > n_a, \; m_1 > m_2 > \cdots > m_b, \; t_1 > t_2 > \cdots > t_c.$$

Recall that the 1-norm of a vector $v = (v_1, v_2, ..., v_k)$ is denoted by $|v|_1$, that is, $|v|_1 = \sum_{i=1}^k |v_i|$. In this section, for the sake of convenience, we will use $|v|$ instead of $|v|_1$ and use

$$|v|_j := \sum_{i=1}^j |v_i|$$

for the 1-norm of the subvector $(v_1, v_2, ..., v_j)$ of v. The inner product of two vectors $u = (u_1, ..., u_k)$ and $v = (v_1, ..., v_k)$ is

$$u \bullet v = u_1 v_1 + \cdots + u_k v_k.$$

Consider iterative functional differential equations of the form [31]

$$\left(x^{(n_1)}(p_1 z)\right)^{N_1} \cdots \left(x^{(n_a)}(p_a z)\right)^{N_a} = A \frac{\left(x^{[t_1]}(r_1 z)\right)^{T_1} \cdots \left(x^{[t_c]}(r_c z)\right)^{T_c}}{\left(x^{[m_1]}(q_1 z)\right)^{M_1} \cdots \left(x^{[m_b]}(q_b z)\right)^{M_b}} \tag{6.218}$$

where $A, p_1, \ldots, p_a, r_1, \ldots, r_c, q_1, \ldots, q_b$ are nonzero complex numbers.

In order that (6.218) is a true differential equation, it is natural to assume $N \bullet n > 0$ (so that $|N| > 0$ and some $n_i N_i > 0$). We will make such an assumption for the moment and comment on the other cases later. We remark that if $|T| = |M| = 0$, the right hand side of (6.218) becomes A. The resulting equation does not contain any iterates of the unknown function and perhaps it is not appropriate to call it an iterative functional equation. We will, however, consider such a possibility as well.

We will seek nontrivial power function solutions of (6.218). By substituting $x(z) = \lambda z^\mu$, where $\lambda \neq 0$, into (6.218), we obtain

$$AF_\mu \lambda^{E_\mu} z^{r+f}$$
$$= P_\mu Q_\mu \lambda^{|N|+C_\mu} \left(\lfloor \mu \rfloor_{n_a}\right)^{|N|_a}$$
$$\times (\lfloor \mu - n_a \rfloor_{n_{a-1}-n_a})^{|N|_{a-1}} \cdots (\lfloor \mu - n_2 \rfloor_{n_1 - n_2})^{|N|_1} z^{|N|\mu - N\bullet n}, \quad (6.219)$$

where

$$P_\mu := \prod_{l=1}^{a} p_l^{N_l \mu}, \quad Q_\mu := \prod_{l=1}^{b} q_l^{M_l \mu^{m_l}}, \quad F_\mu := \prod_{l=1}^{c} r_l^{T_l \mu^{t_l}},$$

$$C_\mu := \sum_{l=1}^{b} M_l(1 + \mu + \cdots + \mu^{m_l - 1}), \quad E_\mu := \sum_{l=1}^{c} T_l(1 + \mu + \cdots + \mu^{t_l - 1}),$$

and

$$r := -\sum_{l=1}^{b} M_l \mu^{m_l}, \quad f := \sum_{l=1}^{c} T_l \mu^{t_l},$$

This leads to two requirements

$$AF_\mu = P_\mu Q_\mu \lambda^{|N|+C_\mu - E_\mu} \left(\lfloor \mu \rfloor_{n_a}\right)^{|N|_a}$$
$$\times (\lfloor \mu - n_a \rfloor_{n_{a-1}-n_a})^{|N|_{a-1}} \cdots (\lfloor \mu - n_2 \rfloor_{n_1 - n_2})^{|N|_1} \quad (6.220)$$

and

$$|N|\mu - N\bullet n = \sum_{l=1}^{c} T_l \mu^{t_l} - \sum_{l=1}^{b} M_l \mu^{m_l}. \quad (6.221)$$

Let

$$\Phi(\mu) := \sum_{l=1}^{b} M_l \mu^{m_l} - \sum_{l=1}^{c} T_l \mu^{t_l} + |N|\mu - N\bullet n, \quad (6.222)$$

which may be called a 'characteristic polynomial' of (6.218). First of all, it is possible that $\Phi(\mu)$ is a constant polynomial. Indeed, this is true only if

$$|N|\mu + M_b \mu^{m_b} + T_c \mu^{t_c} = 0.$$

In such a case, the only possible root of Φ is 0. However, substituting $x(z) = \lambda z^0 = \lambda$ into (6.218), we obtain $\lambda^{|T|} = 0$, which shows that the only possible power function solution is trivial. In general, $\Phi(\mu)$ may be a nonconstant polynomial.

Theorem 6.45. *Suppose Φ is not a constant polynomial and μ is a root which does not belong to the set $\{0, 1, \ldots, \max\{1, n_1 - 1\}\}$. Suppose further that $|T| + |M| > 0$ and*

$$|N| + C_\mu - E_\mu \neq 0. \quad (6.223)$$

Then (6.218) has at least one single-valued, nontrivial, analytic power solution of the form
$$x(z) = \lambda z^\mu, \ z \in \mathbf{C}\setminus(-\infty, 0],$$
where
$$\lambda^{|N|+C_\mu - E_\mu} = \frac{AF_\mu}{P_\mu Q_\mu} \frac{1}{(\lfloor\mu\rfloor_{n_a})^{|N|_a} (\lfloor\mu - n_a\rfloor_{n_{a-1}-n_a})^{|N|_{a-1}} \cdots (\lfloor\mu - n_2\rfloor_{n_1-n_2})^{|N|_1}}. \tag{6.224}$$

Indeed, if Φ is not a constant polynomial with root $\mu \notin \{0, 1, \ldots, \max\{1, n_1 - 1\}\}$, then
$$(\lfloor\mu_i\rfloor_{n_a})^{|N|_a} (\lfloor\mu_i - n_a\rfloor_{n_{a-1}-n_a})^{|N|_{a-1}} \cdots (\lfloor\mu_i - n_2\rfloor_{n_1-n_2})^{|N|_1} \neq 0. \tag{6.225}$$
We may thus substitute μ into (6.220), solve for λ as given by (6.224), show λ is well defined and then check directly that λz^μ is a desired solution.

Observe that for the above arguments to hold, three points should be noted:

1. the exponent of $\lambda^{|N|+C_\mu - E_\mu}$ in (6.220) must be non-zero,
2. the condition that $\Phi(\mu)$ does indeed determine the values of μ, which is equivalent to $\Phi(\mu)$ being non-constant, and
3. the condition (6.225) must hold, which is true if $\mu \notin \{0, 1, \ldots, \max\{1, n_1 - 1\}\}$.

The restriction $\mu \notin \{0, 1, \ldots, \max\{1, n_1 - 1\}\}$, however, can sometimes be relaxed as shown in the following examples.

Example 6.8. Consider the iterative functional differential equation
$$x'(z) = \frac{x(z)}{x^{[2]}(z)}.$$
Here $N = (N_1) = (1)$, $T = (T_1) = (1)$, $M = (M_1) = (1)$, and $n = (n_1) = (1)$, $t = (t_1) = (1)$, $m = (m_1) = (2)$. Hence (6.220) becomes
$$\lambda^{1+\mu}\mu = 1 \tag{6.226}$$
and
$$\Phi(\mu) = \mu^2 - 1,$$
which has roots $\mu_\pm = \pm 1$. Since
$$|N| + C_{\mu_+} - E_{\mu_+} = 1 \neq 0,$$
in view of Theorem 6.45, we see that our equation has a nontrivial solution. In fact, if we substitute $\lambda = \lambda_+$ into (6.226), we see that $\lambda = 1$ or -1, and we can find solutions $x(z) = z$ and $x(z) = -z$. On the other hand,
$$|N| + C_{\mu_-} - E_{\mu_-} = 0,$$
thus our Theorem 6.45 does not guarantee a nontrivial solution. Indeed, substituting $\mu = \mu_-$ into (6.226), we see that $-1 = 1$, which is impossible.

Example 6.9. If $\Phi(\mu)$ has a root at $\mu = 1$, then for (6.225) to hold under the condition $|N|_a > 0$, there are only two possibilities:

1. $a = 1$, $n_1 = 1$, $N_1 > 0$ and (6.218) takes the form

$$\left(x^{(1)}(p_1 z)\right)^{N_1} = A \frac{\left(x^{[t_1]}(r_1 z)\right)^{T_1} \cdots \left(x^{[t_c]}(r_c z)\right)^{T_c}}{\left(x^{[m_1]}(q_1 z)\right)^{M_1} \cdots \left(x^{[m_b]}(q_b z)\right)^{M_b}},$$

which has the solution $x(z) = \lambda z^1$ where

$$\lambda^{N_1 + M \bullet m - T \bullet t} = \frac{A \prod_{l=1}^{c} r_l^{T_l}}{p_1^{N_1} \prod_{l=1}^{b} q_l^{M_l} (\mu)_1^{N_1}},$$

or

2. $N_1 = \cdots = N_{a-1} = 0$, $N_a > 0$, $n_a = 1$ and (6.218) takes the form

$$\left(x^{(1)}(p_a z)\right)^{N_a} = A \frac{\left(x^{[t_1]}(r_1 z)\right)^{T_1} \cdots \left(x^{[t_c]}(r_c z)\right)^{T_c}}{\left(x^{[m_1]}(q_1 z)\right)^{M_1} \cdots \left(x^{[m_b]}(q_b z)\right)^{M_b}},$$

which has the solution $x(z) = \tilde{\lambda} z^1$ where

$$\tilde{\lambda}^{N_a + M \bullet m - T \bullet t} = \frac{A \prod_{l=1}^{c} r_l^{T_l}}{p_a^{N_a} \prod_{l=1}^{b} q_l^{M_l} (\mu)_1^{N_a}}.$$

Example 6.10. In (6.218), let $m_b > 0$, $t = (t_1) = (0)$ and $T = (T_1) = (j)$ where $j \geq 0$. Then we are considering

$$\left(x^{(n_1)}(p_1 z)\right)^{N_1} \cdots \left(x^{(n_a)}(p_a z)\right)^{N_a} = A \frac{z^j}{\left(x^{[m_1]}(q_1 z)\right)^{M_1} \cdots \left(x^{[m_b]}(q_b z)\right)^{M_b}}. \quad (6.227)$$

If $|M| > N \bullet n - |N| + j$, then the characteristic polynomial

$$\Phi(\mu) = \sum_{l=1}^{b} M_l \mu^{m_l} - j + |N|\mu - N \bullet n$$

does not have any roots in $\{0\} \cup [1, \infty)$ (since $\Phi(0) = -N \bullet n - j \leq -N \bullet n < 0$, and since $\Phi(\mu) \geq |M| - j + |N| - N \bullet n > 0$ for $\mu \geq 1$). Thus, under the conditions stated above, Theorem 6.45 asserts the existence of nontrivial power function solutions for (6.227).

If, in addition, all $m_1, m_2, ..., m_b$ are even, or all odd, then (6.227) will have at least m_1 distinct nontrivial solutions. Indeed, we have already shown that $\Phi(\mu) > 0$ for $\mu \geq 1$. If $m_1, m_2, ..., m_b$ are even, then Descarte's rule of sign (see, e.g. page 171 of [14]), tells us that $\Phi(\mu)$ has at most one negative real root and at most one positive real root in $(0, 1)$, while if $m_1, m_2, ..., m_b$ are all odd, then $\Phi(\mu)$ has no negative real root and at most one positive real root. In either cases, $\Phi(\mu)$ cannot have repeated roots, other roots being complex conjugates. Hence all m_1 roots of $\Phi(\mu)$ are distinct and they yield at least m_1 distinct nontrivial solutions.

Example 6.11. In (6.218), let $|T| = 0$, $m_b = 0$ and $M_b = k \geq 0$. Then we are considering

$$\left(x^{(n_1)}(p_1 z)\right)^{N_1} \cdots \left(x^{(n_a)}(p_a z)\right)^{N_a} = A \frac{1}{z^k \left(x^{[m_1]}(q_1 z)\right)^{M_1} \cdots \left(x^{[m_{b-1}]}(q_{b-1} z)\right)^{M_{b-1}}}. \tag{6.228}$$

As in the previous example, if $\left|\sum_{l=1}^{b-1} M_l\right| > N \bullet n - |N| - k$ and if $N \bullet n - k \neq 0$, then the corresponding characteristic polynomial

$$\Phi(\mu) = \sum_{l=1}^{b-1} M_l \mu^{m_l} + k + |N|\mu - N \bullet n$$

does not have any roots in $\{0\} \cup [1, \infty)$. Thus Theorem 6.45 will imply the existence of nontrivial power function solutions for (6.228). Furthermore, a statement similar to that in Example 6.10 can also be made when $m_1, ..., m_{b-1}$ are all even, or, all odd.

Example 6.12. Consider an iterative functional differential equation of the form

$$\left(x^{(n_1)}(p_1 z)\right)^{N_1} \cdots \left(x^{(n_a)}(p_a z)\right)^{N_a} = A z^j \left(x^{[t_1]}(q_1 z)\right)^{T_1} \cdots \left(x^{[t_b]}(q_b z)\right)^{T_b}$$

where $a, b, N_1, \ldots, N_a, T_1, \ldots, T_b$ and $n_1, \ldots, n_a, t_1, \ldots, t_b$ are positive integers such that $n_1 > n_2 > \cdots > n_a$, $t_1 > t_2 > \cdots > t_b$. The number j is an integer and $A, p_1, \ldots, p_a, q_1, \ldots, q_b$ are nonzero complex numbers. By reasons explained in Example 6.10, if $|N| \leq |T|$, $N \bullet n + j \neq 0$ and

$$\Phi(\mu) = \sum_{l=1}^{b} T_l \mu^{t_l} - |N|\mu + N \bullet n + j \tag{6.229}$$

has a root μ, then Theorem 6.45 asserts the existence of at least one nontrivial solution of the form $x(z) = \lambda z^\mu$ in $\mathbf{C} \setminus (-\infty, 0]$.

The case $|T| = |M| = 0$ is treated in our next result.

Theorem 6.46. Suppose $|T| = |M| = 0$ and $\frac{N \bullet n}{|N|} \notin \{0, 1, \ldots, \max\{1, n_1 - 1\}\}$. Then (6.218) has at least one single-valued, nonzero, analytic power solutions of the form

$$x(z) = \lambda z^\mu, \ z \in \mathbf{C} \setminus (-\infty, 0],$$

where

$$\mu = \frac{N \bullet n}{|N|}$$

and

$$\lambda^{|N|} = \frac{A}{P_\mu} \frac{1}{(\lfloor \mu \rfloor_{n_a})^{|N|_a} (\lfloor \mu - n_a \rfloor_{n_{a-1} - n_a})^{|N|_{a-1}} \cdots (\lfloor \mu - n_2 \rfloor_{n_1 - n_2})^{|N|_1}}. \tag{6.230}$$

Indeed, since $|T| = |M| = 0$, thus $Q_\mu = F_\mu = 1$, $C_\mu = E_\mu = 0$,
$$\Phi(\mu) = |N|\mu - N\bullet n,$$
and (6.219) becomes
$$P_\mu \lambda^{|N|} (\lfloor \mu \rfloor_{n_a})^{|N|_a} (\lfloor \mu - n_a \rfloor_{n_{a-1}-n_a})^{|N|_{a-1}} \cdots (\lfloor \mu - n_2 \rfloor_{n_1-n_2})^{|N|_1} = A. \quad (6.231)$$
Since Φ now has the unique root $\mu = N\bullet n/|N|$, and since it does not belong to $\{0, 1, \ldots, \max\{1, n_1 - 1\}\}$, thus λ is a well defined root of the equation (6.231) and we may check directly that λz^μ is a desired solution.

Again, we may find $|N|$ distinct roots of (6.231) and they yield $|N|$ distinct solutions of (6.218).

As in Example 6.9, the restriction $\mu \notin \{0, 1, \ldots, \max\{1, n_1 - 1\}\}$ in Theorem 6.46 can also be relaxed.

Example 6.13. If $\Phi(\mu)$ has a root at $\mu = 1$, then for (6.225) to hold under the condition $|N|_a > 0$, there are only two possibilities:

1. $a = 1, n_1 = 1, N_1 > 0$ and (6.218) takes the form
$$\left(x^{(1)}(p_1 z) \right)^{N_1} = A,$$
which has the solution $x(z) = \lambda z^1$ where
$$\lambda^{N_1} = \frac{A}{p_l^{N_l}(\lfloor \mu \rfloor_1)^{N_1}},$$
or

2. $N_1 = \cdots = N_{a-1} = 0, N_a > 0, n_a = 1$ and (6.218) takes the form
$$\left(x^{(1)}(p_a z) \right)^{N_a} = A,$$
which has the solution $x(z) = \tilde{\lambda} z^1$ where
$$\tilde{\lambda}^{N_a} = \frac{A}{p_a^{N_a}(\lfloor \mu \rfloor_1)^{N_a}}.$$

We have assumed that $N\bullet n > 0$. The reason is that otherwise (6.218) reduces to
$$1 = A \frac{\left(x^{[t_1]}(r_1 z) \right)^{T_1} \cdots \left(x^{[t_c]}(r_c z) \right)^{T_c}}{\left(x^{[m_1]}(q_1 z) \right)^{M_1} \cdots \left(x^{[m_b]}(q_b z) \right)^{M_b}}, \quad (6.232)$$
or to
$$(x(p_1 z))^{N_1} = A \frac{\left(x^{[t_1]}(r_1 z) \right)^{T_1} \cdots \left(x^{[t_c]}(r_c z) \right)^{T_c}}{\left(x^{[m_1]}(q_1 z) \right)^{M_1} \cdots \left(x^{[m_b]}(q_b z) \right)^{M_b}}. \quad (6.233)$$
Since derivatives are missing in both cases, it is not appropriate to call them differential equations.

However, we may try to find solutions of the form $x(z) = \lambda z^\mu$ anyway.

For instance, suppose $|N| = 0$ and Φ is not a constant polynomial with a root μ. Note that $\Phi(\mu)$ is then given by

$$\Phi(\mu) = \sum_{l=1}^{b} M_l \mu^{m_l} - \sum_{l=1}^{c} T_l \mu^{t_l},$$

and (6.219) becomes

$$Q_\mu \lambda^{C_\mu - E_\mu} = AF_\mu.$$

Substituting the root μ into (6.234), we obtain

$$\lambda^{C_\mu - E_\mu} = \frac{AF_\mu}{Q_\mu}. \qquad (6.234)$$

Thus if $|T| \geq 0$ and if $C_\mu - E_\mu \neq 0$, then we may find a solution λ from (6.234) which yields a nontrivial solution of (6.232).

As for equation (6.233), the development leading to Theorems 6.45 and 6.46 still apply.

Theorem 6.47. *Suppose*

$$\Phi(\mu) = N_1 \mu + \sum_{l=1}^{b} M_l \mu^{m_l} - \sum_{l=1}^{c} T_l \mu^{t_l}$$

is not a constant polynomial and μ is a root. Suppose further that $|T| \geq 0$ and

$$N_1 + C_\mu - E_\mu \neq 0.$$

Then (6.218) has at least one single-valued, nontrivial, analytic power solution of the form

$$x(z) = \lambda z^\mu, \; z \in \mathbf{C} \setminus (-\infty, 0],$$

where

$$\lambda^{N_1 + C_\mu - E_\mu} = \frac{AF_\mu}{p_1^{N_1 \mu} Q_\mu}.$$

We remark that if $|T| = |M| = 0$, then equation (6.233) becomes

$$(x(p_1 z))^{N_1} = A.$$

Assuming $x(z) = \lambda z^\mu$ where $\lambda \neq 0$, we easily find that $\mu = 0$ and $\lambda^{N_1} = A$. Thus, $x(z) = A^{1/N_1}$ is a solution.

6.5 Notes

Equation (6.11) and the corresponding Theorems 6.1 and 6.2 are in Li and Cheng [120]. Equation (6.22) and the corresponding Theorems 6.3, 6.4 and 6.5 are in Si [186]. Besides equations (6.11) and (6.22), there are several other Babbage type equations which have been studied. See for exampe Bessis et al. [17], Pfeiffer [156], Issacs [83], Wagner [219], Myberg [144], Sarkovskii [168], Rice [162], Rice et al. [163], Si [175], Mai and Liu [133], Si and Zhang [196], Liu [129, 130], Liu and Mai [131]. In particular, Sarkovski in [168] considered the equation

$$x(f(x(t))) = g(x(t)),$$

where $x(t)$ is the unknown function; Rice in [162] considered the iterative square roots of Chebysev polynomials; Rice et al. [163] show that a quadratic polynomial does not allow iterative square roots.

The invariant curve equation (6.41) and the corresponding existence theorems are contained in Si [176], while equation (6.49) and the corresponding results are in Si and Zhang [195]. For other invariant curve equations, the readers may consult Nitecki [148], Anosov [4], Brydak [20, 21], Dhombres [46, 47], Ng and Zhang [147], Si et al. [194], Li et al. [124].

Equation (6.72) and the corresponding existence results are contained in Si et al. [185]; equation (6.80) and the corresponding existence results are contained in Si and Cheng [181]; equation (6.86) and the corresponding existence results are contained in Si and Cheng [180]; equation (6.95) and the corresponding existence results are contained in McKiernan [140]; and equation (6.105) and the corresponding existence results are contained in Si et al. [192]. Two other equations similar to equation (6.80) are studied recently by Wang and Si [222] and Xu et al. [229].

The equation (6.95) (studied by McKiernan [140]) is associated with the asymptotic behavior of Golomb's sequence $\{F(n)\}_{n=1}^{\infty}$: $1, 2, 2, 3, 3, 4, 4, 4, 5, 5, 5, 6, 6, 6, 6, \ldots$. Golomb's self-describing sequence is a monotone nondecreasing sequence of positive integers with the property that for each $n \geq 1$, $F(n)$ is equal to the number of (not necessarily consecutive) occurences of the integer n in the sequence. Golomb's sequence is proposed by Golomb in Problem 5407 in [70] who asked for an asymptotic formula for the sequence. Marcus [134] showed that as $n \to \infty$, the n-th term of the Golomb's sequence tends to the solution of (6.95). That is $F(n) \sim x(n) (n \to \infty)$ where x is a positive solution of (6.95). He also gave $x_+(z) = \varphi^{2-\varphi} z^{\varphi-1}$, where $\varphi = (\sqrt{5}+1)/2$ is the golden number, as a positive solution to (6.95). The asymptotic behavior of Golomb's sequence is also studied in Petermann [152, 153] and Petermann and Remy [155, 154]. Two other similar equations are studied recently by Si and Zhang [197] and Si et al. [202].

For additional motivation and results related to iterative differential equations of the form

$$x'(t) = f(t, x(t), x(\alpha(t, x(t))), x'(\beta(t, x(t)))),$$

the readers may consult Driver [49–51, 53, 54], Dunkel [56], Eder [58], Grimm [71], Gusarenko [72], Hsing [80], Hoag [79], Jackiewicz [86, 87], Oberg [149], Shi and Li [171, 172], Tavernini [215], Wang [217], Wu [226, 227], Xiang [228].

Equation (6.125) and the corresponding existence results are contained in Li [119]; equation (6.152) and the corresponding existence results are contained in Si and Wang [189]; equation (6.167) and the corresponding existence results are contained in Si and Wang [187]; and equation (6.186) and the corresponding existence results are contained in Si and Wang [188]. Other equations involving second order derivatives of the unknown function can be found in Si and Wang [190], Si and Zhang [199], Liu and Li [128] and Li and Liu [123].

Power functions solutions for iterative functional equations were noted by T. T. Lu in a note to one of the authors (Cheng) before [122] was written. Since then several papers [121, 214, 31] were written, all based on similar ideas. In particular, equation (6.218) and the corresponding results are contained in Cheng et al. [31]. The same idea is also applied by Si and Zhang [199] to obtain power functions solutions of

$$x''(z) = \sigma \left(x^{[m]}\right)^\tau + \gamma z^\beta$$

where $\sigma, \gamma, \tau, \beta \in \mathbf{C}\backslash\{0\}$ and $m \in \mathbf{Z}^+$. It is expected similar solutions can be found for several other iterative functional equations.

Appendix A

Univariate Sequences and Properties

We summarize some of the notations and facts related to basic univariate sequences and their associated operations.

The set of real numbers is denoted by $\mathbf{R}$, the set of all complex numbers by $\mathbf{C}$, the set of integers by $\mathbf{Z}$, the set of positive integers by $\mathbf{Z}^+$, and the set of nonnegative integers by $\mathbf{N}$. The imaginary unit is denoted by $\mathbf{i}$. $l^{\mathbf{N}}$ is the set of all (real or complex) sequences of the form $\{f_k\}_{k \in \mathbf{N}} = \{f_0, f_1, f_2, ...\}$. Therefore, sequences $a, b, ...$ in $l^{\mathbf{N}}$ are assumed to have the form $\{a_k\}_{k \in e\mathbf{N}}$, $\{b_k\}_{k \in \mathbf{N}}$, ..., respectively.

A.1 Common Sequences

- Let $\alpha \in \mathbf{C}$, the sequence $\{\alpha, 0, 0, ...\}$ is denoted by $\overline{\alpha}$ and is called a *scalar sequence*.
- The sequences $\{0, 0, ...\}$ and $\{1, 0, ...\}$ is denoted by $\overline{0}$ and $\overline{1}$ respectively.
- For any number $\lambda \in \mathbf{C}$, the *geometric sequence* $\{\lambda^n\}_{n \in \mathbf{N}}$, where $\lambda \in \mathbf{C}$, is denoted by $\underline{\lambda}$.

- $\underline{-1} = \{+1, -1, +1, -1, ...\}$, $\underline{1} = \{1, 1, 1, ...\}$ which is also denoted by $\boldsymbol{\sigma}$.
- $\underline{\mathbf{i}} = \{1, \mathbf{i}, -1, -\mathbf{i}, 1, \mathbf{i}, -1, -\mathbf{i}, ...\}$ and $\underline{-\mathbf{i}} = \{1, -\mathbf{i}, -1, \mathbf{i}, 1, -\mathbf{i}, -1, \mathbf{i}, ...\}$.
- The *arithmetic sequence* $\{0, 1, 2, 3, ...\}$ is denoted by $\boldsymbol{\eta}$.
- The *difference sequence* $\{1, -1, 0, 0, ...\}$ is denoted by $\boldsymbol{\delta}$.
- The *exponential sequence* $\{1/0!, 1/1!, 1/2!, 1/3!, 1/4!, ...\}$ is denoted by $\boldsymbol{\varpi}$.
- The *Dirac sequence* $\hbar^{\langle m \rangle} \in l^{\mathbf{N}}$, where $m \in \mathbf{N}$, is defined by

$$\hbar_k^{\langle m \rangle} = \begin{cases} 1 & k = m \\ 0 & k \neq m \end{cases}.$$

  The sequence $\hbar^{\langle 1 \rangle}$ is also written as $\hbar$.
- The *jump (or Heaviside) sequence* $\mathbf{H}^{(m)}$, where $m \in \mathbf{N}$, is defined by

$$\mathbf{H}_k^{(m)} = \begin{cases} 0 & 0 \leq k < m \\ 1 & k \geq m \end{cases}.$$

- The *factorial sequence* $\lfloor z \rfloor \in l^{\mathbf{N}}$, where $z \in \mathbf{C}$, is defined by
$$\lfloor z \rfloor = \{1, z, z(z-1), z(z-1)(z-2), ...\}.$$
Thus $\lfloor z \rfloor_0 = 1$ and
$$\lfloor z \rfloor_m = z(z-1)(z-2)\cdots(z-m+1), \ m \in \mathbf{Z}^+.$$
Note that $\lfloor 0 \rfloor_0 = 0! = 1$, $\lfloor n \rfloor_n = n!$ and $0 = \lfloor n \rfloor_{n+1} = \lfloor n \rfloor_{n+2} = \cdots$ for $n \in \mathbf{N}$.

- The sequence $\lceil z \rceil \in l^{\mathbf{N}}$, where $z \in \mathbf{C}\setminus\{-1, -2, -3, ...\}$, is defined by
$$\lceil z \rceil = \left\{1, \frac{1}{z+1}, \frac{1}{(z+1)(z+2)}, ...\right\}.$$
Thus $\lceil z \rceil_0 = 1$ and
$$\lceil z \rceil_m = \frac{1}{(z+1)(z+2)\cdots(z+m)}, \ m \in \mathbf{Z}^+.$$
Note that the fraction $1/(z+1)(z+2)\cdots(z+m)$, when $m \in \mathbf{Z}^+$, is defined if $z \neq -1, -2, ..., -m$.

- The *binomial sequence* $C^{(z)} \in l^{\mathbf{N}}$, for any $z \in \mathbf{C}$, is defined by
$$C^{(z)} = \left\{\frac{1}{0!}, \frac{z}{1!}, \frac{z(z-1)}{2!}, \frac{z(z-1)(z-2)}{3!}, ...\right\},$$
so that $C_0^{(z)} = 1$ and
$$C_m^{(z)} = \frac{z(z-1)\cdots(z-m-1)}{m!}, \ m \in \mathbf{N}, z \in \mathbf{C}.$$
In particular, for $i, j \in \mathbf{N}$ such that $j \leq i$, $C_j^{(i)}$ is the usual binomial coefficient.

A.2 Sums and Products

- Let $f, g \in l^{\mathbf{N}}$. Their (termwise) sum is
$$f + g = \{f_0 + g_0, f_1 + g_1, f_2 + g_2, ...\}.$$

- Let $\alpha \in \mathbf{C}$ and $f \in l^{\mathbf{N}}$. The α multiple of f is
$$\alpha f = \{\alpha f_0, \alpha f_1, \alpha_2 f_2, ..., \}.$$

- Let $f, g \in l^{\mathbf{N}}$. The termwise product of f and g is
$$f \cdot g = \{f_0 g_0, f_1 g_1, f_2 g_2, ...\}_{k \in \mathbf{N}}.$$

The products $f \cdot f$, $f \cdot f \cdot f$, ... will be denoted by $f^2, f^3, ...$ respectively. We define $f^1 = f$ and $f^0 = \boldsymbol{\sigma}$. The sequence f^p is called the p-th termwise (product) power of f. The k-th term of the sequence f^p is $(f^p)_k$, which is written as f_k^p.

- Let $f, g \in l^{\mathbf{N}}$. The *convolution product* of f and g is

$$(f * g)_k = \sum_{i=0}^{k} f_{k-i} g_i, \quad k \in \mathbf{N}.$$

The products $f*f$, $f*f*f$, ... will be denoted by $f^{\langle 2 \rangle}, f^{\langle 3 \rangle}, \ldots$ respectively. We define $f^{\langle 1 \rangle} = f$ and $f^{\langle 0 \rangle} = \bar{1}$. The sequence $f^{\langle p \rangle}$ is called the p-th convolution (product) power of f. The k-th term of the sequence $f^{\langle p \rangle}$ is $\left(f^{\langle p \rangle}\right)_k$, which is written as $f_k^{\langle p \rangle}$.

- Let $f, g \in l^{\mathbf{N}}$. If

$$\lim_{k \to \infty} \sum_{i=0}^{k} f_i g_n^{\langle i \rangle} = \sum_{i=0}^{\infty} f_i g_n^{\langle i \rangle} < \infty, \ n \in \mathbf{N},$$

then the the *composition product* of f and g is

$$f \circ g = \left\{ \sum_{i=0}^{\infty} f_i g_n^{\langle i \rangle} \right\}_{n \in \mathbf{N}}.$$

The products $f \circ f$, $f \circ f \circ f$, ..., will be denoted by $f^{[2]}$, $f^{[3]}$, ..., respectively. We also define $f^{[1]} = f$ and $f^{[0]} = \hbar$. The sequence $f^{[p]}$ is called the p-th composition (product) power of f.

A.3 Quotients

- Let $f, g \in l^{\mathbf{N}}$. If $g_0 \neq 0$, then there is a unique sequence $x = \{x_k\} \in l^{\mathbf{N}}$ such that $g * x = f$. This sequence is also denoted by the quotient f/g.
- Let $g \in l^{\mathbf{N}}$. If $g_0 = 0$ and $g_1 \neq 0$, then there is a unique sequence $x \in l^{\mathbf{N}}$ such that $x \circ g = \hbar$. This sequence is also denoted by $g^{[-1]}$.
- Let $g \in l^{\mathbf{N}}$. If $g_i \neq 0$ for $i \in \mathbf{N}$, then there is a unique sequence $x \in l^{\mathbf{N}}$ such that $x \cdot g = \underline{1}$. This sequence is also denoted by g^{-1}.

A.4 Algebraic Derivatives and Integrals

- Given a sequence $f \in l^{\mathbf{N}}$, we define the *algebraic derivative* of f by

$$Df = \{(k+1)f_{k+1}\}_{k=0}^{\infty}.$$

The higher algebraic derivatives $D^n f$ are defined recursively by $D^n f = D(D^{n-1}f)$. Thus we have

$$D\{f_0, f_1, f_2, \ldots\} = \{f_1, 2f_2, 3f_3, \ldots\},$$

and

$$D^n f = \{(k+1) \cdots (k+n) f_{k+n}\}$$

for $n \in \mathbf{Z}^+$. It is natural to define $D^0 f = f$ and $D^1 f = Df$.

- Let ϕ be a sequence. If there is a sequence ψ such that $D\psi = \phi$, then ψ is called the primitive of ϕ. In particular, given $\phi = \{\phi_0, \phi_1, \phi_2, ...\}$, the primitive

$$\left\{0, \frac{\phi_0}{1}, \frac{\phi_1}{2}, \frac{\phi_2}{3}, \frac{\phi_3}{4}, ...\right\}$$

is called the *algebraic integral* of ϕ and is denoted by $\int \phi$. Hence

$$\int \phi = \hbar * \left\{\frac{\phi_k}{k+1}\right\}_{k \in \mathbf{N}}.$$

A.5 Tranformations

- Given $f \in l^\mathbf{N}$, the first difference of f is $\Delta f = \{f_1 - f_0, f_2 - f_1, ...\}$. The higher differences $\Delta^m f$, $m = 2, 3, ...$, are defined recursively by $\Delta^m f = \Delta\left(\Delta^{m-1} f\right)$. Furthermore, $\Delta^0 f = f$ and $\Delta^1 f = \Delta f$.
- The shifted sequences $E^m f$ and $E^{-m} f$, where $m \geq 1$, are respectively

$$E^m f = \{f_{m+k}\}_{k \in \mathbf{N}},$$

and

$$(E^{-m} f)_k = \begin{cases} f_{-m+k} & k \geq m \\ 0 & 1 \leq k \leq m \end{cases}.$$

We also define $E^0 f = f$ and $Ef = E^1 f$. Note that $\hbar^{\langle m \rangle} * f = E^{-m} f$ for $m \in \mathbf{N}$.

- Given $f \in l^\mathbf{N}$, the partial sum sequence generated from f is

$$\left\{\sum_{i=0}^{k} f_i\right\}_{k \in \mathbf{N}} = \sigma * f.$$

- Given $f \in l^\mathbf{N}$, the absolute sequence of f is

$$|f| = \{|f_0|, |f_1|, |f_2|, ...\}.$$

- Given $f \in l^\mathbf{N}$, the positive and negative parts of f are

$$f^+ = \frac{1}{2}(|f| + f)$$

and

$$f^- = \frac{1}{2}(|f| - f).$$

Similarly, other transformations can be obtained by termwise operation, e.g.

$$\ln f = \{\ln f_k\}_{k \in \mathbf{N}}$$

provided the transformed sequence is in $l^\mathbf{N}$.

A.6 Limiting Operations

- For each $j \in \mathbf{N}$, let $f^{(j)} \in l^{\mathbf{N}}$. The sequence $\{f^{(j)}\}_{j \in \mathbf{N}}$ (of sequences in $l^{\mathbf{N}}$) is said to converge (pointwise) to the limit sequence $f \in l^{\mathbf{N}}$ if
$$\lim_{j \to \infty} f_k^{(j)} = f_k \in \mathbf{C},\ k \in \mathbf{N}.$$

- The infinite sum of a sequence $\{f^{(j)}\}_{j \in \mathbf{N}}$ of sequences is the limit sequence of the partial sum sequence $\left\{\sum_{j=0}^n f^{(j)}\right\}_{n \in \mathbf{N}}$:
$$\sum_{j=0}^{\infty} f^{(j)} = \lim_{n \to \infty} \sum_{j=0}^n f^{(j)}.$$

If such a limiting sequence exists, we say that the series $\sum_{j=0}^{\infty} f^{(j)}$ converges. Note that $\sum_{j=0}^{\infty} f^{(j)}$ converges if, and only if, $\sum_{j=0}^{\infty} f_n^{(j)}$ converges for each $n \in \mathbf{N}$; furthermore,
$$\sum_{j=0}^{\infty} f^{(j)} = \sum_{j=0}^{\infty} \left\{f_n^{(j)}\right\}_{n \in \mathbf{N}} = \left\{\sum_{j=0}^{\infty} f_n^{(j)}\right\}_{n \in \mathbf{N}},$$

that is, the k-th term of the series is obtained by 'adding' all the k-th terms of the individual sequences.

A.7 Operational Rules

- Equipped with the termwise addition and the convolution product, $l^{\mathbf{N}}$ is a commutative ring with no zero divisor, i.e. $f * g = \overline{0}$ implies $f = \overline{0}$ or $g = \overline{0}$, and the additive and multiplicative identities are $\overline{0}$ and $\overline{1}$ respectively.
- For any $f, g \in l^{\mathbf{N}}$,
$$(f + g)^{\langle k \rangle} = \sum_{i=0}^{k} C_i^{(k)} f^{\langle i \rangle} * g^{\langle k-i \rangle},\ k \in \mathbf{N}.$$

- Let $f, g \in l^{\mathbf{N}}$ and $\lambda, \mu \in \mathbf{C}$. Then $f \cdot g = g \cdot f$, $\underline{\lambda} \cdot \underline{\mu} = \underline{\lambda \mu}$, $\underline{0} \cdot f = \overline{f_0}$, $\underline{1} \cdot f = f$, and $\underline{\lambda} \cdot (f + g) = \underline{\lambda} \cdot f + \underline{\lambda} \cdot g$.
- Let $f = \{f_k\} \in l^{\mathbf{N}}$. If $f_0 = 0$, then the first n, where $n \geq 1$, terms of the n-th convolution power $f^{\langle n \rangle}$ are equal to zero, that is, $f_k^{\langle n \rangle} = 0$ for $k = 0, 1, ..., n-1$; and $f_n^{\langle n \rangle} = f_1^n$. Furthermore, since
$$f_i^{\langle n \rangle} = \sum_{v_1 + \cdots + v_n = i; v_1, ..., v_n \in \mathbf{N}} f_{v_1} f_{v_2} \cdots f_{v_n} = \sum_{l_1 + \cdots + l_n = i; l_1, ..., l_n \in \mathbf{Z}^+} f_{l_1} \cdots f_{l_n},$$

for each $j \in \{0, ..., k\}$, the term $f_k^{(j)}$ involves $f_1, ..., f_{k-1}$ only and can be expressed as
$$f_k^{(j)} = P(f_1, ..., f_{k-1}),\ k \geq 2, 0 \leq j \leq k,$$

where P depends on j and k and is a $(n-1)$-variate polynomial with positive coefficients. Hence the conditions $f_0 = 0$, $f_1 = \mu$ and the iteration formula

$$f_k = F\left(f_k^{\langle 2\rangle}, ..., f_k^{\langle k\rangle}\right), \quad k \geq 2,$$

will define f in a unique manner.

- Let $f \in l^{\mathbf{N}}$. If $f_0 = f_1 = 0$, then the first $2n$ terms of the convolution product $f^{\langle n\rangle}$ are equal to zero.
- Let $f, g \in l^{\mathbf{N}}$. Then $\underline{\lambda} \cdot (f * g) = (\underline{\lambda} \cdot f) * (\underline{\lambda} \cdot g)$ for $\lambda \in \mathbf{C}$. Hence $(f \cdot \underline{\lambda})^{\langle m\rangle} = \underline{\lambda} \cdot f^{\langle m\rangle}$, $m \in \mathbf{N}$.
- Let $f, g, p, q \in l^{\mathbf{N}}$ such that $g_0 \neq 0$ and $q_0 \neq 0$. Then

$$\frac{f * p}{g * q} = \frac{f}{g} * \frac{p}{q}.$$

- Let $f, g \in l^{\mathbf{N}}$ such that $g_0 \neq 0$. Then $(f/g)^{\langle n\rangle} = f^{\langle n\rangle}/g^{\langle n\rangle}$ for $n \in \mathbf{N}$.
- For $\alpha, \beta \in \mathbf{C}$ and $f, g \in l^{\mathbf{N}}$,

$$D(\alpha f + \beta g) = \alpha Df + \beta Dg,$$

$$D(f * g) = f * Dg + g * Df,$$

$$D(f \cdot g) = (Df) \cdot Eg = (Ef) \cdot Dg$$

and

$$D\left(\frac{f}{g}\right) = \frac{g * Df - f * Dg}{g^{\langle 2\rangle}},$$

where we recall that f/g is only defined when the zeroth term of g is not 0.

- For $f \in l^{\mathbf{N}}$,

$$D^n f = \{(k+1)\cdots(k+n)f_{k+n}\}, \quad n \in \mathbf{Z}^+.$$

$$\hbar * Df = \{kf_k\},$$

$$\hbar^{\langle m\rangle} * D^n f = \{\lfloor k+n-m\rfloor_n f_{k+n-m}\}, \quad n \geq m \geq 0,$$

$$Df^{\langle n\rangle} = f^{\langle n-1\rangle} * Df + f * Df^{\langle n-1\rangle} = \cdots = \overline{n} * f^{\langle n-1\rangle} * Df, \quad n \in \mathbf{Z}^+,$$

$$D^n\left(\hbar^{\langle m\rangle} * f\right) = \hbar^{\langle m-n\rangle} * \{\lfloor m+k\rfloor_n f_k\}, \quad m \geq n \geq 1.$$

- If $D\phi = 0$, then ϕ is a scalar sequence.
- For any $f, g \in l^{\mathbf{N}}$ and any $\alpha, \beta \in \mathbf{C}$,

$$\int (\alpha f + \beta g) = \alpha \int f + \beta \int g.$$

- Let f be a sequence in $l^{\mathbf{N}}$, Δf its first difference, and $\overline{f_0}$ the sequence $\{f_0, 0, 0, ...\}$. Then it is easily checked that $\boldsymbol{\delta} * f = \Delta f + \overline{f_0} - \boldsymbol{\delta} * (\Delta f) = \overline{f_0} + \hbar * (\Delta f)$.

- Let $a, b \in l^{\mathbf{N}}$ such that $b_0 = 0$. Then $b_n^{\langle i \rangle} = 0$ for $n \in \{0, ..., i\}$. Thus
$$\sum_{i=0}^{\infty} a_i b_n^{\langle i \rangle} = \sum_{i=0}^{n} a_i b_n^{\langle i \rangle} = \begin{cases} a_0 & n = 0 \\ \sum_{i=1}^{n} a_i b_n^{\langle i \rangle} & n \geq 1 \end{cases}.$$
- Let $g \in L^{\mathbf{N}}$ such that $g_0 = 0$. Then
$$\mathbf{H}^{(m)} \circ g = \left\{ \sum_{i=m}^{n} g_n^{\langle i \rangle} \right\}_{n \in \mathbf{N}}.$$
- If $\{f^{(j)}\}_{j \in \mathbf{N}}$ and $\{g^{(j)}\}_{j \in \mathbf{N}}$ are two sequences of sequences which converge to f and g respectively, then
$$\lim_{j \to \infty} \left(\alpha f^{(j)} + \beta g^{(j)} \right) = \alpha f + \beta g, \quad \alpha, \beta \in \mathbf{C},$$
$$\lim_{j \to \infty} \left(f^{(j)} \cdot g^{(j)} \right) = f \cdot g,$$
$$\lim_{j \to \infty} f^{(j)} * g^{(j)} = f * g,$$
$$\lim_{j \to \infty} D f^{(j)} = D \left(\lim_{j \to \infty} f^{(j)} \right)$$

and

$$\lim_{j \to \infty} \int f^{(j)} = \int \lim_{j \to \infty} f^{(j)}.$$
- If the composition product $f \circ g$ of $f, g \in l^{\mathbf{N}}$ is defined, then
$$f \circ g = \left\{ \sum_{i=0}^{\infty} f_i g_n^{\langle i \rangle} \right\}_{n \in \mathbf{N}} = \sum_{i=0}^{\infty} f_i g^{\langle i \rangle}.$$
- If the infinite sums $\sum_{j=0}^{\infty} f^{(j)}$ and $\sum_{j=0}^{\infty} g^{(j)}$ of two respective sequences $\{f^{(j)}\}_{j \in \mathbf{N}}$ and $\{g^{(j)}\}_{j \in \mathbf{N}}$ of sequences in $l^{\mathbf{N}}$ converge, then
$$\sum_{j=0}^{\infty} \left(\alpha f^{(j)} + \beta g^{(j)} \right) = \alpha \sum_{j=0}^{\infty} f^{(j)} + \beta \sum_{j=0}^{\infty} g^{(j)}, \quad \alpha, \beta \in \mathbf{C},$$
$$\sum_{j=0}^{\infty} D f^{(j)} = D \left(\sum_{j=0}^{\infty} f^{(j)} \right),$$

and

$$\sum_{j=0}^{\infty} \int f^{(j)} = \int \left(\sum_{j=0}^{\infty} f^{(j)} \right).$$
- If the infinite sum $\sum_{j=0}^{\infty} f^{(j)}$ of the sequence $\{f^{(j)}\}_{j \in \mathbf{N}}$ of sequences in $l^{\mathbf{N}}$ converge, then for any $g \in l^{\mathbf{N}}$,
$$\sum_{j=0}^{\infty} f^{(j)} \cdot g = \left(\sum_{j=0}^{\infty} f^{(j)} \right) \cdot g$$

and

$$\sum_{j=0}^{\infty} f^{(j)} * g = \left(\sum_{j=0}^{\infty} f^{(j)} \right) * g.$$

A.8 Knowledge Base

- $\hbar^{\langle 0 \rangle} = \{1, 0, 0, ...\}$, $\hbar^{\langle 1 \rangle} = \{0, 1, 0, 0, ...\}$, $\hbar^{\langle 2 \rangle} = \{0, 0, 1, 0, 0, ...\} = \hbar * \hbar$, $\hbar^{\langle 3 \rangle} = \{0, 0, 0, 1, 0, 0, ...\} = \hbar * \hbar * \hbar$, etc.
- $\frac{1}{2}(\underline{\mathbf{i}} + \underline{-\mathbf{i}}) = \{1, 0, -1, 0, 1, 0, -1, 0, ...\}$.
- $\frac{1}{2}(\underline{\mathbf{i}} - \underline{-\mathbf{i}}) = \{0, \mathbf{i}, 0, -\mathbf{i}, 0, \mathbf{i}, 0, -\mathbf{i}, ...\}$.
- $\frac{1}{2}(\underline{\mathbf{1}} + \underline{-\mathbf{1}}) = \{1, 0, 1, 0, 1, 0, 1, 0, ...\}$.
- $\frac{1}{2}(\underline{\mathbf{1}} - \underline{-\mathbf{1}}) = \{0, 1, 0, 1, 0, 1, 0, 1, ...\}$.
- $\boldsymbol{\sigma}^{\langle 2 \rangle} = \left\{ \sum_{i=0}^{k} 1 \right\} = \{k+1\}_{k \in \mathbf{N}} = \{\lfloor k+1 \rfloor_1\}_{k \in \mathbf{N}}$.
- $\boldsymbol{\sigma}^{\langle n \rangle} = \left\{ \frac{\lfloor k+n-1 \rfloor_{n-1}}{(n-1)!} \right\}_{k \in \mathbf{N}}$ for $n \in \mathbf{N}$.
- $\boldsymbol{\delta}^{\langle 2 \rangle} = \{1, -2, 1, 0, ...\}$.
- $\boldsymbol{\delta}^{\langle n \rangle} = \{(-1)^k \lfloor n \rfloor_k / k!\}_{k \in \mathbf{N}}$ for $n \in \mathbf{Z}^+$.
- For $\alpha \in \mathbf{C}$, $D\overline{\alpha} = \overline{0}$.
- $D\hbar^{\langle n \rangle} = n\hbar^{\langle n-1 \rangle}$ for $n \in \mathbf{Z}^+$.
- $D\boldsymbol{\delta}^{\langle n \rangle} = -n\boldsymbol{\delta}^{\langle n-1 \rangle}$ for $n \in \mathbf{Z}^+$.
- $D^n \boldsymbol{\sigma} = \{(k+1)(k+2) \cdots (k+n)\}_{k \in \mathbf{N}}$ for $n \in \mathbf{Z}^+$.
- For any $\alpha \in \mathbf{C}$,
$$\underline{\alpha} = \frac{\overline{1}}{\overline{\alpha} * \boldsymbol{\delta} - \overline{\alpha} + \overline{1}} = \frac{\overline{1}}{\overline{1} - \overline{\alpha} * \hbar}.$$

- For any $\beta \neq 1$,
$$\frac{\overline{1}}{\boldsymbol{\delta} - \overline{\beta}} = \left\{ \left(\frac{1}{1 - \beta} \right)^{k+1} \right\}.$$

- For any $c \in \mathbf{C}$,
$$(\underline{c})^{\langle 2 \rangle} = (\underline{c} \cdot \boldsymbol{\sigma}) * (\underline{c} \cdot \boldsymbol{\sigma}) = \underline{c} \cdot \boldsymbol{\sigma}^2 = \left\{ c^k \lfloor k+1 \rfloor_1 \right\}_{k \in \mathbf{N}}.$$

- For any $\beta \neq 1$,
$$\frac{\overline{1}}{(\boldsymbol{\delta} - \overline{\beta})^{\langle n \rangle}} = \left\{ \frac{\lfloor k+n-1 \rfloor_{n-1}}{(n-1)!} (1 - \beta)^{-k-n} \right\}_{k \in \mathbf{N}}, \ n \in \mathbf{Z}^+.$$

- For any $\gamma \neq 0$,
$$\frac{\overline{1}}{(\overline{\gamma} - \hbar)^{\langle n \rangle}} = \left\{ \frac{\lfloor k+n-1 \rfloor_{n-1}}{(n-1)!} \gamma^{-k-n} \right\}, \ \gamma \neq 0, \ n \in \mathbf{Z}^+.$$

- For any $\gamma \neq 1$,
$$\frac{\overline{1}}{(\overline{1} - \overline{\gamma}\hbar)^{\langle n \rangle}} = \left\{ \frac{\lfloor k+n-1 \rfloor_{n-1}}{(n-1)!} \gamma^k \right\}, \ n \in \mathbf{Z}^+.$$

- $\eta = \{k\}_{k \in \mathbf{N}} = \hbar * \boldsymbol{\sigma}^{\langle 2 \rangle}$.
- $D\eta = \{(k+1)^2\}_{k \in \mathbf{N}}$ implies
$$\{(k+1)^2\} = D(\hbar * \boldsymbol{\sigma}^{\langle 2 \rangle}) = \hbar * D\boldsymbol{\sigma}^{\langle 2 \rangle} + \boldsymbol{\sigma}^{\langle 2 \rangle} = 2\hbar * \boldsymbol{\sigma}^{\langle 3 \rangle} + \boldsymbol{\sigma}^{\langle 2 \rangle},$$
$$\{(k+1)^3\} = D\left(\hbar * D\left(\hbar * \boldsymbol{\sigma}^{\langle 2 \rangle} \right) \right) = 6\hbar^{\langle 2 \rangle} * \boldsymbol{\sigma}^{\langle 4 \rangle} + 6\hbar * \boldsymbol{\sigma}^{\langle 3 \rangle} + \boldsymbol{\sigma}^{\langle 2 \rangle},$$
etc.

- Since $\bar{c}_0^{\langle i \rangle} = c^i$ and $\bar{c}_n^{\langle i \rangle} = 0$ for $i \in N$ and $n \in \mathbf{Z}^+$, we see that

$$(\varpi \circ \bar{c})_0 = 1 + c + \frac{1}{2!}c^2 + \cdots = e^c$$

and

$$(\varpi \circ \bar{c})_n = 0, \ n \in \mathbf{Z}^+.$$

A.9 Analytic Functions

Let $a \in l^{\mathbf{N}}$ with positive (or infinite) radius of convergence

$$\rho(a) = \frac{1}{\limsup_{k \to \infty} |a_k|^{1/k}}.$$

Then the function $\hat{a}$ defined by

$$\hat{a}(\lambda) = \sum_{k=0}^{\infty} a_k \lambda^k = \lim_{n \to \infty} \sum_{k=0}^{n} (\underline{\lambda} \cdot a)_k, \ |\lambda| < \rho(a),$$

is called the power series function generated by a. A function $f = f(\lambda)$ is said to be analytic at 0 if it is (equal to) a power series function (near 0) generated by a sequence with positive radius of convergence. A function g is said to be analytic at c if it is the 'translation' of an analytic function at 0.

Some analytic functions that are analytic at 0 and generated by common sequences are:

- $\widehat{\underline{\alpha}}(\lambda) = \alpha$, $\alpha \in \mathbf{C}$.
- $\hat{\sigma}(\lambda) = 1 + \lambda + \lambda^2 + \lambda^3 + \cdots = \frac{1}{1-\lambda}$.
- $\hat{\delta}(\lambda) = 1 - \lambda$.
- $\hat{\varpi}(\lambda) = 1 + \lambda + \frac{1}{2!}\lambda^2 + \frac{1}{3!}\lambda^3 + \cdots = e^{\lambda}$.
- $\widehat{h^{\langle m \rangle}}(\lambda) = \lambda^m$ for $m \in \mathbf{N}$.
- $\widehat{\mathbf{H}^{(m)}}(\lambda) = \lambda^m + \lambda^{m+1} + \lambda^{m+2} + \cdots = \lambda^m \left(1 + \lambda + \lambda^2 + \cdots\right) = \lambda^m/(1-\lambda)$ for $m \in \mathbf{N}$.
- $\widehat{C^{(n)}}(\lambda) = 1 + \frac{n}{1!}\lambda + \frac{n(n-1)}{2!}\lambda^2 + \cdots + \lambda^n = (1+\lambda)^n$ for $n \in \mathbf{N}$.
- Let $a = \frac{1}{2}(\underline{\mathbf{i}} + \underline{-\mathbf{i}})$. Then $\widehat{a \cdot \varpi}(z) = \cos z$.
- Let $a = \frac{1}{2}(\underline{\mathbf{i}} - \underline{-\mathbf{i}})$. Then $\widehat{a \cdot \varpi}(z) = \sin z$.
- Let $a = \frac{1}{2}(\underline{1} + \underline{-1})$. Then $\widehat{a \cdot \varpi}(z) = \cosh z$.
- Let $a = \frac{1}{2}(\underline{1} - \underline{-1})$. Then $\widehat{a \cdot \varpi}(z) = \sinh z$.

A.10 Operations for Analytic Functions

Analytic functions can be combined, decomposed, or transformed.

- Let $\widehat{a}$ and $\widehat{b}$ be analytic at 0. For any $\alpha, \beta \in \mathbf{C}$, the linear combination $\alpha\widehat{a}+\beta\widehat{b}$ and the product $\widehat{a}\widehat{b}$ are analytic at 0. Furthermore, for λ sufficiently near to 0,
$$\left(\alpha\widehat{a} + \beta\widehat{b}\right)(\lambda) = (\widehat{\alpha a + \beta b})(\lambda).$$

- Let $\widehat{a}$ and $\widehat{b}$ be analytic at 0. Then the product function $\widehat{a}\widehat{b}$ is analytic at 0. Furthermore, for λ sufficiently near to 0,
$$\left(\widehat{a}\widehat{b}\right)(\lambda) = \widehat{a}(\lambda)\widehat{b}(\lambda) = \widehat{a*b}(\lambda)$$

- Let $\widehat{a}$ be analytic at 0. Then $\widehat{a}^m$, where $m \in \mathbf{N}$, is analytic at 0. Furthermore, for λ sufficiently near to 0,
$$\widehat{a}^m(\lambda) = \widehat{a^{\langle m \rangle}}(\lambda).$$

- Let $\widehat{a}$ be analytic at 0. Then $\widehat{a}^{(m)}$, where $m \in \mathbf{N}$, is analytic at 0. Furthermore, for λ sufficiently near to 0,
$$\widehat{a}^{(m)}(\lambda) = \widehat{D^m a}(\lambda)$$

- Let $\widehat{a}$ be analytic at 0. Then the Cauchy integral $\int_0^\lambda \widehat{a}(w)dw$, as a function of λ, is analytic at 0. Furthermore, for λ sufficiently near to 0,
$$\int_0^\lambda \widehat{a}(w)dw = \widehat{\int a}.$$

- Let $\widehat{a}$ be analytic at 0 and $a_0 \neq 0$. Then the quotient function $1/\widehat{a}$, is analytic at 0. Furthermore, for λ sufficiently near to 0,
$$(1/\widehat{a})(\lambda) = \widehat{1/a}(\lambda).$$

- Let $\widehat{a}$ be analytic at 0 such that $\widehat{a}(0) = 0$ and $\widehat{a}'(0) \neq 0$. Then its inverse function $\widehat{a}^{-1}$ is analytic at 0. Furthermore, for λ sufficiently near to 0,
$$\widehat{a}^{-1}(\lambda) = \widehat{a^{[-1]}}(\lambda).$$

- Let $\widehat{a}$ and $\widehat{b}$ be analytic at 0 and $\sum_{n=0}^\infty |b_n \lambda^n| < \rho(a)$ for $|\lambda| < \rho(b)$. Then the composite function $\widehat{a} \circ \widehat{b}$ is analytic at 0. Furthermore, for λ sufficiently near to 0,
$$\widehat{a}(\widehat{b}(\lambda)) = \widehat{a \circ b}(\lambda).$$

where we recall that $a \circ b$ is the composition product defined by $(a \circ b)_n = \sum_{i=0}^\infty a_i b_n^{\langle i \rangle}$ for $n \in N$.

- Let $\widehat{a}$ be analytic at 0 and $\beta \in \mathbf{C}$. Then for λ sufficiently near to 0,
$$\widehat{a}(\beta\lambda) = \widehat{\underline{\beta} \cdot a}(\lambda).$$

- Let $\widehat{a}$ be analytic at 0. Then for nonzero μ sufficiently near 0, the sequence
$$b = \left\{ \sum_{n=0}^{\infty} C_k^{(n)} a_n \mu^{n-k} \right\}_{k \in \mathbf{N}}$$
is well defined and for λ sufficiently near 0,
$$\widehat{a}(\lambda) = \widehat{b}(\lambda - \mu).$$

- Let $\widehat{a}$ and $\widehat{b}$ be analytic at 0 such that $\widehat{a}(\lambda) = \widehat{b}(\lambda)$ for all λ in an open ball sufficiently near 0, then $a = b$.

In case each term a_k in the sequence $a \in l^{\mathbf{N}}$ is not zero, the ratio test for series also yields
$$\liminf_{n \to \infty} \left| \frac{a_n}{a_{n+1}} \right| \leq \rho(a) \leq \limsup_{n \to \infty} \left| \frac{a_n}{a_{n+1}} \right|.$$

We remark that the series $\underline{\lambda} \cdot (a \circ b)$ is the power series which arises by substituting $w = \widehat{b}(\lambda)$ into $\widehat{a}(w)$ and then formally expand the resulting expression and rearranging terms in increasing powers of λ.

Bibliography

[1] J. Aczel, Lectures on Functional Equations and Their Applications, New York, London, 1966.
[2] J. Aczel and J. Dhombres, Functional Equations in Several Variables, Cambridge University Press, 1989.
[3] J. Aczel and D. Gronau, Some differential equations related to iteration theory, Canad. J. Math. 40(3)(1988), 695–717.
[4] V. Anosov, Geodesic flows on closed Riemannian manifolds of a negative curvature (in Russian), Trudy Math. Inst. im. V. A. Steklova, 90, 1967.
[5] T. M. Apostol, Mathematical Analysis, Second Edition, Addison Wesley, 1974.
[6] A. Arikoglu and I. Ozkol, Solution of boundary value problems for integro-differential equations by using differential transform method, Appl. Math. Comput., 168(2005), 1145–1158.
[7] A. Augustynowicz and H. Leszczynski, On x-analytic solutions to the Cauchy problem for partial differential equations with retardedd variables, Z. Anal. Anwendungen, 15(2)(1996), 345–356.
[8] A. Augustynowicz and H. Leszczynski, On the existence of analytic solutions of the Cauchy problem for first-order partial differential equations with retarded variables, Comment. Math. Prace Mat., 36(1996), 11–15.
[9] A. Augustynowicz, Analytic solutions to the first order partial differential equations with time delays at the derivative, Funct. Differ. Equ., 6(1-2)(1997), 19–29.
[10] F. Ayaz, Solutions of the systems of differential equations by differential transform method, Appl. Math. Comput., 147(2004), 547–567.
[11] C. Babbage, Essay towards the calculus of functions, Philosophical Trans., 1815, 389–423.
[12] C. Babbage, Essay towards the calculus of functions II, Philosophical Trans., 1816, 175–256.
[13] W. Balser, Formal Power Series and Linear Systems of Meromorphic Ordinary Differential Equations, Springer-Verlag, New York, 200.
[14] E. J. Barbeau, Polynomials, Springer, 1989.
[15] K. Baron, R. Ger and J. Matkowski, Analytic solutions of a system of functional equations, Publ. Math. Debrecen 22(3-4)(1975), 189–194.
[16] R. Bellman and K. Cooke, Differential Difference Equations, Academic Press, 1963.
[17] D. Bessis, S. Marmi and G. Turchetti, On the singularities of divergent majorant series arising from normal form theory, Rend. Mat. Appl. Serie VII, 9(4)(1989), 645-659.
[18] L. E. Bottcher, Beitrage zu der Theorie der Iterationsrechnung, Dissertation,

Leipzig, 1898.

[19] L. E. Bottcher, Principles of the iteration calculus (Polish), I, II, Prac. Math. Fiz. 10(1899), 65–101.

[20] D. Brydak, Sur une e'quation fonctionnelle I (in French), Ann. Polon. Math., 15(1964), 237–251.

[21] D. Brydak, Sur une e'quation fonctionnelle II (in French), Ann. Polon. Math., 21(1968), 1–13.

[22] L. Carleson and T. W. Gamelin, Complex Dynamics, Springer-Verlag, New York, 1993.

[23] J. Carr and J. Dyson, The functional differential equation $y'(x) = ay(\lambda x) + by(x)$, Proc. Roy. Soc. Edinburgh Sect. A, 74(1974/75), 165–174 (1976).

[24] C. K. Chen and S. H. Ho, Application of differential transformation to eigenvalue problems, Appl. Math. Comput., 79(1996), 173–188.

[25] C. K. Chen and S. H. Ho, Solving partial differential equations by two-dimensional differential transform method, Appl. Math. Comput., 106(1999), 171–179.

[26] C. L. Chen and Y. C. Liu, Solution of two-point boundary value problems using the differential transformation method, J. Optimization Theory Appl., 99(1)(1998), 23–35.

[27] F. Y. Cheng, Nonexistence of concave analytic solutions to Feigenbaum functional equations (in Chinese), Kexue Tongbao, 35(3)(1990), 171–172.

[28] S. S. Cheng, Partial Difference Equations, Taylor and Francis, 2003.

[29] S. S. Cheng, Smooth solutions of iterative functional differential equations, 2004-Dynamical Systems and Applications, GBS Publishers and Distributions, 2005, pp. 228-252.

[30] S. S. Cheng, J. G. Si. and X. P. Wang, An existence theorem for iterative functional differential equations, Acta Math. Hungarica, 94(1-2)(2002), 1–17.

[31] S. S. Cheng, S. Talwong and V. Laohakosol, Exact solutions of iterative functional differential equations, Computing, 76(2006), 67-76.

[32] V. B. Cherepennikov, Analytic solutions of a class of functional-differential equations (in Russian), Mathematical Physics, 74–77, Leningrad. Gos. Ped. Inst., Leningrad, 1987.

[33] V. B. Cherepennikov, Analytic solutions of some systems of functional-differential equations (in Russian), Partial differential equations, 17–21, Leningrad. Gos. Ped. Inst., Leningrad, 1987.

[34] V. B. Cherepennikov, Analytic solutions of the Cauchy problem for some linear systems of functional-differential equations of neutral type (in Russian), Izv. Vyssh. Uchebn. Zaved. Mat., 6(1994), 90–98.

[35] V. B. Cherepennikov, Analytic solutions of some linear systems of functional-differential equations in a neighborhood of a regular singular point, Translation in Differential Equations, 30 (1994), no. 12, 2020–2022 (1995).

[36] V. B. Cherepennikov, Analytic solutions of some linear systems of functional-differential equations of neutral type. Funct. Differ. Equ. 3(1-2)(1995), 69–82.

[37] V. B. Cherepennikov, Singular analytic solutions of some linear systems of functional-differential equations in a neighborhood of a regular singular point, Siberian Math. J. 37(1)(1996), 171–183.

[38] V. B. Cherepennikov, Analytic solutions of some functional-differential equations linear systems. Proceedings of the Second World Congress of Nonlinear Analysts, Part 5 (Athens, 1996). Nonlinear Anal., 30(5)(1997), 2641–2651.

[39] C. H. Choi, Time-varying Riccati differential equations with known analytic solutions, IEEE Trans. Automat. Control 37(5)(1992), 642–645.

[40] E. A. Coddington, N. Levinson, Theory of ordinary differential equations, McGraw Hill, 1955.

[41] K. L. Cooke, Functional differential systems: some models and perturbation problems, Inter. Symp. Diff. Eqs. Dynamical Systems, Puerto Rico, 1965.

[42] K. L. Cooke and J. Wiener, Retarded differential equations with piecewise constant delays, J. Math. Anal, Appl., 99(1984), 265–297.

[43] C. C. Cowen, Analytic solutions of Bottcher's functional equation in the unit disk, Aequationes Math., 24(1982), 187–194.

[44] C. C. Cowen and B. d. MacCluer, Omposition Operators on Spaces of Analytic Functions, Studies in Advanced Mathematics, CRC Press, Boca Raton, Fl., 1995.

[45] H. Cremer, Uber die Haufigkeit der Nichtzentrem, Math. Ann., 115(1938), 573–580.

[46] J. G. Dhombres, Iteration lineaire dordre 2. C. R. Acad. Sci. Paris, 280 (1975), A275-277.

[47] J. G. Dhombres, Iteration lineaire dordre deux, Publ. Math. Debrecen, 24(1977), 277–287.

[48] R. Driver, Delay-Differential Equations with an Application to a Two-body Problem of Classical Electrodynamics, Tech. Report, University of Minnesota, Minnespolis, 1960.

[49] R. D. Driver, A two-body problem of classical electrodynamics: the one-dimensional case, Ann. Physics, 21(1963), 122-142.

[50] R. D. Driver, A functional-differential system of neutral type arising in a two-body problem of classical electrodynamics, International Symposium on Nonlinear Differential Equations and Nonlinear Mechanics, pp. 474-484, Academic Press, New York, 1963.

[51] R. D. Driver, A "backwards" two-body problem of classical relativistic electrodynamics, Phys. Rev. (2), 178(1969), 2051–2057.

[52] R. D. Driver, Ordinary and Delay Differential Equatins, Springer-Verlag, New York, 1977.

[53] R. D. Driver, Can the future influence the present? Phy. Rev. D., 19(4)(1979), 1098-1107.

[54] R. D. Driver, A neutral system with state dependent delay, J. Diff. Eq., 54(1)(1984), 73–86.

[55] J. M. Dubbey, The Mathemtical Works of Charles Babbage, Cambridge University Press, 1978.

[56] G. M. Dunkel, On nested functional differential equations, SIAM J. Appl. Math., 18(2)(1970), 514-525.

[57] E. Eder, Existence, uniqueness and iterative construction of motions of charged particles with retarded interactions, Ann. Inst. H. Poincare, 39(1)(1983), 1-27.

[58] E. Eder, The functional differential equation $x'(t) = x(x(t))$, J. Diff. Eq., 54(1984), 390-400.

[59] A. Elbert, Asymptotic behaviour of the analytic solution of the differential equation $y'(t) + y(qt) = 0$ as $q \to 1^-$, J. Comput. Appl. Math., 41(1992), 5-22.

[60] M. J. Feigenbaum, Quantitative universality for a class of nonlinear transformations, J. Statist. Phys., 19(1)(1978), 25–52.

[61] Y. A. Fiagbedzi and M. A. El-Gebeily, Existence and uniqueness of the solution of a class of nonlinear functional differential equations, Ann. of Diff. Eq. 16(2000), 381-390.

[62] G. M. Fichtenholz, Infinite Series: Ramifications, Gordon and Breach, 1970.

[63] G. M. Fichtenholz, Functional Series, Gordon and Breach, 1970.

[64] A. Feldstein and Z. Jackiewicz, Unstable neutral functional differential equations,

Canad. Math. Bull., 33(4)(1990), 428–433.
[65] S. W. Golomb, D. Marcus and N. J. Fine, Solution to problem 5407, Amer. Math. Monthly 74(1967), 740-743.
[66] K. O. Friedrichs, Lectures on Advanced Ordinary Differential Equations, Gordon and Breach, 1965.
[67] W. G. Ge, A transform theorem for differential-iterative equations and its application (in Chinese), Acta Math. Sinica, 40(6)(1997), 881–888.
[68] W. G. Ge and Y. R. Mo, Lyapunov stability of an iterative functional differential equation, Proc. Fifth National Conference on Stability Theory of Differential Equations and its Applications, Dailien, 1996.
[69] W. G. Ge and Y. R. Mo, Existence and bifurcation of solutions to differential-iterative equations (in Chinese), Acta Math. Appl. Sinica, 21(1)(1998), 113–122.
[70] S. W. Golomb, Problem 5407, Amer. Math. Monthly 73(1966), 674.
[71] L. J. Grimm, Existence and continuous dependence for a class of neutral nonlinear equations, Proc. Amer. Math. Soc., 29(1971), 467-473.
[72] S. A. Gusarenko, Solvability of equations of neutral type with a locally Volterra operator, Functional-Differential Equations (Russian), 26-29, ii-iii, Perm. Politekhn. Inst. Perm, 1985.
[73] J. Hale, Theory of Functional Differential Equations, Springer Verlag, 1977.
[74] J. K. Hale and M. A. Cruz, Existence, uniqueness, and continuous dependence for hereditary systems, Ann. Mat. Pura Appl., 85(1970), 63–81.
[75] I. H. A. Hassan, Differential transformation technique for solving higher-order initial value problems, Appl. Math. Comput., 154(2004), 299–311.
[76] Z. M. He and W. G. Ge, Monotone iterative technique and periodic boundary value problem for first order impulsive functional differential equations, Acta Math. Sinica, (Engl. Ser.) 18(2)(2002), 253–262.
[77] I. H. Herron, The radius of convergence of power series solutions to linear differential equations. Amer. Math. Monthly 96(9)(1989), 824–827.
[78] E. Hille, Lectures on Ordinary Differential Equations, Addison. Wesley, 1968.
[79] J. T. Hoag and R. D. Driver, A delayed-advanced model for the electrodynamics two-body problem, Nonlinear Anal., 15(2)(1990), 165–184.
[80] D. P. K. Hsing, Existence and uniqueness theorem for the one dimensional backwards two-body problem of electrodynamics, Phy. Rev. D., 16(4)(1977), 974-982.
[81] E. L. Ince and I. N. Sneddon, The Solutions of Ordinary Differential Equations, Second Edition, Longman Scientific, 1987.
[82] E. K. Ifantis, An existence theory for functional-differential equations and functional-differential systems, J. Differential Equations 29(1)(1978), 86–104.
[83] R. Isaacs, Iterates of fractional order, Canad. J. Math., 2(1950), 409–416.
[84] A. Iserles and Y. K. Liu, On neutral functional-differential equations with proportional delays. J. Math. Anal. Appl. 207(1)(1997), 73–95.
[85] E. Jabotinsky, Analytic iteration, Trans. Amer. Math. Soc. 108(1963), 457-477.
[86] Z. Jackiewicz, Existence and uniqueness of solutions of neutral delay-differential equations with state dependent delays, Funk. Ekv., 30(1987), 9-17.
[87] Z. Jackiewicz, A note on existence and uniqueness of solutions of neutral functional differential equations with state dependent delays, Comment. Math. Univ. Carolin., 36(1)(1995), 15–17.
[88] M. J. Jang, C. L. Chen and Y. C. Liu, On solving the initial-value problems using the differential transformation method, Appl. Math. Comput., 115(2000), 145–160.
[89] M. J. Jang, C. L. Chen and Y. C. Liu, Two-dimensional differential transform method for partial differential equations, Appl. Math. Comput., 121(2001), 261–

270.

[90] C. Jordan, Calculus of Finite Differences, Chelsea, New York, 1965.

[91] M. Jia, W. G. Ge and X. P. Liu, Existence and continuation of solutions for a class of nonautonomous differential-iterative equations (in Chinese), J. Beijing Inst. Technol. (Chinese Ed.), 18(1)(1998), 5–10.

[92] E. Kamke, Differentialgleichungen reeller funktionen, Akademische Verlag, 1930.

[93] A. Kaneko, On extension of analytic solutions of linear partial differential equations, Proceedings of the St. Petersburg Mathematical Society, Vol. III, 105-110. Amer. Math. Soc. Transl. ser. 2, 188, Amer. Math. Soc., Providence, RI, 1999.

[94] W. Kaplan, Introduction to Analytic Functions, Addison-Wesley, 1966.

[95] T. Kato and J. B. McLeod, The functional differential equation $y'(x) = ay(\lambda x) + by(x)$, Bull. Amer. Math. Soc., 77(1971), 891–937.

[96] G. Koenigs, Recherches sur les integrales de certaines equations fonctionnelles, Ann. Sci. Ecole Norm. Sup. (3), 1(1884), Supplement, 3–41.

[97] G. Koenigs, Nouvelles recherches sur les equations fonctionnelles, Ann. Sci. Ecode Norm. Sup. (3), 2(1885), 385–404.

[98] J. Kovats, Real analytic solutions of parabolic equations with time-measurable coefficients, Proc. Amer. Math. Soc., 130(4)(2002), 1055-1064.

[99] S. G. Krantz and H. R. Parks, A Primer of Real Analytic Functions, Birkaüser Verlag, 1992.

[100] S. G. Krantz, Function Theory of Several Complex Variables, Wadsworth & Brooks, 1992.

[101] S. G. Krantz and H. R. Parks, The Implicit Function Theorem, Birkaüser Boston, 2002.

[102] M. Kuczma, A Survey of the Theory of Functional Equations, Univ. Beograg, Publ. Elektrotehn. Fak. Ser. Math. Fiz., No. 130, 1964.

[103] M. Kuczma, Une remargue sur les solutions analytiques d'une equation fonctionnelle, Colloq. Math., 16(1967), 93–99.

[104] M. Kuczma, Functional Equation in a Single Variable, Polish Scientific Publishers, Warsaw, 1968.

[105] M. Kuczma, Analytic solutions of a linear functional equation, Ann. Polon. Math., 21(1969), 297–303.

[106] M. Kuczma, On a functional equation with divergent solutions, Ann. Polon. Math., 22(1979), 173–178.

[107] M. Kuczma, B. Choczewshi and R. Ger, Iterative Functional Equations, Cambridge University Press, 1990.

[108] M. Kuczma and W. Smajdor, Analytic solutions of some functional equations, J. London Math. Soc. (2), 4(1971/72), 418–424.

[109] M. Kuczma and W. Smajdor, On the radius of series solutions of a functional equation, Ann. Polon. Math., 24(1970/71), 233–240.

[110] C. Z. Li and W. G. Ge, Existence of periodic solutions to a type of differential-iterative equations (in Chinese) J. Beijing Inst. Technol. (Chin. Ed.) 20(5)(2000), 534–538.

[111] T. Y. Li and J. A. Yorke, Period three implies chaos, Amer. Math. Monthly, 82(1974), 985–992.

[112] W. R. Li, Existence and uniqueness of analytic solutions to linear functional equations (in Chinese), Qufu Shiyuan Xuebao, 10(2)(1984), 16–21.

[113] W. R. Li, The binomial theorem and functional equation (in Chinese), Shuxue Tongbao, 12(7)(1985), 43–44.

[114] W. R. Li, The existence and uniqueness of continuous solutions of functional equa-

tions of the form $f(x) = G(x, f(qx))$ in Banach spaces (in Chinese), Qufu Shifan Daxue Xuebao Ziran Kexue Ban, 12(1)(1986), 33–37.

[115] W. R. Li, The existence of a analytic solution to a second order linear equation (in Chinese), Advan. Math., 2(1)1986, 19–23.

[116] W. R. Li, A proof, using real analysis, of the existence of power series solutions to second-order linear ordinary differential equations (in Chinese), Qufu Shifan Daxue Xuebao Ziran Kexue Ban, 13(3)(1987), 50–53.

[117] W. R. Li, Differentiable solutions to two classes of functional equations with many unknown functions (in Chinese), Qufu Shifan Daxue Xuebao Ziran Kexue Ban, 19(1)(1993), 44–50.

[118] W. R. Li, The analytic solutions of two Jabotinsky's differential equations, J. Math. Study 29(3)(1996), 12–17.

[119] W. R. Li, Analytic solutions for a class of second-order iterative functional-differential equations (in Chinese), Acta Math. Sinica, 41(1)(1998), 167–176.

[120] W. R. Li and S. S. Cheng, Analytic solutions of an iterative functional equation, Aequationes Math., 68(1-2)(2004), 21–27.

[121] W. R. Li, S. S. Cheng and L. J. Cheng, Analytic solution of an iterative differential equation, Math. Practice Theory, 32(6)(2002), 994-998 (in Chinese).

[122] W. R. Li, S. S. Cheng and T. T. Lu, Closed form solutions of iterative functional differential equations, Appl. Math. E-Notes, 1(2001), 1-4.

[123] W. R. Li and H. Z. Liu, Discussion on the analytic solutions of the second order iterated differential equation, Bull. Korean Math. Soc., 43(2006), 1–14.

[124] W. R. Li, H. Z. Liu and S. S. Cheng, Analytic solutions of an equation of invariant curves, Dyn. Contin. Discrete Impuls. Syst., Ser. A, Math. Anal., 13A(2006), Part 2, suppl., 1014-1021.

[125] W. R. Li and J. G. Si, Continuous solutions and analytic solutions of two classes of functional equations (in Chinese), J. Math. Res. Exposition, 9(2)(1989), 231–237.

[126] W. R. Li and J. G. Si, The theory, applications, and development of functional equations in one variable, (in Chinese) Qufu Shifan Daxue Xuebao Ziran Kexue Ban, 21(1)(1995), 34–44.

[127] S. X. Liang and J. Z. Zhang, A complete discrimination system for polynomials with compelx coefficients and its automatic generation, Sci. china Ser. E, 42(2)(1999), 113-128.

[128] H. Z. Liu and W. R. Li, Analytic solutions of an iterative equation with second order derivatives, Dyn. Contin. Discrete Impuls. Syst. Ser. A Math. Anal., 13A(2006), Part 2, suppl., 1030–1037.

[129] X. H. Liu, Existence and uniqueness of analytic solutions of systems of iterative functional equations, Northeast Math. J., 16(4)(2000), 428–438.

[130] X. H. Liu, Analytic solutions of systems of functional equations, Appl. Math. J. Chinese Univ. Ser. B, 18(2)(2003), 129–137.

[131] X. H. Liu and J. H. Mai, Analytic solutions of iterative functional equations, J. Math. Anal. Appl., 270(1)(2002), 200–209.

[132] X. P. Liu, M. Jia and W. G. Ge, A class of nonautonomous functional-differential iterative equations (in Chinese), Gaoxiao Yingyong Shuxue Xuebao Ser. A, 13(3)(1998), 249–256.

[133] J. H. Mai and X. H. Liu, On a property of roots of polynomials, J. Math. Res. Exposition, 21(1)(2001), 17–20.

[134] D. Marcus, Solution to problem 5407, Amer. Math. Monthly, 74(1967), 740.

[135] J. Matkowski, Note on a functional equation, Zeszyty Nauk. Uniw. Jagiello. Prace Mat. No. 15, (1971), 109–111.

[136] J. Matkowski, On the continuous dependence of local analytic solutions of a functional equation on given functions, Ann. Polon. Math., 24(1970/1971), 21–26.

[137] J. Matkowski, On the continuous dependence of local analytic solutions of the functional equation in the non-uniqueness case, Ann. Polon. Math., 24 (1970/71), 319–326.

[138] J. Matkowski, Correction to the paper: "On the continuous dependence of local analytic solutions of a functional equation on given functions". Ann. Polon. Math., 26 (1972), 219.

[139] M. McKiernan, On the n-th derivative of composite functions, Amer. Math. Monthly, 63(5)(1956), 331-333.

[140] M. A. McKiernan, The functional differential equation $Df = 1/ff$, Proc. Amer. Math. Soc. 8(1957), 230-233.

[141] J. Moser, A rapidly convergent iteration method and non-linear partial differential equations I, Ann. Scuola Norm. Sup. Pisa (3), 20(1966), 265–315.

[142] J. Moser, A rapidly convergent iteration method and non-linear differential equations II, Ann. Scuola Norm. Sup. Pisa (3), 20(1966), 499–535.

[143] B. Muckenhoupt, Some results on analytic iteration and conjugacy, Amer. J. Math., 84(1962), 161–169.

[144] P. J. Myrberg, Inversion der Iteration fur rationale Funktionen (in German), Ann. Acad. Sci. Fenn. Ser. A I No. 292 1960.

[145] P. J. Myrberg, Eine verallgemeinerung der abelschen funktionalgleichung, Ann. Acad. Sci. Fenn, Ser. A. I. 327(1962), 18.

[146] P. J. Myrberg, Uber die analytischen Losungen der Gleichung $f(R(z)) = f(z)$ bei rationalem $R(z)$ (in German), Ann. Acad. Sci. Fenn. Ser. A I No. 308 1962.

[147] C. T. Ng and W. N. Zhang, Invariant curves for planar mappings, J. Differ. Equations Appl. 3(1997), 147–168.

[148] Z. Nitecki, Differentiable Dynamics - An Intoduction to the Orbit Structure of Diffeomorphisms, The MIT Press, Cambridge, Mass, and London, England, 1971.

[149] R. J. Oberg, On the local existence of solutions of certain functional differential equations, Proc. Amer. Math. Soc., 20(1969), 295-302.

[150] L. Pandolfi, Some observations on the asymptotic behaviour of the solutions of the equation $\dot{x} = A(t)x(\lambda t) + B(t)x(t), \lambda > 0$, J. Math. Anal. Appl., 67(2)(1979), 483–489.

[151] V. R. Petuhov, On a boundary value problem (in Russian) Trudy Sem. Teor. Differencial. Uravnenii Otklon. Argumentom Univ. Druzby Narodov Patrisa Lumumby, 3(1965), 252–255.

[152] Y. F. S. Petermann, On Golomb's self describing sequence, J. Number Theory, 53(1995), 13–24.

[153] Y. F. S. Petermann, On Golomb's self describing sequence II, Arch. Math., 67(1996), 4473-477.

[154] Y. F. S. Petermann and J. L. Remy, Golomb's self-described sequence and functional differential equations, Illinois J. Math., 42(3)(1998), 420–440.

[155] Y. F. S. Petermann and J. L. Remy, Increasing self-described sequences, Ann. Comb., 8(3)(2004), 325–346.

[156] G. A. Pfeiffer, The functional equation $f[f(x)] = g(x)$, Ann. Math., 20(2)(1918), 13–22.

[157] H. Poincare, Sur une classe estendue de transcendantes uniformes, C. R. Acad. Sci. Paris 103(1886), 862–864.

[158] O. Rausenberger, Theorie der allgemeinen Periodiziat, Math. Ann. 18(1881), 379–410.

[159] L. Reich, On a differential equation arising in iteration theory in rings of formal power series in one variable, Iteration Theory and its Functional Equation, Lecture Notes in Mathmatics, 1163, Springer Verlag, Berlin, Heidelberg, New York, Tokyo, 1984, 135–148.
[160] L. Reich, Holomorphic solutions of the differential equation of E. Jabotinsky, Österreich. Akad. Wiss. Math. Natur. Kl. Sitzungsber II, 195(1-3)(1986), 157–166.
[161] L. Reich, Generalized Böttcher equations in the complex domain, Symposium on Complex Differential and Functional Equations, 135–147, Univ. Joensuu Dept. Math. Rep. Ser., 6, Univ. Joensuu, Joensuu, 2004.
[162] R. E. Rice, Iterative square roots of Chebyshev polynomials, Stochastica, 3(2)(1979), 1–14.
[163] R. E. Rice, B. Schweizer and A. Sklar, When is $f(f(z)) = az^2 + bz + c$? Amer. Math. Monthly, 87(1980), 252–262.
[164] S. Roman, The formula of Faà di Bruno, Amer. Math. Monthly, 87(1980), 805–809.
[165] H. Rüssmann, Über die Iteration analytischer funktionen, J. Math. Mech., 17(1967), 523–532.
[166] H. Rüssmann, Über die Normalform analytischer Hamiltonscher Differentialgleichungen in der Nähe einer Gleichgewichtslösung (German), Math. Ann., 169(1967), 55–72.
[167] G. Sansone, Equazioni differenziali nel campo reale I, II, Nicola Zanichelli Editore, 1948, 1949.
[168] A. N. Sarkovskii, Functional and functional-differential equations in which the deviation of the argument depends on the unknown function, Functional and Differential-Difference Equations (Russian), pp. 148-155, Izdanie Inst. Mat. Akad. Nauk Ukrain. SSR, Kiev, 1974.
[169] I. M. Sheffer, Convergence of multiply-infinite series, Amer. Math. Monthly, 52(7)(1945), 365–376.
[170] I. M. Sheffer, Note on multiply infinite series, Bull. Amer. Math, Soc., 52(1946), 1036–1041.
[171] B. Shi and Z. X. Li, Existence of solutions and bifurcation of a class of first order functional differential equations, Acta Math. Appl. Sinica, 18(1)(1995), 83-89.
[172] B. Shi and Z. X. Li, An investigation of a new class of functional differential equations, J. Central China Normal Univ. Natur. Sci., 29(1)(1995), 12–15.
[173] E. Schröder, Ueber iterirte Functionen, Math. Ann., 3(2)(1871), 296–322.
[174] J. G. Si, Existence of analytic solutions for two classes of linear functional differential equations, Acta Math. Appl. Sinica, 14(2)(1991), 262–276.
[175] J. G. Si, The existence of local analytic solutions of the iterated equations $\sum_{i=1}^{n} \lambda_i f^i(z) = F(z)$, Acta Math. Sinica, 37(5)(1994), 590–600.
[176] J. G. Si, On analytic solutions of the equation of invariant curves, C. R. Math. Rep. Acad. Sci. Canada, 17(1)(1995), 49–52.
[177] J. G. Si, The existence of analytic solutions for two classes of linear functional differential equations of neutral type, J. Qingdao Univ. Nat. Sci. Ed., 9(1)(1996), 20–33.
[178] J. G. Si, Analytic solutions of a nonlinear functional differential equation with proportional delays, Demonstratio Math., 33(4)(2000), 747–752.
[179] J. G. Si, Analytic solutions of linear neutral functional differential-difference equations, Appl. Math. E-Notes, 1(2001), 11–17.
[180] J. G. Si and S. S. Cheng, Analytic solutions of a functional-differential equation with state dependent argument, Taiwanese J. Math., 1(4)(1997), 471–480.
[181] J. G. Si and S. S. Cheng, Note on an iterative functional-differential equation,

Demonstratio Math., 31(3)(1998), 609–614.

[182] J. G. Si and S. S. Cheng, Smooth solutions of a nonhomogeneous iterative functional-differential equation, Proc. Roy. Soc. Edinburgh Sect. A, 128(4)(1998), 821–831.

[183] J. G. Si and S. S. Cheng, Analytic solutions of a functional differential equation with proportional delays, Bull. Korean Math. Soc. 39(2)(2002), 225–236.

[184] J. G. Si and W. R. Li, Analytic solutions of some functional differential equations (in Chinese), J. Qindao Univ. nat. Sci. Ed., 11(1)(1998), 24–32.

[185] J. G. Si, W. R. Li and S. S. Cheng, Analytic solutions of an iterative functional-differential equation, Comput. Math. Appl., 33(6)(1997), 47–51.

[186] J. G. Si and X. P. Wang, Analytic solutions of a polynomial-like iterative functional equation, Demonstratio Math. 32(1)(1999), 95–103.

[187] J. G. Si and X. P. Wang, Analytic solutions of a second-order functional-differential equation with a state derivative dependent delay, Colloq. Math., 79(2)(1999), 273–281.

[188] J. G. Si and X. P. Wang, Analytic solutions of a second-order iterative functional differential equation, J. Comput. Appl. Math., 126(1-2)(2000), 277–285.

[189] J. G. Si and X. P. Wang, Analytic solutions of a second-order functional differential equation with a state dependent delay, Results Math., 39(3-4)(2001), 345–352.

[190] J. G. Si and X. P. Wang, Analytic solutions of a second-order iterative functional differential equation, Comput. Math. Appl. 43(1-2)(2002), 81–90.

[191] J. G. Si and X. P. Wang, Erratum to: "Analytic solutions of a second-order iterative functional differential Equation" [Comput. Math. Appl. 43 (2002), no. 1-2, 81–90; MR1873243 (2002i:34153)], Comput. Math. Appl. 44(5-6)(2002), 833.

[192] J. G. Si, X. P. Wang and S. S. Cheng, Analytic solutions of a functional-differential equation with a state derivative dependent delay, Aequationes Math., 57(1)(1999), 75–86.

[193] J. G. Si, X. P. Wang and S. S. Cheng, Nondecreasing and convex $C^{(2)}$-solutions of an iterative functional-differential equation, Aequationes Math. 60(1-2)(2000), 38–56.

[194] J. G. Si, X. P. Wang and W. N. Zhang, Analytic invariant curves for a planar map, Appl. Math. Lett. 15(5)(2002), 567–573.

[195] J. G. Si and W. N. Zhang, Analytic solutions of a functional equation for invariant curves, J. Math. Anal. Appl. 259(1)(2001), 83–93.

[196] J. G. Si and W. N. Zhang, Analytic solutions of a nonlinear iterative equation near neutral fixed points and poles, J. Math. Anal. Appl. 284(1)(2003), 373–388.

[197] J. G. Si and W. N. Zhang, Analytic solutions of a class of iterative functional differential equations, J. Comput. Appl. Math. 162(2)(2004), 467–481.

[198] J. G. Si and W. N. Zhang, Analytic solutions of a q-difference equation and applications to iterative equations, J. Difference Equ. Appl. 10(11)(2004), 955–962.

[199] J. G. Si and W. N. Zhang, Analytic solutions of a second-order nonautonomous iterative functional differential equation, J. Math. Anal. Appl. 306(2)(2005), 398–412.

[200] J. G. Si, W. N. Zhang and S. S. Cheng, Smooth solutions of an iterative functional equation, Functional Equations and Inequalities, 221–232, Math. Appl., 518, (ed. Th. M. Rassias) Kluwer Acad. Publ., Dordrecht, 2000.

[201] J. G. Si, W. N. Zhang and S. S. Cheng, Continuous solutions of an iterative functional inequality on compact interval, Nonlinear Stud., 7(1)(2000), 105–108.

[202] J. G. Si, W. N. Zhang and G. H. Kim, Analytic solutions of an iterative functional differential equation, Appl. Math. Comput. 150(3)(2004), 647–659.

[203] C. L. Siegel, Iteration of analytic functions, Ann. Math., 43(2)(1942), 607–612.

[204] C. L. Siegel, Vorlesungen uber Himmelsmechanik, Springer Verlag, Berlin-

Gottingen-Heidelberg, 1956.

[205] W. Smajdor, On the extience and uniqueness of analytic solutions of the functional equation $\varphi(z) = h(z, \varphi[f(z)])$, Ann. Polon. Math., 19(1967), 37–45.

[206] W. Smajdor, Local analytic solutions of the functional equation $\varphi(z) = h(z, \varphi[f(z)])$ in multidimensional spaces, Aequationes Math., 1(1968), 20–36.

[207] W. Smajdor, Analytic solutions of the equation $\varphi(z) = h(z, \varphi[f(z)])$ with right side contraction, Aequationes. Math., 2(1968), 30–38.

[208] W. Smajdor, Local analytic solutions of the functional equation $\Phi(z) = H(z, \Phi[f_1(z)], ..., \Phi[f_m(z)])$, Ann. Polon. Math., 24(1970), 39–43.

[209] W. Smajdor, Solutions of the Schröder equation, Ann. Polon. Math., 27(1972/73), 61–65.

[210] A. Smajdor and W. Smajdor, On the existence and uniqueness of analytic solutions of a linear functional equation, Math. Zeitschr., 98(1967), 235–243.

[211] K. T. Smith, Power Series from a Computational Point of View, Springer Verlag, 1987.

[212] I. N. Sneddon, Special Functions of Mathematical Physics and Chemistry, Oliver and Boyd, 1956.

[213] B. H. Stephan, On the existence of periodic solutions of $z'(t) = -az(t - r + \mu k(t, z(t))) + F(t)$, J. Diff. Eq., 6(1969), 408-419.

[214] S. Talwong, V. Laohakosol and S. S. Cheng, Power function solutions of iterative functional differential equations, Appl. Math. E-Notes, 4(2004), 160-163.

[215] L. Tavernini, The approximate solution of Volterra differential systems with state dependent time lags, SIAM J. Numer. Anal., 15(1978), 1039-1052.

[216] G. Valiron, Fonctions analytigues, Presses Universitaires de France, Paris, 1954.

[217] K. Wang, On the equation $x'(t) = f(x(x(t)))$, Funckcialaj Ekvacioj, 33(1990), 405-425.

[218] R. W. Wagner, An extension of analytic functions to matrices, Amer. J. Math., 62(1940), 380–390.

[219] R. Wagner, Eindeutige Losungen der funktionalgleichung $f(x + f(x)) = f(x)$, Elem. Math., 14(1957), 73–78.

[220] X. P. Wang and J. G. Si, Continuous solutions of an iterative functional equation (in Chinese), Acta Math. Sinica, 42(5)(1999), 945–950.

[221] X. P. Wang and J. G. Si, Differentiable solutions of an iterative functional equation, Aequationes Math., 61(1-2)(2001), 79–96.

[222] X. P. Wang and J. G. Si, Analytic solutions of an iterative functional differential equation, J. Math. Anal. Appl., 262(2)(2001), 490–498.

[223] G. N. Waston, A Treatise on the Theory of Bessel Functions, 2nd ed., Cambridge, 1944.

[224] E. T. Whittaker and G. N. Waston, Modern Analysis, Cambridge, 1958.

[225] A. Wilansky, On the convergence of double series, Bull. Amer. Math. Soc., 53(1947), 793-799.

[226] H. Z. Wu, On the existences and asymptotic behavior of strong solutions of equation $x'(t) = f(x(x(t)))$, Ann. Diff. Eqs., 9(1993), 336-351.

[227] H. Z. Wu, A class of functional differential equations with deviating arguments depending on the state, Acta Math. Sinica, 38(6)(1995), 803-809.

[228] X. F. Xiang and W. G. Ge, On the periodic solution of the differential-iterative equation $x'(t) = \omega(t)(ax(t) - bx(x(t)))$ ($a > b > 0$) (in Chinese), J. Systems Sci. Math. Sci., 19(4)(1999), 457–464.

[229] B. Xu, W. N. Zhang and J. G. Si, Analytic solutions of an iterative differential equation which may violate the Diophantine condition, J. Difference Equ. Appl.,

10(2)(2004), 201–211.

[230] W. X. Yang and W. G. Ge, Periodic solutions for the differential-iterative equation $\dot{x} + g(x(x)) = p(t)$ (in Chinese), J. Beijing Inst. Technol. (Chinese Ed.) 22(5)(2002), 537–539.

[231] J. Z. Zhang and Y. Lu, On criterion of existence and uniqueness of real iterative groups in a single variable (in Chinese), Acta. Sci. Nat. Univ. Peking, 6(1982), 23–45.

[232] J. Z. Zhang, L. Yang and W. N. Zhang, Some advances on functional equations, Adv. Math. China, 24(5)(1995), 385–405.

[233] J. Z. Zhang, L. Yang and W. N. Zhang, Iterative equations and embedding flow, Shanghai Scientific and Technological Education Publishing House, China, 1998.

[234] J. Z. Zhang and Y. Lu, Iterative roots of a piecewise monotone continuous self-mapping (in Chinese), Acta. Math. Sinica, 26(4)1983, 398–412.

[235] W. N. Zhang, Discussion on the differentiable solutions of the iterated equation $\sum_{i=1}^{n} \lambda_i f^i(x) = F(x)$, Nonlinear Analysis, 15(4)(1990), 387–398.

[236] Z. X. Zheng, J. H. Xu and H. S. Ren, A class of FDE with deviating argument which depends on state and rate of change of the state, Annals Diff. Eq., 14(2)(1998), 454–459.

[237] J. K. Zhou, Differential Transformation and Its Applications for Electrical Circuits, Huazhong Univ. Press, Wuhan, China, 1986 (in Chinese).

Index

bi-index, 63

Cauchy's estimate, 72
Cauchy's estimation, 72, 101, 102, 112, 114, 115, 125, 134, 140, 142, 144, 150, 153–155, 157, 159, 162, 168, 179, 181, 183, 186, 191–193, 195, 209, 220, 227, 228, 233, 241
contraction mapping, 9, 107

derivative, 3
    algebraic, 32, 50, 53, 261
    partial, 8

enumeration, 15
equation
    Airy's, 128
    Böttcher, 121
    Babbage, 176
    Chebyshev's, 128
    conjugacy, 121
    Euler, 190
    functional, 83, 86
        linear, 100
    functional differential, 123
    invariant curve, 176, 190
    iterative functional, 175
    nonlinear differential, 133
    Poincaré, 110, 114
    polynomial functional, 90
    rational functional, 90
    Schröder, 110, 176

Faa di Bruno formula, 39, 106, 212
function
    analytic

        bivariate, 67
        matrix, 71
        multivariate, 68
        univariate, 56
    exponential, 5, 89
    generating, 49
    logarithm, 5
    power, 89, 244, 246–249, 251–254, 256, 258
    rational, 5

integral
    algebraic, 34, 50, 54, 262
    Cauchy, 8
invariant curve, 176
iterate, 3, 175, 182

majorant, 72, 87, 101, 125, 136
multi-index, 6, 68

Newton binomial expansion, 59, 89, 90

ordinary point, 124, 127

polycylinder, 8
polynomial, 4, 57
power series, 1, 49
    bivariate, 63, 64
    formal, 2
    matrix, 71
    multivariate, 68
product
    composition, 34, 261
    convolution, 26, 43, 261
    inner, 250
    termwise, 11, 260

quotient, 29, 261

radius of convergence, 50, 127
Reinhart domain, 65

sequence
  absolute, 262
  absolutely summable, 19
  arithmetic, 259
  attenuated, 26, 49
  binomial, 7, 260
  bivariate, 6, 15, 42
  Cauchy, 9
  difference, 25, 259
  difference of, 262
  Dirac, 6, 11, 25, 259
  exponential, 25, 259
  exponential of, 36
  factorial, 260
  geometric, 6, 259
  Heaviside, 6, 25, 93, 259
  Lebesgue summable, 12
  limit, 263
  multivariate, 11
  partial sum, 13, 262
  relatively summable, 19
  scalar, 6, 25, 259
  shifted, 26, 262
  summable, 18
  translated, 43
  univariate, 6
Siegel number, 77, 112, 180, 190, 194, 195, 198, 200–202, 204, 206, 208–210, 218, 220, 221, 223, 226, 227, 229–233, 235, 236, 238, 240–242

Siegel's lemma, 77
space
  Banach, 107
  metric, 9, 119
sum
  termwise, 260
support, 11
system
  linear differential, 124
  neutral differential, 128

Taylor series, 49, 64
theorem
  Able's limit, 52
  Banach contraction, 121
  Cauchy-Kowalewski, 139
  dominated convergence, 17
  Fubini, 15
  implicit function, 86, 92–98
  inverse function, 88, 176
  inversion, 54
  Merten, 27
  monotone convergence, 13
  representation, 53
  substitution, 55, 67
  unique continuation, 58
  unique representation, 54, 84, 92, 100, 102, 104, 113, 124, 135
transformation, 262